AF388941

Discrete Multiphysics

Discrete Multiphysics

Modelling Complex Systems with Particle Methods

Editor

Alessio Alexiadis

MDPI • Basel • Beijing • Wuhan • Barcelona • Belgrade • Manchester • Tokyo • Cluj • Tianjin

Editor
Alessio Alexiadis
School of Chemical Engineering,
University of Birmingham
UK

Editorial Office
MDPI
St. Alban-Anlage 66
4052 Basel, Switzerland

This is a reprint of articles from the Special Issue published online in the open access journal *ChemEngineering* (ISSN 2305-7084) (available at: https://www.mdpi.com/journal/ChemEngineering/special_issues/Particle_Methods).

For citation purposes, cite each article independently as indicated on the article page online and as indicated below:

LastName, A.A.; LastName, B.B.; LastName, C.C. Article Title. *Journal Name* **Year**, *Volume Number*, Page Range.

ISBN 978-3-0365-2213-5 (Hbk)
ISBN 978-3-0365-2214-2 (PDF)

Contents

About the Editor

Alessio Alexiadis is a Reader at the School of Chemical Engineering at the University of Birmingham in UK. He received his Ph.D. in 2001 at the Politecnico di Torino (Italy). Since then, he has worked at the Ecole Nationale Superieure des Mines de Paris (2001–2003), Max-Planck-Institut fur Kohlenforschung (2003–2004), University of New South Wales (2004–2006), University of Cyprus (2006–2008), Washington University in St. Louis (2008–2010), KTH—Royal Institute of Technology (2010–2011) and University of Warwick (2012–2013). He has been PI or co-I in various research projects supported by research councils and other funding bodies in the UK, Australia, USA and EU. In 2008, he was awarded the Marie Curie OIF fellowship (EU), and between 2008 and 2009 he was part of the Plasma–wall Interaction EURATOM Task force. His current research mainly focuses on mathematical modelling and computer simulations. Past research activities include photocatalysis, computational fluid dynamics, bubbly flows, multiphase reactors, polymer dynamics, membrane science, global warming modelling, electrochemistry, micro- and nano-fluidics, and both classical and ab initio molecular simulations.

Preface to "Discrete Multiphysics"

This book is based on the ChemEngineering special issue "Discrete Multiphysics: Modelling Complex Systems with Particle Methods" and presents various computational studies, characterized by the use of particle-based methods such as Smooth Particle Hydrodynamics (SPH), Discrete Element Method (DEM), Lattice Spring Model (LSM) and several others. Particle methods can be applied to a variety of different problems that, on the surface, look very different from each other. For example, SPH is often used to model hydrodynamics problems, LSM solid mechanics and DEM real particles. However, despite these differences, they all follow the same algorithm. The computational domain is subdivided into computational particles that represent real particles (e.g., DEM), or discrete portions of liquid (e.g., SPH) or solid (e.g., LSM) matter. These particles move because they exchange forces with each other. What characterizes different particle methods is the type of physical phenomena represented by the forces. For instance, in SPH, they represent the pressure and viscous forces occurring in fluids; in LSM, the elastic forces occurring in solids; in DEM, Hertzian contact forces between real particles.

The focus of this book is not on single-particle methods, but on the coupling of different-particle methods. As particle methods follow the same algorithm, they can easily be coupled within Discrete Multiphysics (DMP) simulations. By coupling multiple particle methods, we can simulate a variety of phenomena occurring in solid–liquid flows, fluid–structure interaction, solidification, melting, breakage and agglomeration. Generally, DMP has proven itself more versatile than traditional, mesh-based, multiphysics, and researchers familiar with this technique will have a very effective tool for modelling complex physical phenomena at their disposal.

While there are many books on traditional mesh-based multiphysics and on single particle-based methods such as SPH, LSM or DEM, to the best of our knowledge, no book on Discrete Multiphysics is available. The publisher and I felt this was an unjustifiable gap in the current literature that needed to be filled. This is even more important at present, because the ability of particle methods to be coupled with other particle methods seems to go beyond the traditional boundaries of mathematical modelling and cross over into research fields such as machine learning and artificial intelligence. This specific aspect is not part of this book (the interested reader will find some material by googling "particle-neuron duality"), but it shows the growing importance of particle methods in the scientific community.

Some chapters, such as "Mass Spring Models of Amorphous Solids" by Maciej Kot, deal with specific particle methods from a theoretical perspective. They do introduce new ideas to the field, but they can also be used as an introduction to the subject. Other chapters, such as "Modelling Complex Particle–Fluid Flow with a Discrete Element Method Coupled with Lattice Boltzmann Methods (DEM-LBM)" by Liu and Wu, focus on the coupling of different particle methods. The advantages of DMP with respect to traditional multiphysics are particularly clear in biological and medical problems (e.g., "Numerical Simulations of Red-Blood Cells in Fluid Flow: A Discrete Multiphysics Study" by Rahmat et al. or "Using Discrete Multiphysics Modelling to Assess the Effect of Calcification on Hemodynamic and Mechanical Deformation of Aortic Valve" by Mohammed et al.). The ability of DMP to easily model solid–liquid flows accounting for particle agglomeration is highlighted in "Modelling Particle Agglomeration on through Elastic Valves under Flow" by Baksamawi et al. Finally, DMP can also be used to improve the mathematical modelling of viscoelastic materials: both viscoelastic solids (e.g., "A Coarse Grained Model for Viscoelastic Solids in Discrete

Multiphysics Simulations" by Sahputra et al.) and viscoelastic fluids (e.g., "A Simplified Framework for Modelling Viscoelastic Fluids in Discrete Multiphysics" by Duque-Daza and Alexiadis).

A feature that makes this book unusual and, hopefully, worth reading is its practical, hands-on approach. We wanted to go beyond the theory, and even beyond the application, of DMP. Several of the authors that contributed to the book have many years of experience with particle methods and we wanted to take advantage of this fact to give to the book a unique practical value. There are many tricks-of-the-trade and rules of good-practice that experienced researchers have developed over the years. This knowledge is often passed from supervisor to student, but is rarely shared in the traditional open literature. Certain chapters of this book (e.g., "How to Modify LAMMPS: From the Prospective of a Particle Method Researcher" by Albano et al.) are specifically based on this idea, while others (e.g., "Fortran Coarray Implementation of Semi-Lagrangian Convected Air Particles within an Atmospheric Model" by Rasmussen et al.) focus on the programming aspects from the perspective of the computer scientist. On this matter, we are very happy for the feedback we are received so far and, in particular, for comments of the type "I wished this article was available when I was a PhD student". This was an aspect I really wanted to highlight in the book and I am grateful to the publisher for giving me the freedom to do so.

I believe that both researchers that are new to particle methods and those that have years of experience will be interested in and benefit from this book. The former will find an introduction to several particle methods and how they can be coupled together. The latter will find several advanced applications of DMP that refer to research topics such as biology, atmospheric research or cavitation.

Another interesting aspect of the book, which is worth mentioning, is the number of authors and co-authors involved in the different chapters. They come from institutions distributed around the word, covering all five continents. It was a pleasure working with each one of them and having the opportunity to build collaborations that go beyond the publication of this book.

Alessio Alexiadis

Editor

 *chemengineering*

Article

A Coarse Grained Model for Viscoelastic Solids in Discrete Multiphysics Simulations

Iwan H. Sahputra [1,2,*], Alessio Alexiadis [1] and Michael J. Adams [1]

[1]   School of Chemical Engineering, University of Birmingham, Birmingham B15 2TT, UK;
      a.alexiadis@bham.ac.uk (A.A.); m.j.adams@bham.ac.uk (M.J.A.)
[2]   Industrial Engineering Department, Petra Christian University, Surabaya 60236, Indonesia
*   Correspondence: iwanh@petra.ac.id

Received: 10 March 2020; Accepted: 25 April 2020; Published: 1 May 2020

Abstract: Viscoelastic bonds intended for Discrete Multiphysics (DMP) models are developed to allow the study of viscoelastic particles with arbitrary shape and mechanical inhomogeneity that are relevant to the pharmaceutical sector and that have not been addressed by the Discrete Element Method (DEM). The model is applied to encapsulate particles with a soft outer shell due, for example, to the partial ingress of moisture. This was validated by the simulation of spherical homogeneous linear elastic and viscoelastic particles. The method is based on forming a particle from an assembly of beads connected by springs or springs and dashpots that allow the sub-surface stress fields to be computed, and hence an accurate description of the gross deformation. It is computationally more expensive than DEM, but could be used to define more effective interaction laws.

Keywords: Kelvin–Voigt viscoelastic bonds; coarse grained model; particle method; viscoelastic particles; inhomogeneous particles

---

1. Introduction

The Discrete Element Method (DEM) has been employed to study a range of pharmaceutical manufacturing processes and products including powder mixing [1], agglomeration with and without a liquid binder [2], and the release of Active Pharmaceutical Ingredients (APIs) from powder inhalation products [3]. Invariably, this has not involved inhomogeneous particles, and those of arbitrary shape have been simulated by gluing primary particles together such that the interior is essentially rigid in order to minimise the computational cost, which is not representative of real particles [4]. An important example of mechanical inhomogeneity is the softening of particles due the presence of moisture during agglomeration or dispersion/dissolution. In such cases, a gradient of moisture content is developed with a corresponding gradient in the mechanical properties. Another example is the encapsulation of APIs for which there is commonly a hard shell and a softer core. For particles formed from an organic polymer such as microcrystalline cellulose, the ingress of moisture will cause them to become viscoelastic.

Mesh-free methods and, in particular, particle methods such as DEM are increasingly popular in the scientific community due to their ability to overcome some drawbacks of the conventional, mesh-based, numerical methods; see [5] for a review. Particle methods can also be coupled together within a Discrete Multiphysics (DMP) framework that, unlike conventional multiphysics techniques, is based on "computational particles" rather than on computational meshes [6,7]. In fact, there is a range of systems for which DMP can address problems that would be very difficult, if not impossible, for traditional multiphysics approaches. Examples are cardiovascular valves [8,9], blood clotting [10], phase transitions [11], capsules' breakup [12,13], and fuzzy boundaries (e.g., a tablets' dissolution) [14]. In many of the above examples, the solid phase is often represented by a Lattice Spring Model (LMS) and involves both linear and non-linear springs for modelling elastic materials. In the current study, the method is extended to viscoelastic materials by implementing the Kelvin–Voigt (KV) viscoelastic model that involves springs and also dashpots to represent the viscous friction.

KV bonds have been proposed in the LSM literature, but only to model wave propagation in viscoelastic media (e.g., seismic wave propagation [15]), where the media are treated as homogenous and no external forces are applied to the system. KV bonds have never been implemented to study the strain field of solid objects under the effect of external loads. Achieving this objective would provide particle-based multiphysics techniques (e.g., DMP) with the ability to model viscoelastic materials, which is currently not possible.

The current study addresses the above shortcoming in the literature. For benchmark and validation purposes, the diametric compression of homogeneous spherical particles between parallel platens is described, which may be considered as a special case of indentation. A flat indenter or platen is widely used especially for the diametric compression of single particles [16] and microcapsules [17]. Generally, they are loaded at a constant velocity to a specified displacement and unloaded, or alternatively held in position, to measure the stress relaxation.

A quasistatic model based on Hertz's contact theory has been employed to describe the interaction between the particles that are packed together to represent unconsolidated porous media [18]. The evolution of the permeability with the deformation was computed by the lattice-Boltzmann approach. Here, the approach is that macroscopic bodies (such as particles) are sub-divided into computational beads. Each bead is connected to the nearest neighbours by linear springs or by KV bonds. It will be shown that for the spherical particle represented by beads connected by linear springs model, under diametric compression simulation, the relationship between force and displacement is nearly identical to the Hertz contact theory.

In the current work, the KV model is compared initially with the theoretical results for a single viscoelastic bond. Then, elastic and viscoelastic spherical particle models including multiple bonds are developed and simulated under diametrical loading. Finally, applications of DMP to spherical particles composed of core and shell regions with different properties are also presented to demonstrate the potential for inhomogeneous systems.

2. Materials and Methods

2.1. Theoretical Background

2.1.1. Hertz Theory for Elastic Normal Contact Force

Hertz proposed a theory to analyse the contact of two elastic isotropic spherical solids by assuming linear elasticity and frictionless boundary conditions [19]. For diametric compression, a spherical body is in contact with two flat surfaces, and the radius of curvature of the flat surfaces is set to infinity. Since the total deformation is evaluated, it is divided by two [20], and therefore, the relationship between the force, F_H, and the relative displacement of the plates, δ, is as follows:

$$F_H = \frac{E\sqrt{2R}}{3(1-v^2)}\delta^{3/2},\tag{1}$$

where E, R, and v are the Young's modulus, radius, and Poisson's ratio of the particle, respectively.

2.1.2. Viscoelastic Normal Contact Force

For the diametric compression of a spherical viscoelastic particle, the force may be partitioned between the elastic deformation and the viscoelastic dissipation, thus [21,22]:

$$F_{VE} = F_{elastic} + F_{dissipative} = A\delta^{3/2} + B\delta^{1/2}\dot{\delta},\tag{2}$$

where δ and $\dot{\delta}$ are the displacement and the rate of displacement, respectively. The elastic term is the Hertzian contact force where A is the constant in the Hertz theory. The dissipative part has a dissipative constant B that was derived independently in [21,23,24].

2.1.3. Mass-Spring-Dashpot Models

Figure 1 depicts two particles of mass m connected by a KV model, which is defined as a "KV" bond and implemented numerically as described in the next section. When a KV bond is displaced by a distance X from its equilibrium position, the resulting force is given by the following relationship:

$$F_{KV} = kX + b\frac{dX}{dt},\tag{3}$$

where k is the spring constant and b is the dashpot constant. If such a force is applied to the model, the displacement will be a function of time, t, as follows:

$$X(t) = \frac{F}{k}\left(1 - e^{-\frac{k \cdot t}{b}}\right),\tag{4}$$

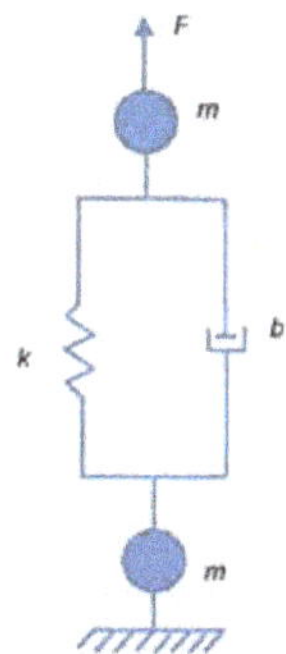

Figure 1. Two particles connected by a spring and a dashpot in parallel.

2.2. Model and Simulation

In this section, we initially compare the numerical implementation of the spring and dashpot model with the theoretical results for a single viscoelastic bond. Then, we extend the study to a large geometry (spherical) including multiple bonds.

2.2.1. Validation of a Single KV Bond

The KV bond was implemented numerically in LAMMPS [25] following the standard Hooke's law and Newton's law of fluid flow for the spring and dashpot, respectively, as shown in Equation (3). To validate the numerical implementation of the mass-spring-dashpot model, a simple system was created as shown in Figure 1, and the displacement was calculated from the simulations and compared to the analytical solution of Equation (4). The following parameter values were employed for the simulations: $F = 1$ N, $m = 0.00001$ kg, $k = 0.2$ N/m, and a range of values of b, as shown in Figure 2, where m is the mass of the beads. The simulated displacements are in close agreement with the analytical solution.

A second validation was performed by comparing the creep and recovery responses of the system in Figure 1 to that calculated using Simulink (Version 9.1, The MathWorks Inc., Natick, MA, USA). As depicted in Figure 3, the simulated displacements were in close agreement with the calculated values from Simulink. After the force was applied, the displacement increased rapidly until it reached a steady state. When the force was removed, the displacement decreased rapidly, and as time increased, it approached asymptotically to zero.

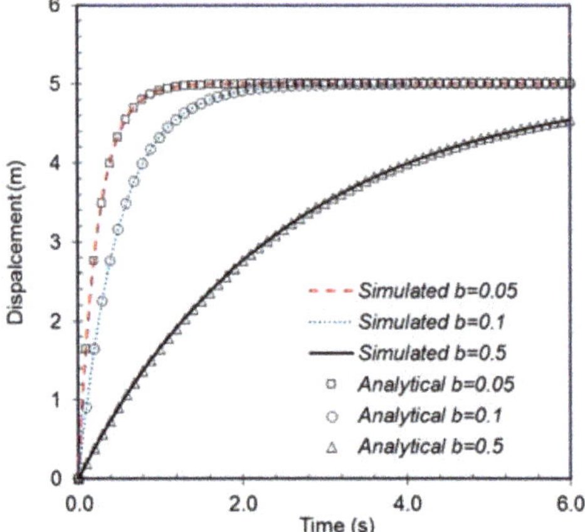

Figure 2. Comparison of the displacement calculated from the simulation and the analytical solution, i.e., Equation (4).

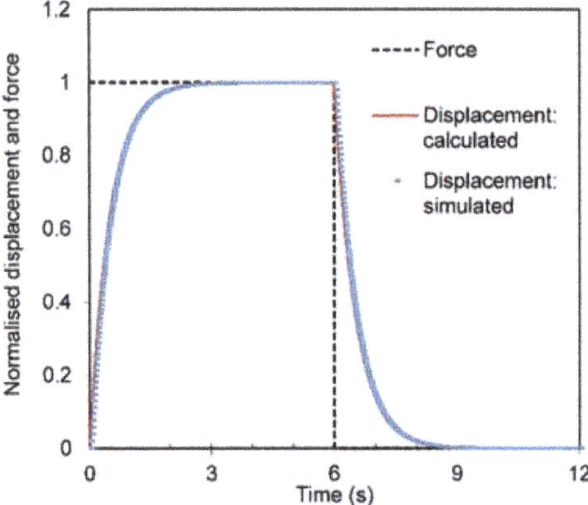

Figure 3. Response of the system depicted in Figure 1 to a constant force, which is removed after 6 s. The blue line is the displacement calculated from the simulation, and the red line is calculated using Simulink.

A third validation was performed by comparing the displacement response of the system in Figure 1 to a sinusoidal load (dynamic force), to the value calculated using Simulink (Version 9.1, The MathWorks Inc. Natick, MA, USA), as shown in Figure 4. Furthermore, in this case, the simulated displacements were in close agreement with the calculated values.

2.2.2. Modelling the Diametric Compression of a Spherical Particle

In DMP models, macroscopic bodies are sub-divided into computational particles (beads). Since in this work, we study KV bonds that can be used in DMP (or other particle-based multiphysics methods), we extended the validation to macroscopic spheres that accounted for multiple KV bonds. A sphere could be sub-divided into computational beads in different ways. Here, we employed two approaches: the beads were arranged on (a) a regular cubic lattice and (b) an irregular tetrahedral lattice.

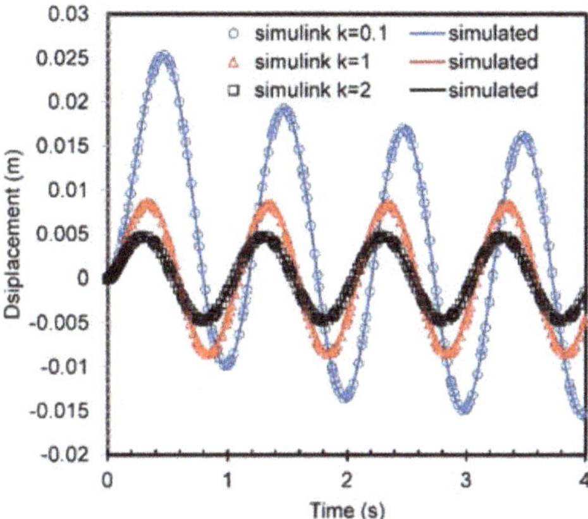

Figure 4. Response of the system depicted in Figure 1 to a sinusoidal loading force. The lines are the displacements calculated from the simulation, and the points are calculated using Simulink.

In the first case, the spherical particle (Figure 5a) was constructed from cubic lattice cells (Figure 5b). It contained 137,059 beads with each connected to the nearest neighbours and along face diagonals by linear springs or by KV bonds. This work focuses on viscoelasticity (KV bonds), and the case of purely elastic spheres (linear springs) was computed for comparison. The case of linear springs, in fact, has been already studied, and mass-spring cubic lattice cell models are known to represent (purely) elastic homogenous isotropic materials if the connection between the masses and the stiffness of the springs are selected appropriately [26]. For a cubic lattice cell with nearest neighbour and next nearest-neighbour linear springs, the Poisson's ratio is predicted by the theory to be 0.25 [26], and Young's modulus is given by the following relationship [27]:

$$E = 2.5 \, k/l \tag{5}$$

where l is the length of an edge of the cell. In the "Results and Discussion" (Section 3), our model will be initially validated against these theoretical values for a perfectly elastic sphere by only accounting for linear springs and, later, it will be extended to a viscoelastic sphere by substituting the springs with KV bonds.

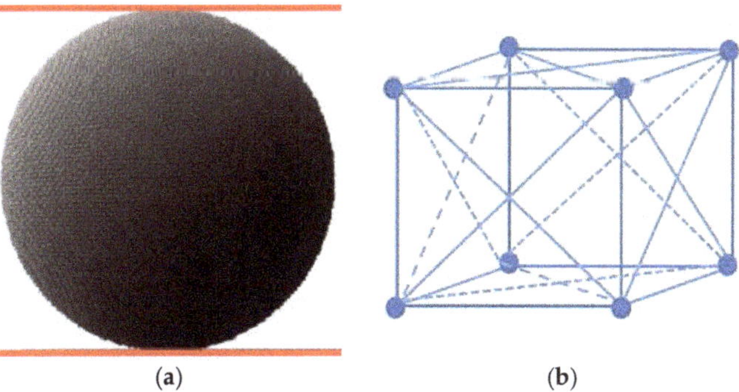

(a)   (b)

Figure 5. (**a**) Visualisation of a spherical particle between two parallel compression planes, which are represented by red lines; (**b**) an elementary cell of a cubic lattice.

The tetrahedral cells were created by discretising the sphere with a finite-element mesh generator. In this case, the distance between the beads was not perfectly uniform, and for this reason, we called it a disordered model. Using this approach, less beads were required, but the calculation of the elastic modulus a priori (Equation (5)) was less accurate [26]. A spherical particle based on a disorder model with 5921 beads was created using an open-source 3D finite element grid generator [28]. As in the case of the cubic lattice, the beads were connected to their neighbours either with linear springs or KV bonds to model, respectively, elastic and viscoelastic materials.

Two parallel solid planes were applied to the particle in order to simulate diametric compression. They exerted a force to compress the particle, where the magnitude of the force, $F(r)$, is given by [29]:

$$F(r) = S(r_b - R_i)^2, \tag{6}$$

where S is the specified force constant, R_i is the position of the plane and $r_b - R_i$ is the distance from the bead to the plane. The force is repulsive, and $F(r) = 0$ for $r_b > R_i$. The force constant was set to be 10^{10} Nm^{-2} for all simulations in order to represent rigid compression planes. During the compression loading simulations, one plane compressed the particle with a constant velocity for both the elastic and viscoelastic particle models, while the other was maintained static. For the viscoelastic particle, the displaced plane was held at its final position after the loading to allow for relaxation. The force and particle displacement were recorded during the simulations, and a time step of 10^{-11} s was used to integrate Newton's equations of motion.

It is well known that the KV model can produce the creep and recovery responses of a two-bead system, as shown in Figure 3, but cannot model stress relaxation behaviour. However, as will be shown in the next section, for the many-bead spherical particle models connected with KV bonds, stress relaxation behaviour could be observed. This is because a many-bead particle model connected with KV bonds is similar to a generalized KV model, i.e., a viscoelastic material model composed of N Kelvin–Voigt units assembled in series. The generalized KV model has been employed, for example to study the viscoelastic properties of micro-cracked materials [30].

3. Result and Discussion

3.1. Perfectly Elastic Spherical Particles

3.1.1. Cubic Lattice Cell Model

Figure 6a presents the simulated force as a function of displacement for an elastic spherical particle based on the cubic lattice cell with a spring constant of 200 Nm^{-1}. The data were compared against the Hertz theory predictions (Equation (1)) and the comparison depicted in Figure 6b. The force and displacement calculated from the simulations were nearly identical to the Hertz theory. The fluctuating behaviour in Figure 6 was due to slight numerical inaccuracies that artificially perturbed the total energy of the system. Since the particle was perfectly elastic, this energy was never dissipated and manifested itself as a high frequency perturbation. This is a known issue with the LSM, which, in the literature, is usually solved by adding a small artificial dissipative term that damps these high frequencies [31]. In this study, since the focus was on validation, we did not implement any artificial dissipation.

The calculated Young's modulus for the spherical particle was 39.1 MPa, which was in a close agreement with the Young's modulus of the elementary cubic lattice cell of 40 MPa calculated using Equation (5). The small discrepancy arose because, due to the cubic cell internal structure, the bead model was not a perfect spherical shape, so that it did not fully comply with the Hertzian contact model.

The bulk shear stress, which is the sub-surface principal stress difference, i.e., $|(\sigma_1 - \sigma_3)|/2$, may be calculated for each bead. The principal stresses (σ_1 and σ_3) were calculated from the virial stress and kinetic energy contributions [32] for each bead. The contours of the calculated shear stress are presented in Figure 7, where a is the contact radius and r is the particle radius. The shape of the contours was similar to that calculated theoretically [33,34]. The maximum value was found at a depth of 0.5a. This is in a close agreement with theoretical value of 0.48a [33].

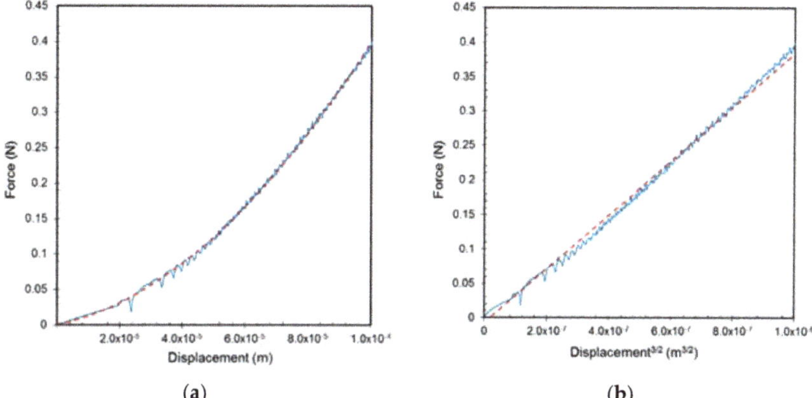

Figure 6. (**a**) Contact force as a function of displacement; (**b**) contact force as a function of displacement$^{3/2}$. Both were calculated from the simulations of an elastic spherical particle based on the cubic lattice cell model. The dashed lines are the best fits of the data to the Hertz theory.

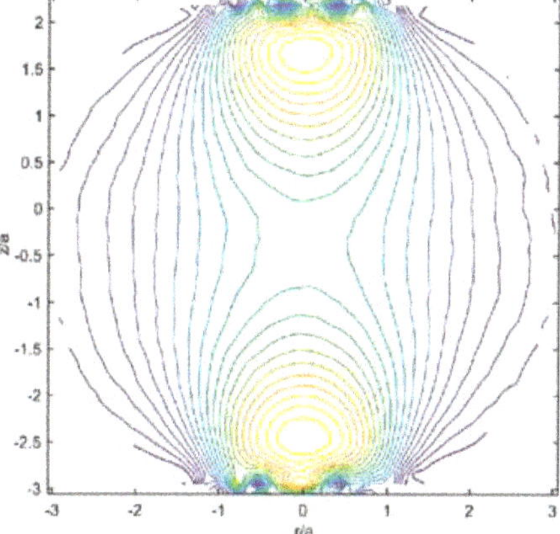

Figure 7. Contours of the sub-surface shear stresses estimated from the simulations using the cubic lattice cell.

3.1.2. Disorder Model

Figure 8a presents the normal contact force as a function of displacement calculated from the simulations using the disorder model with a spring constant of 200 Nm^{-1}. Although this involved a smaller number of beads, the data were a closer fit to a Hertzian response (Figure 8b), but with greater fluctuations of the force. As mentioned above, these fluctuations normally would be removed with an artificial dissipation term, but in this validation example, it is noteworthy that, as expected, disordered structures increased the amplitude of the perturbation. Using Equation (1) and assuming that $v = 0.25$, the Young's modulus was calculated to be 12.4 kPa.

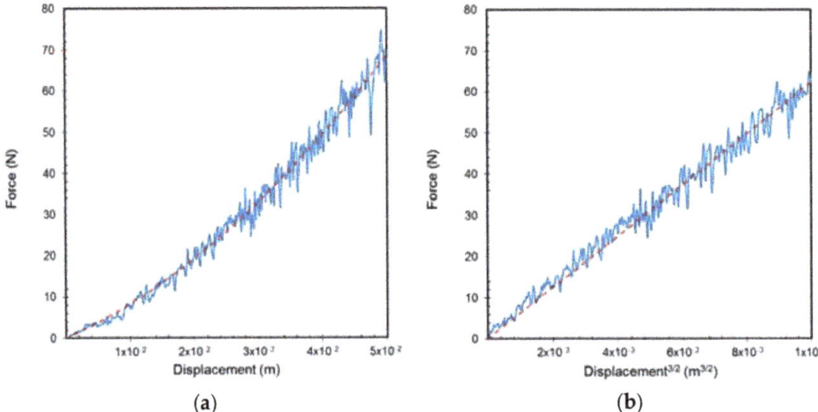

Figure 8. (**a**) Contact force as a function of displacement; (**b**) contact force as a function of displacement$^{3/2}$. Both were calculated from the simulations of an elastic spherical particle based on the disorder model. The dashed lines are the fits of the simulated data to the Hertz theory.

Figure 9 presents the contours of the calculated sub-surface shear stresses, for which due to the random location of the beads, the pattern was not similar in form to that of the cubic lattice cell model. However, the maximum value was also found at a depth of about $0.5a$ below the surface, which was similar to the cubic lattice model.

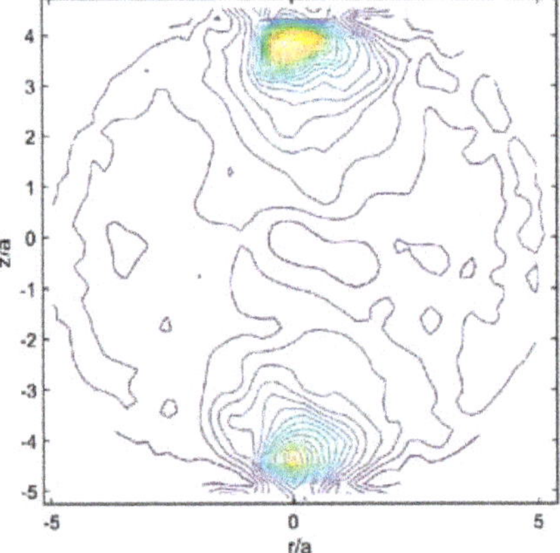

Figure 9. Contours of maximum shear stress beneath the particle surface estimated from the simulations of the particle based on the disorder model.

3.2. Viscoelastic Spherical Particles

3.2.1. Cubic Lattice Cell Model

Figure 10 presents the contact force as a function of time during compression and relaxation calculated from the simulations of a viscoelastic spherical model based on the cubic lattice cell during compression with a spring constant of 200 Nm^{-1} and a dashpot constant of 10^{-6} Nm^{-1} s. The force and displacement data were fitted to Equation (2) in order to obtain the value of A, and hence Young's modulus using Equation (1). It was found to be 36.8 MPa, which was slightly less than that for the elastic particle.

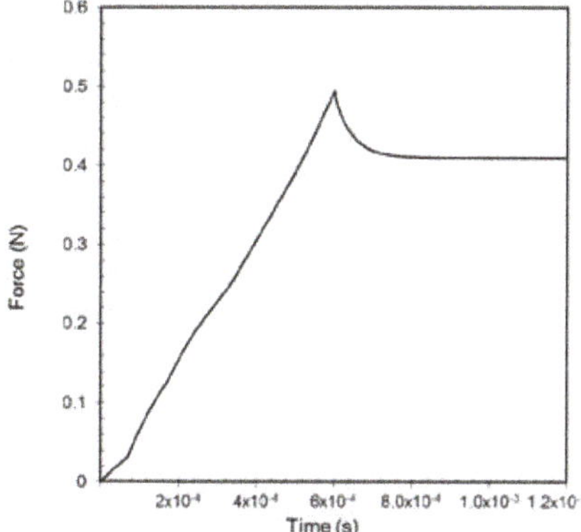

Figure 10. Contact force as a function of time during compression and relaxation calculated from the simulations of the viscoelastic spherical particle based on the cubic lattice cell model.

An analysis of the force relaxation after compression was performed using a previous method for experimental compression of an agarose micro-particle [35]. Instantaneous (E_0, corresponding to $t = 0$) and long-time (E_∞, corresponding to $t = \infty$) elastic moduli were then calculated. The values were found to be $E_0 = 54$ MPa and $E_\infty = 34$ MPa. The Hertzian Young's modulus was close to the calculated relaxed value. The relaxation times are $t_1 = 0.49$ s and $t_2 = 4.6 \times 10^{-5}$ s.

3.2.2. Disorder Model

Figure 11 presents the simulated contact force as a function of time for a viscoelastic spherical particle based on the disorder model with a spring constant of 200 Nm^{-1} and a dashpot constant of 10^{-6} Nm^{-1} s. The force and displacement data were fitted to Equation (2) in order to obtain the value of A, from which the Young's modulus was obtained using Equation (1) as 112 KPa, which was greater than the calculated value for the elastic particle.

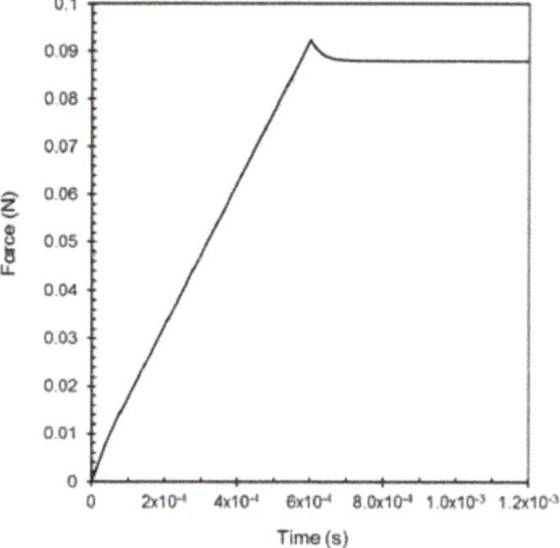

Figure 11. Contact force as a function of time during compression and relaxation calculated from the simulations of the viscoelastic spherical particle based on the disorder model.

An analysis of the force relaxation after compression was performed using the procedure used for the cubic lattice model. The elastic modulus values were found to be $E_0 = 203$ KPa and $E_\infty = 200$ KPa. The Hertzian Young's modulus was less than the calculated relaxed value, indicating that the force was not fully relaxed after compression. The relaxation time $t_1 = 0.82$ s and $t_2 = 2.83 \times 10^{-5}$ s. It is noteworthy that since the dashpot accounted for the physical viscosity of the material, artificial dissipation was not necessary in these examples.

3.3. Application of the Elastic Disorder Model: Hard Core-Soft Shell and Soft Core-Hard Shell Spherical Particles under Compression

In this section, we consider spherical inhomogeneous particles composed of a hard core-softer shell (HC-SS) and a soft core-harder shell (SC-HS). The models were based on a disorder lattice with 6065 beads. The beads were connected with springs to their neighbours. The hard regions of the particles were modelled using a larger spring constant than that for the soft part. The ratios of shell thickness, h, and the particle radius, r, were set to be 0.5, 0.2, and 0.05. A visualization of the shell thickness and particle radius is presented in Figure 12. A small artificial damping force (1^{-10} Nm^{-1}.s) was added to the beads, which was proportional to the relative velocity of the beads, in order to damp the kinetic energy from the system and obtain a smoother compression force. Two parallel solid planes were positioned on the surface of the particle in order to simulate diametric compression, as described in the previous section. Using Equation (1) and assuming that $v = 0.25$, the Young's modulus was calculated for each case.

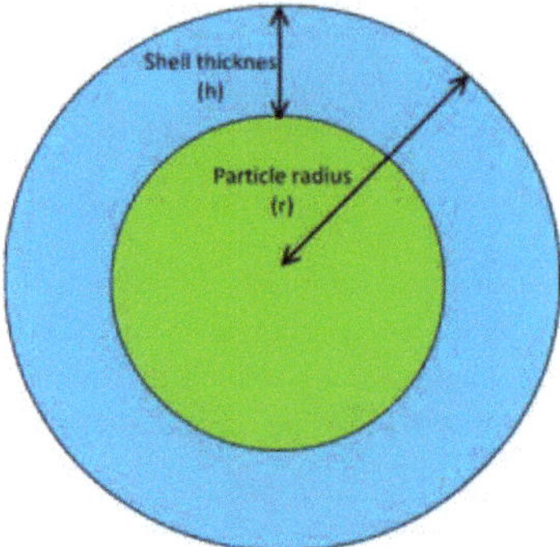

Figure 12. Illustration of shell thickness (h) and particle radius (r).

Figure 13 presents the force as a function of fractional deformation ($\delta/2r$) of HC-SS particles with different values of k_{core} and h/r. Figure 14 presents the force as a function of fractional deformation ($\delta/2r$) of SC-HS particles with different values of k_{shell} and h/r. The force profiles in Figures 13 and 14 can be compared with experimental data to simplify theoretical equations for small deformation e.g. for microcapsules [36].

By comparing Figures 13a and 14a, for the deformation up to 5%, it may be seen that the change in k_{shell} affected the deformation force more significantly than the change in k_{core}. Figure 13a shows that by reducing k_{shell} from 200 to 20 Nm^{-1}, the deformation force is now just about 11% of the initial value; while in Figure 14a, by reducing I_{core} from 200 to 0.2 Nm^{-1}, the force now is about 67% of the initial value. Thus, at small deformations, the compression load is mainly absorbed by the shell.

Increasing the h/r ratio from 0.05 to 0.2 changed the deformation force more significantly than increasing the h/r ratio from 0.2 to 1, as presented in Figures 13b and 14b for the deformation up to 5%. Increasing the h/r ratio from 0.5 to 1 did not change the force significantly. Moreover, it may be seen from Figures 15b and 16b that changing the h/r ratio from 0.5 to 1 did not change the lumped Young's modulus significantly. In these cases, the h/r ratio of 0.5 could be considered as a cut-off ratio where there would be no significant change to the modulus.

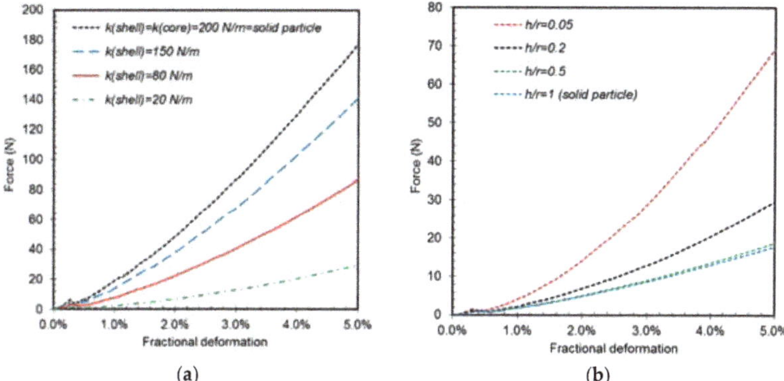

Figure 13. (**a**) The force as a function of fractional deformation ($\delta/2r$) of HC-SS particles ($h/r = 0.2$ and different k_{shell}) and a hard solid particle ($k_{shell} = k_{core} = 200\ \mathrm{Nm}^{-1}$). (**b**) The force as a function of fractional deformation ($\delta/2r$) of HC-SS particles ($k_{core} = 200\ \mathrm{Nm}^{-1}$ and $k_{shell} = 20\ \mathrm{Nm}^{-1}$) with different values of h/r and a soft solid particle ($h/r = 1$, $k_{core} = k_{shell} = 20\ \mathrm{Nm}^{-1}$).

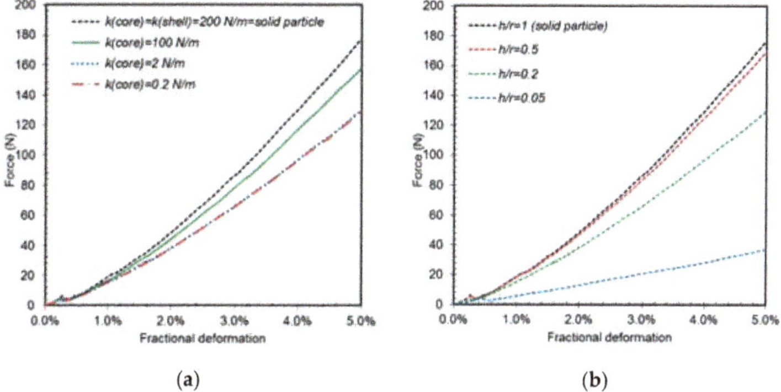

Figure 14. (**a**) The force as a function of fractional deformation ($\delta/2r$) of SC-HS particles ($h/r = 0.2$ and different k_{core}) and a hard solid particle ($k_{core} = k_{shell}$). (**b**) The force as a function of fractional deformation of SC-HS particles ($k_{core} = 2\ \mathrm{Nm}^{-1}$ $k_{shell} = 200\ \mathrm{Nm}^{-1}$) with different values of h/r and a hard solid particle ($h/r = 1$, $k_{core} = k_{shell} = 200\ \mathrm{Nm}^{-1}$).

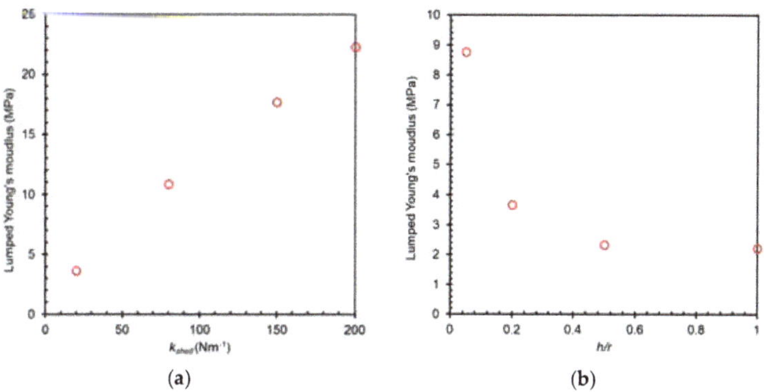

Figure 15. (**a**) Lumped Young's modulus of HC-SS particles of Figure 12a as a function of k_{shell}. (**b**) Lumped Young's modulus of HC-SS particles of Figure 12b as a function of h/r.

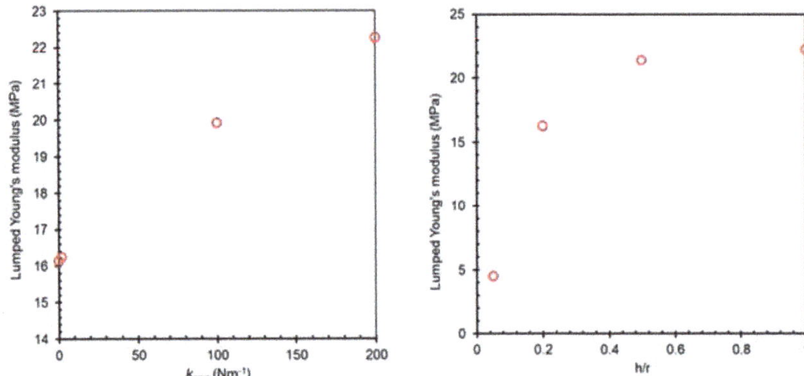

Figure 16. (**a**) Lumped Young's modulus of SC-HS particles of Figure 13a as a function of k_{core}. (**b**). Lumped Young's modulus of SC-HS particles of Figure 13b as a function of h/r.

Based on the results, a lumped Young's modulus could be calculated using Equation (1), which represented the equivalent Young's modulus the particle would have if it were homogeneous. Figures 15 and 16 present the lumped Young's moduli of HC-SS and SC-HS particles as a function of k_{shell} and k_{core}, respectively, and as a function of the h/r ratio. As expected, with increasing values of k_{shell} or k_{core}, the Young's modulus increases for HC-SS and SC-HS particles, respectively. For the HC-SS particle, increasing the h/r ratio decreased the Young's modulus, while for the SC-HS particle, it increased the value.

4. Conclusions

It has been demonstrated that the LSM could accurately represent the deformation, including the associated sub-surface stress fields, not only for elastic particles, but also for viscoelastic particles when linear springs were substituted with KV bonds. The disorder model was computationally more efficient than that based on a cubic lattice cell and led to a more refined definition of particle shape. Although only spherical particles were investigated in the current study, the approach is readily applicable to more complex shapes of the type that are often encountered in the pharmaceutical sector. The proposed technique could be employed, within a particle-based multiphysics model such as DMP, to model mechanical inhomogeneity, for example the softening of a particle immersed in water could be modelled by coupling the Young's modulus with the diffusion coefficient. It could also be extended to non-linear elastic deformation, plastic deformation, and fracture by introducing non-linear springs, friction elements, and springs of limited extensibility. For example, the fracture strength is of particular interest for encapsulates.

If compared with gluing DEM particles together to model different shapes, the proposed technique provided not only accurate contact stresses, but also the stresses within the particle. The disadvantage, however, was in the greater computational cost. However, it was considerably more convenient to implement than discrete finite elements, which require more complex material models. The proposed approach could also be adopted to develop effective interaction laws for inhomogeneous systems in DEM simulations to reduce the computational cost. It would also be possible to incorporate LSM particles in a DEM simulation.

Author Contributions: Conceptualization, I.H.S., A.A., and M.J.A.; methodology, I.H.S., A.A., and M.J.A.; software, I.H.S.; validation, I.H.S.; writing, original draft preparation, I.H.S; writing, review and editing, I.H.S., A.A. and M.J.A.; funding acquisition, A.A. and M.J.A. All authors read and agreed to the published version of the manuscript.

Funding: This research was funded by Engineering and Physical Sciences Research Council, Grant Number EP/M02959X/1.

Acknowledgments: The computations described in this paper were performed using the University of Birmingham's BlueBEAR HPC service, which provides a High Performance Computing service to the University's research community. See http://www.birmingham.ac.uk/bear for more details.

Conflicts of Interest: The authors declare no conflict of interest.

Nomenclature

a	contact radius
API	Active Pharmaceutical Ingredients
b	dashpot constant
δ	relative displacement
$\dot{\delta}$	rate of displacement
DEM	Discrete Element Method
DMP	Discrete Multiphysics
E	Young's modulus
F	force
$F(r)$	force (diametric compression)
F_E	viscoelastic normal contact force
F_H	force (Hertz theory)
F_{KV}	force (Kelvin–Voigt model)
h	shell thickness
HC-SS	hard core-softer shell
k	spring constant
KV	Kelvin–Voigt
l	length of an edge of the cell
LSM	Lattice Spring Model
m	mass
r	radius
r_b-R_i	distance from the bead to the compression plane
R_i	position of the compression plane
σ_1, σ_3	principal stresses
S	specified force constant
SC-HS	soft core-harder shell
υ	Poisson's ratio
X	distance from equilibrium position

References

1. Alian, M.; Ein-Mozaffari, F.; Upreti, S.R. Analysis of the mixing of solid particles in a plowshare mixer via discrete element method (DEM). *Powder Technol.* **2015**, *274*, 77–87. [CrossRef]
2. Ketterhagen, W.R.; am Ende, M.T.; Hancock, B.C. Process modeling in the pharmaceutical industry using the discrete element method. *J. Pharm. Sci.* **2009**, *98*, 442–470. [CrossRef] [PubMed]
3. Yang, J.; Wu, C.-Y.; Adams, M. DEM analysis of the effect of particle–wall impact on the dispersion performance in carrier-based dry powder inhalers. *Int. J. Pharm.* **2015**, *487*, 32–38. [CrossRef] [PubMed]
4. Lu, G.; Third, J.; Müller, C. Discrete element models for non-spherical particle systems: From theoretical developments to applications. *Chem. Eng. Sci.* **2015**, *127*, 425–465. [CrossRef]
5. Garg, S.; Pant, M. Meshfree methods: A comprehensive review of applications. *Int. J. Comput. Methods* **2018**, *15*, 1830001. [CrossRef]
6. Alexiadis, A. A smoothed particle hydrodynamics and coarse-grained molecular dynamics hybrid technique for modelling elastic particles and breakable capsules under various flow conditions. *Int. J. Numer. Methods Eng.* **2014**, *100*, 713–719. [CrossRef]
7. Alexiadis, A. The discrete multi-hybrid system for the simulation of solid-liquid flows. *PLoS ONE* **2015**, *10*, e0124678. [CrossRef]

8. Ariane, M.; Allouche, M.H.; Bussone, M.; Giacosa, F.; Bernard, F.; Barigou, M.; Alexiadis, A. Discrete multi-physics: A mesh-free model of blood flow in flexible biological valve including solid aggregate formation. *PLoS ONE* **2017**, *12*, e0174795. [CrossRef]

9. Ariane, M.; Allouche, M.H.; Bussone, M.; Giacosa, F.; Bernard, F.; Barigou, M.; Alexiadis, A. Modelling and simulation of flow and agglomeration in deep veins valves using discrete multi physics. *Comput. Biol. Med.* **2017**, *89*, 96–103. [CrossRef]

10. Ariane, M.; Vigolo, D.; Brill, A.; Nash, F.G.B.; Barigou, M.; Alexiadis, A. Using Discrete Multi-Physics for studying the dynamics of emboli in flexible venous valves. *Comput. Fluids* **2018**, *166*, 57–63. [CrossRef]

11. Alexiadis, A.; Ghraybeh, S.; Qiao, G. Natural convection and solidification of phase-change materials in circular pipes: A SPH approach. *Comput. Mater. Sci.* **2018**, *150*, 475–483. [CrossRef]

12. Alexiadis, A. A new framework for modelling the dynamics and the breakage of capsules, vesicles and cells in fluid flow. *Procedia IUTAM* **2015**, *16*, 80–88. [CrossRef]

13. Rahmat, A.; Barigou, M.; Alexiadis, A. Deformation and rupture of compound cells under shear: A discrete multiphysics study. *Phys. Fluids* **2019**, *31*, 051903. [CrossRef]

14. Rahmat, A.; Barigou, M.; Alexiadis, A. Numerical simulation of dissolution of solid particles in fluid flow using the SPH method. *Int. J. Numer. Methods Heath Fluid Flow* **2019**. [CrossRef]

15. O'Brien, G.S. Discrete visco-elastic lattice methods for seismic wave propagation. *Geophys. Res. Lett.* **2008**, *35*. [CrossRef]

16. Paul, J.; Romeis, S.; Tomas, J.; Peukert, W. A review of models for single particle compression and their application to silica microspheres. *Adv. Powder Technol.* **2014**, *25*, 136–153. [CrossRef]

17. Kuo-Kang, L. Deformation behaviour of soft particles: A review. *J. Phys. D Appl. Phys.* **2006**, *39*, R189.

18. Bakhshian, S.; Sahimi, M. Computer simulation of the effect of deformation on the morphology and flow properties of porous media. *Phys. Rev. E* **2016**, *94*, 042903.

19. Hertz, H. Ueber die Berührung fester elastischer Körper. *J. Reine Angew. Math.* **1882**, *1882*, 156–171.

20. Mook, W.M.; Nowak, J.D.; Perrey, C.R.; Carter, C.B.; Mukherjee, R.; Girshick, S.L.; Gerberich, W.W. Compressive stress effects on nanoparticle modulus and fracture. *Phys. Rev. B* **2007**, *75*, 214112. [CrossRef]

21. Brilliantov, N.V.; Spahn, F.; Hertzsch, J.M.; Pöschel, T. Model for collisions in granular gases. *Phys. Rev. E* **1996**, *53*, 5382–5392. [CrossRef] [PubMed]

22. Schwager, T.; Pöschel, T. Coefficient of restitution for viscoelastic spheres: The effect of delayed recovery. *Phys. Rev. E* **2008**, *78*, 051304. [CrossRef] [PubMed]

23. Zheng, Q.J.; Zhu, H.P.; Yu, A.B. Finite element analysis of the contact forces between a viscoelastic sphere and rigid plane. *Powder Technol.* **2012**, *26*, 130–142. [CrossRef]

24. Kuwabara, G.; Kono, K. Restitution Coefficient in a Collision between Two Spheres. *Jpn. J. Appl. Phys.* **1987**, *26*, 1230–1233. [CrossRef]

25. Plimpton, S. Fast Parallel Algorithms for Short-Range Molecular Dynamics. *J. Comput. Phys.* **1995**, *117*, 1–19. [CrossRef]

26. Ladd, A.J.; Kinney, J.H.; Breunig, T.M. Deformation and failure in cellular materials. *Phys. Rev. E* **1997**, *55*, 3271. [CrossRef]

27. Kot, M.; Nagahashi, H.; Szymczak, P. Elastic moduli of simple mass spring models. *Vis. Comput.* **2015**, *31*, 1339–1350. [CrossRef]

28. Geuzaine, C.; Remacle, J.-F. Gmsh: A 3-D finite element mesh generator with built-in pre- and post-processing facilities. *Int. J. Numer. Methods Eng.* **2009**, *79*, 1309–1331. [CrossRef]

29. Kelchner, C.L.; Plimpton, S.; Hamilton, J. Dislocation nucleation and defect structure during surface indentation. *Phys. Rev. B* **1998**, *58*, 11085. [CrossRef]

30. Nguyen, S. Generalized Kelvin model for micro-cracked viscoelastic materials. *Eng. Fract. Mech.* **2014**, *127*, 226–234. [CrossRef]

31. Chen, H.; Lin, E.; Liu, Y. A novel Volume-Compensated Particle method for 2D elasticity and plasticity analysis. *Int. J. Solids Struct.* **2014**, *51*, 1819–1833. [CrossRef]

32. Thompson, A.P.; Plimpton, S.J.; Mattson, W. General formulation of pressure and stress tensor for arbitrary many-body interaction potentials under periodic boundary conditions. *J. Chem. Phys.* **2009**, *131*, 154107. [CrossRef] [PubMed]

33. Johnson, K.L. *Contact Mechanics*; Cambridge University Press: Cambridge, UK, 1987.

34. Lee, S.C.; Ren, N. The Subsurface Stress Field Created by Three-Dimensionally Rough Bodies in Contact with Traction. *Tribol. Trans.* **1994**, *37*, 615–621. [CrossRef]

35. Yan, Y.; Zhang, Z.; Stokes, J.R.; Zhou, Q.Z.; Ma, G.H.; Adams, M.J. Mechanical characterization of agarose micro-particles with a narrow size distribution. *Powder Technol.* **2009**, *192*, 122–130. [CrossRef]

36. Butt, H.-J.; Cappella, B.; Kappl, M. Force measurements with the atomic force microscope: Technique, interpretation and applications. *Surf. Sci. Rep.* **2005**, *59*, 1–152. [CrossRef]

Article

Using Discrete Multiphysics Modelling to Assess the Effect of Calcification on Hemodynamic and Mechanical Deformation of Aortic Valve

Adamu Musa Mohammed [1,2,*], Mostapha Ariane [3] and Alessio Alexiadis [1]

[1] School of Chemical Engineering, University of Birmingham, Birmingham B15 2TT, UK;
    A.Alexiadis@bham.ac.uk
[2] Department of Chemical Engineering, Faculty of Engineering and Engineering Technology, Abubakar
    Tafawa Balewa University, Bauchi 740272, Nigeria
[3] Department of Materials and Engineering, Sayens—University of Burgundy, 21000 Dijon, France;
    Mostapha.Ariane@u-bourgogne.fr
* Correspondence: amm702@bham.ac.uk or ammohd@atbu.edu.ng; Tel.: +44-(0)776-717-3356

Received: 2 April 2020; Accepted: 29 July 2020; Published: 3 August 2020

Abstract: This study proposes a 3D particle-based (discrete) multiphysics approach for modelling calcification in the aortic valve. Different stages of calcification (from mild to severe) were simulated, and their effects on the cardiac output were assessed. The cardiac flow rate decreases with the level of calcification. In particular, there is a critical level of calcification below which the flow rate decreases dramatically. Mechanical stress on the membrane is also calculated. The results show that, as calcification progresses, spots of high mechanical stress appear. Firstly, they concentrate in the regions connecting two leaflets; when severe calcification is reached, then they extend to the area at the basis of the valve.

Keywords: discrete multiphysics modelling; smoothed particle hydrodynamics; lattice spring model; particle-based method; aortic valve; calcification; stenosis

---

1. Introduction

Aortic valve disease is the malfunction of the aortic valve due to heart malformation at birth (congenital) or developed during a lifetime related to injury, age, or calcification of the valve [1]. Calcification, in particular, may result in calcific aortic valve disease (CAVD) caused by calcium deposits on the valve leaflets, which mainly affects the elderly population with an incidence rate of 2 –7% in the population above 65 years of age [2]. Over time, calcium build-up makes the aortic valve stiffer, preventing full opening (stenosis) and hindering the blood flow from the left ventricle to the aorta. It may also prevent the valve from closing properly (regurgitation), resulting in blood leakages back to the ventricle.

Stenosis starts with the risk of leaflet deformation and progresses from early lesions to valve obstruction, which is initially mild to moderate but eventually becomes severe, with or without clinical symptoms [3] and the patient has high risk of cardiac failure.

In the literature, several studies focus on the dynamics of blood flow and the deformation of the calcified aortic valve leaflets to better understand and assess the severity of calcification. Computational fluid dynamics (CFD) was used to generate patient-specific aortic valve models from patients' medical images in [4,5]. Other studies [6–10] focus on investigating the stresses on the valve leaflets, excluding the fluid flow behaviour in the model. Since the behaviour of the aortic valve depends on both the fluid and the leaflets, a fluid–structure interaction (FSI) approach is recommended to model the complex dynamic of this problem [2,11].

In this paper, we aim to use discrete multiphysics (DMP) [12–14], which enables the fluid–solid interaction, to develop a 3D model representing various stages of calcification of the aortic valve. DMP has been extensively used in in silico medicine for modelling a variety of human systems including the aortic valve [15], the intestine [16], deep venous valves [17,18], the lungs [19], and, in conjunction with machine learning, peristalsis in the oesophagus [20]. Sections 2 and 3 give an overview of the method and model used, respectively.

2. Discrete Multiphysics

Multiphysics simulation allows for tackling problems that involve multiple physical models or simultaneous physical phenomena, and that gives it widespread application in both industry and academia [20]. Discrete multiphysics, therefore, is a method based on particles and that combines various particle methods such as smoothed particle hydrodynamics (SPH), lattice spring model (LSM), and discrete element method [15]. The current model only accounts for SPH and LSM. Both modelling techniques are briefly discussed below. The reader can refer to [21] for a more extensive introduction to SPH, and to [22] for the LSM. Details on how the two models are linked together in DMP can be found in [12,13].

2.1. Smooth Particle Hydrodynamics

Smoothed particle hydrodynamics (SPH) was invented to simulate nonaxisymmetric phenomena in astrophysics by [23,24] and to solve the astrophysical problems in three-dimensional open space [24]. It is a meshless Lagrangian particle method used to solve differential equations which are often found in engineering and scientific problems [25,26]. The method, being easy to work with a reasonable degree of accuracy, could be extended to complicated physics without much trouble [27]. The SPH method is gaining popularity in a variety of fields and has a wide range of applications such as multiphase flow [25,28], biomedicals [15,17,19,29], fluid–solid interaction [12,29–33], and hydrodynamic instability [34,35].

The SPH method represents the state of a system with particles which possess individual material properties and move according to a governing equation [36]. The fundamental idea behind this discrete approximation lies in the mathematical identity

$$f(r) = \iiint f(r')\delta(r - r')dr' \tag{1}$$

which gives a discretized particle approximation within a domain of a smoothing function (kernel) W within a characteristic width h called the smoothing length [12] such that

$$\lim_{h \to 0} W(r,\, h) = \delta(r) \tag{2}$$

This gives rise to the discrete approximation

$$f(r) \approx \iiint f(r')W(r - r',\, h)dr' \tag{3}$$

which can be discretised over a series of particle of mass m_i and density ρ_i to obtain

$$f(r) \approx \sum_i \frac{m_i}{\rho_i} f(r_i)W(r - r_i,\, h) \tag{4}$$

where $f(r)$ is a generic function defined over the volume V, r is the three-dimensional vector point, and $\delta(r)$ is the three-dimensional delta function.

Equation (4) is a representation of the discrete approximation of a generic continuous field. Further simplification of Equation (4) can be done to approximate the Navier–Stokes equation. See detail and complete equations in [14,15]. In addition, more information on the SPH and equation derivation can also be found in [21,35,37].

2.2. Lattice Spring Model (LSM)

The behaviour of deformable objects can be simulated using the LSM approach. This model consists of mass point and linear springs which exert forces at the nodes located at the end point [38], which are placed on either a regular lattice or positioned randomly within the system [38]. The term LSM is interchangeable with mass spring model (MSM), where the latter is mostly used in computer graphics and the former in solid mechanics. Nowadays, LSM is applicable in many areas and disciplines, for instance, cloth simulation, face animation, or soft tissue behaviour in surgery training systems [39,40] are simulated using LSM.

The LSM follows the idea that any material point of the body can be referred to by its position vector $r = (x,y,z)$ [38]. When that body undergoes deformation due to an applied force F, the displacement is proportional to the force applied.

$$F = k(r - r_0) \tag{5}$$

where r_0 is the initial distance between the two particles, r is the distance at time t, and k is a Hookean constant

3. The Model

As mentioned, the multiphysics model used in this study is based on DMP, a computational method that combines various particle methods. In the case under investigation, the computational domain is divided into two parts: (i) a liquid part (blood) where smooth particle hydrodynamics (SPH) is used, and (ii) the solid structure part (valve) where the stress deformation equation is solved using the lattice spring model (LSM). The two parts are coupled together to represent an FSI model that simulates the valve deformation and blood flow dynamics within the valve.

In Figure 1a, the valve's leaflets (tricuspid) are shown; Figure 1b shows the overall geometry. The geometry was firstly designed as a CAD model (nodes and elements) and then transformed to a particle model as explained in the Appendix A. Simulations were run with the open-source software LAMMPS [41].

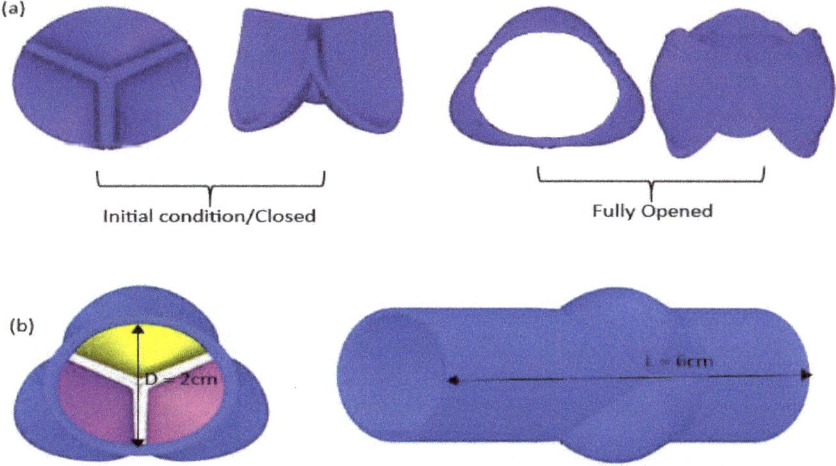

Figure 1. Valve leaflets (**a**) and complete geometry (**b**).

The system is three-dimensional and consists of 418,743 particles: 342,358 particles for the fluid, 19,725 particles for the leaflets, and 56,660 particles for the rigid pipe. The distance between particles (particle spacing) and the number of particles were chosen from the previous work done where DMP was used to model similar FSI problems [29,32]. In this study, the aortic valve geometry of Bavo et al. (2016) was adopted. The leaflets tissue was modelled as linear and elastic, whereas the elasticity of the aorta (rigid pipe) was neglected and the arterial wall was assumed to be rigid [42].

In the simulation, the flow is driven by a periodic acceleration (Figure 2) given by

$$G = G_0 \sin^n(\omega t) \cos(\omega t - \phi) \tag{6}$$

where $G_0 = 400$ m s^{-2}, $\omega = 2\pi f$ is the angular frequency, $n = 13$, and $\phi = \pi/10$ as discussed in Steven et al., 2003 [40]. The value of G_0 is determined to achieve full opening of the valve that gives an average flow rate around 600 mL s^{-1} for valves at normal condition, which is consistent with the literature, e.g., [40,43,44].

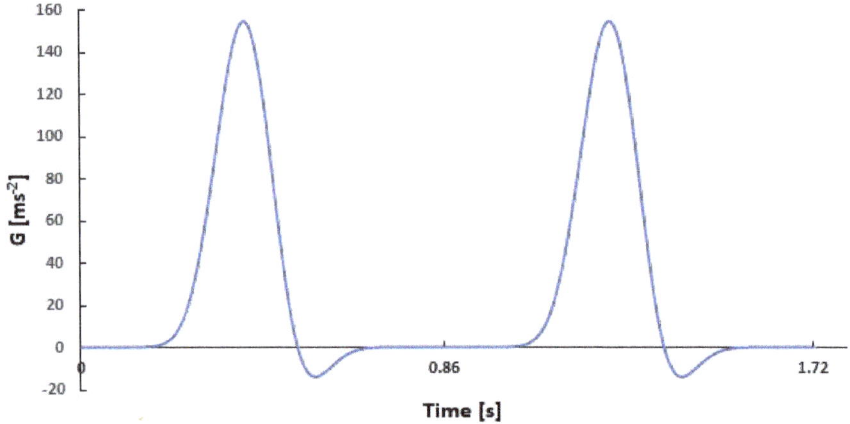

Figure 2. Pulsatile flow function.

As mentioned, the flexible valve is modelled with LSM, which is a common approach for cardiovascular valves [45]. This implies that each computational particle is linked with its neighbour particles by means of a force (in Equation (5)).

In LSM, the value of k is linked to the Young modulus E of the material [38]. In practice, however, it is not always straightforward to calculate k from E in the case of complex geometries, in particular, with irregular particles distribution [39]. For this reason, we use a more practical approach: k is determined, together with G_0, in such a way that the valve opens fully and the flow rate for a healthy and non-calcified valve is 600 mL s^{-1} as discussed above.

The numerical model used in this paper has extensively been tested and validated for a variety of similar fluid flow problems [15–18,32]. In this paper, the value of k in the lattice model, as mentioned earlier, and G_0 in Equation (6) were chosen in such a way to make sure the flow in the valve and its opening are consistent with available observations (e.g., [40,43,44,46,47]). Convergence of the results on the number of particles used to discretise the system was carried out and the numbers reported in Table 1 represent the best compromise between accuracy and computational times. Since we refer to the human body, the available observations of real valves show very scattered results because of individual variability. Therefore, we cannot provide a systematic validation like in the case of experiments carried out in a laboratory under controlled conditions. This is even more true for calcified valves, where, on top of physiological variations, we have additional variations due to the severity and the course of the disease. Given all of the above, the best we can do is to make sure that the results are consistent with real data. In our study, we achieved this by making sure that the flow in the valve and the opening of the leaflets are both within the range of real observations.

Table 1. Model parameters used in the simulation; for the meaning of SPH parameters such as α or h, refer to Liu and Liu, 2003 [21].

Parameters	Values
Number of SPH wall particle	56,660
Number of SPH fluid particle	342,358
Number of SPH leaflets	19,725
Mass of each particle (fluid)	6.7×10^{-8} kg
Mass of each particle (Solid)	14×10^{-8} kg
Smoothing length h	1.0×10^{-3} m
Length L	6×10^{-2} m
Diameter D	2×10^{-2} m
Particle spacing l	0.4×10^{-3} m
Fluid Density ρ	1060 kg m^{-3}
Frequency f	1.167 s^{-1} (70 beats min^{-1})
Pseudo-gravity G_0	400 m s^{-2}
Viscosity μ	0.003 Pa·s
Elastic constant k	10–14,500 N m^{-1}
Sound speed c_0	16 m s^{-1}
Time step Δt	1×10^{-6} s

4. Results and Discussion

4.1. Stages of Calcification

In our model, the value of k was chosen to control the stiffness of the valve and model calcification. The higher the value of k, the stiffer the valve. In our model, higher values of k are used to model a higher degree of calcification. We define the degree of calcification γ as

$$\gamma = \log\left(\frac{k}{k_H}\right) \tag{7}$$

where k is the spring constant used to simulate the calcified valve and k_H is the stiffness of the healthy valve. Figure 3 shows the valve during maximal opening for four different degrees of calcification.

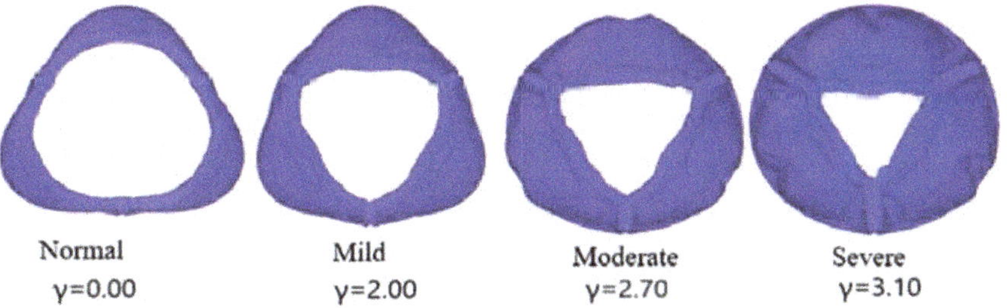

Figure 3. Severity of aortic valve stenosis in terms of orifice opening (stenosis).

Some important parameters like flow velocity, volume flow rate, and stress can be used to ascertain the level of calcification of the valves. Figure 4 shows how the flow velocity in the valve is affected by calcification.

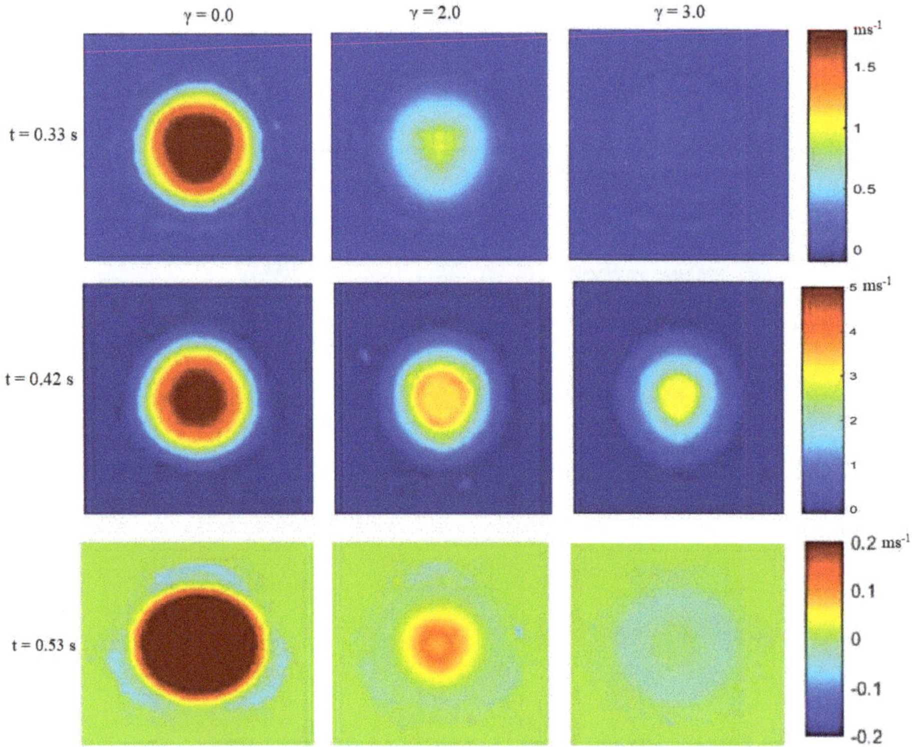

Figure 4. Velocity profile at different time steps.

The flow velocity was measured on a section just above the valve for three different degrees of calcification and three different times. According to the level of calcification, the valve opens and closes at different times. The healthy valve ($\gamma = 0$) starts opening at $t = 0.29$ s, reaches maximal opening (peak systole) at $t = 0.41$ s, and closes again at $t = 0.56$ s. The valve with $\gamma = 2.0$ starts opening at $t = 0.32$ s, reaches maximal opening at $t = 0.43$ s, and closes again at $t = 0.54$ s. The valve with $\gamma = 3.0$ starts opening at $t = 0.37$ s, reaches maximal opening at $t = 0.43$ s, and closes again at $t = 0.53$ s. Figure 4 also relates with Figure 3 in showing how the valve opening is reduced (stenosis) in case of calcification. The time to attain peak velocity varies with the severity of the valve's stenosis, which signifies high mortality risk and the need for a valve replacement [48]. The time at which the valve closes also varies with the severity of calcification; at $\gamma = 3.0$, there is also evidence of regurgitation (back flow), a characteristic of a stenotic aortic valve [2,3,9]. Figure 4 also shows that, as expected, the blood flow in the healthy valve ($\gamma = 0.0$) is higher than in the calcified valves ($\gamma = 2.0$ and $\gamma = 3$).

Transvalvular flow is another important parameter for measuring aortic valve functionality. The flow reduces as γ increases (Figure 5); it decreases almost linearly up to a critical value $\gamma_{CR} = 3$, after which it decreases sharply. In Figure 5, γ_{CR} corresponds to a flow rate of 200 mL s^{-1}. Our calculations are consistent with the medical literature where a flow rate $\geq$250 mL s^{-1} is considered acceptable, whereas <200 mL s^{-1} is associated with an increase of mortality rate in cases of patients with aortic stenosis [44,49]. The maximum orifice diameter and the average stress on the valve (see next section) can also be used to monitor the progression of the valve stenosis. Table 2 summarizes the parameters as the condition worsens from normal to severe.

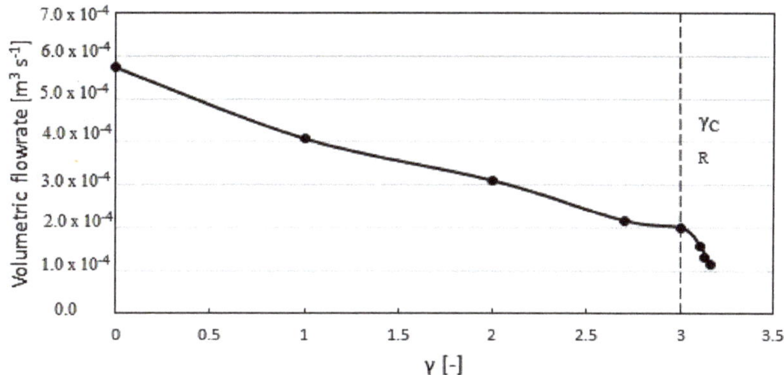

Figure 5. Volumetric blood flow with respect to γ.

Table 2. Parameters for determining severity of calcification.

Calcification	Maximum Orifice Diameter [cm]	Mean Flow $\times 10^{-4}$ [m^{3}s^{-1}]	Average Stress [kPa]
Normal ($\gamma = 0.0$)	1.81	5.72	10.60
Mild ($\gamma = 2.0$)	1.41	3.21	91.68
Moderate ($\gamma = 2.7$)	1.29	2.17	181.61
Severe ($\gamma = 3.1$)	1.17	1.58	324.27

4.2. Stress Distribution on the Membrane

Calcification increases the stiffness of the valve, which results in higher stresses on the membrane. This is particularly important because mechanical stress plays a major role in the calcification of bioprostheses [50]. This, potentially, can create a vicious circle where calcification leads to higher mechanical stress, which, in turn, leads to further calcification. Figure 6 shows how the average stress increases with the degree of calcification. Since stenosis due to calcification can be related to congenital bicuspid valve disease or fused leaflets [51], the numerical values of 0.28 to 0.35 MPa obtained for calcified valves are consistent with those reported by [46] for fused tri-leaflet valves. For stress distribution in a transcatheter aortic valve, the maximum value of 0.35 MPa is also consistent with the report of Auricchio et al. (2014) [47].

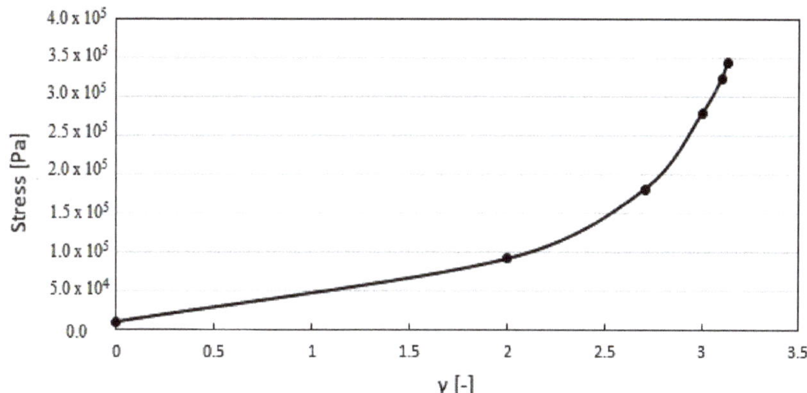

Figure 6. Stress at different degrees of calcification.

The local stress distribution (i.e., the Frobenius norm of the stress tensor) on the leaflets is shown in Figure 7, considering a membrane thickness of 0.6 mm [7]. In this work, calcification is modelled by uniformly increasing the Young modulus of the membrane; Figure 7 suggests that calcification does not progress uniformly.

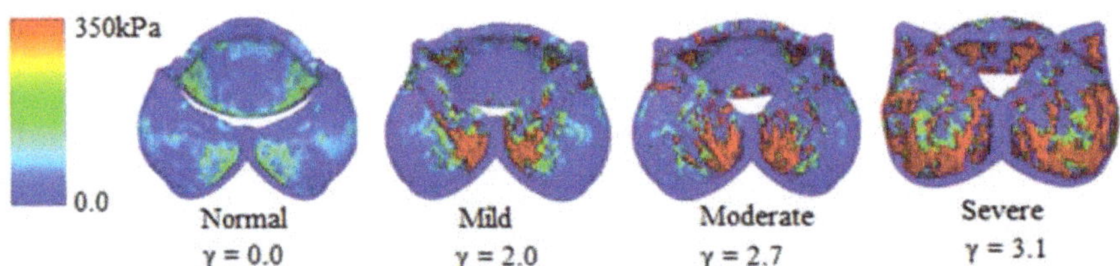

Figure 7. Mechanical stress on the valve leaflets at maximal opening.

In fact, as calcification progresses, spots of high mechanical stress appear. Firstly, they concentrate in the regions connecting two leaflets; when severe calcification is reached, they extend to the area at the basis of the valve.

5. Conclusions

In this study, the effect of calcification in a 3D aortic valve is simulated with discrete multiphysics. The model accounts for both hemodynamics and leaflet deformation, and it can be considered an improvement over a previous 2D model [15]. The results show that the mean transvalvular flow could be used to assess valve calcification, and severe calcification occurs when the flow rate is lower than 200 mL s^{-1}.

The model can also assess the local stress on the membrane. Calcification increases mechanical stress, which, in turn, promotes further calcification. In this work, calcification is modelled by uniformly increasing the Young modulus of the membrane. The results show that, as calcification progresses, spots of high mechanical stress appear. Firstly, they concentrate in the regions connecting two leaflets; when severe calcification is reached, they extend to the area at the basis of the valve. This suggests that the model could be improved by accounting for local changes in stiffness, which depends on the local stress distribution.

Methodologically, this work can also benefit researchers interested in particle methods, but not necessarily to cardiovascular applications. In fact, the Appendix A explains how geometries designed with CAD model can be transformed to particle models.

Author Contributions: Conceptualization, A.A., A.M.M., and M.A.; simulation and visualization, A.M.M.; numerical calculations, A.M.M.; interpretation and analysis of results, A.M.M. and A.A.; writing—original draft preparation, A.M.M.; supervision, A.A.; writing—review and editing, A.A., A.M.M., and M.A.; input script, M.A. and A.A.; model and appendix, M.A. All authors have read and agreed to the published version of the manuscript.

Funding: This research received no external funding.

Acknowledgments: The Nigerian Petroleum Technology Development Fund (PTDF) is acknowledged for the provision of a scholarship to Adamu Musa Mohammed.

Conflicts of Interest: The authors declare no conflict of interest.

Appendix A

This appendix explains how to adapt the 3D CAD model for Discrete Multiphysics. The valve geometry is taken from [42] and was initially created with the help of CAD software. In general, particle distribution can be generated with a programming code or a pre-processing solver. In the first case, the coordinates of the points are created with a separate standard programming code such as C++ and MATLAB, or directly in the processing solver with an integrated algorithm (in the software

LAMMPS for instance). For complex geometries, a pre-processing commercial builder (GAMBIT, ABAQUS, SALOME) can be used for the structure and the associated mesh is then replaced with particles (Figure A1). In the paper, we used the second approach in order to design our tricuspid valve system.

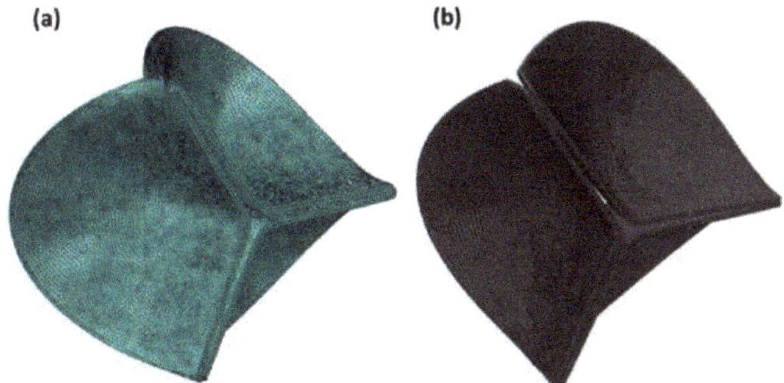

Figure A1. Mesh generation using a pre-processing solver (a) and the generated particle distribution (b).

The procedure of implementation is simple, but it requires a couple of sequences to obtain the final input file. Here, we present the one used for our valve model, but the methodology can be extended to any other applications.

The different steps can be described as follow:

1. Creation or importation of the CAD geometry
2. Writing of the data file
3. Generation of the bond and coefficient files
4. Implementation of the input file

The first step consists of designing the part geometry or importing an existing one to any CAD commercial or open-source software (Figure A2a). The idea is to use the software capability for automatically or manually generating the mesh (elements and nodes, Figure A2b). Next, the nodes information (numbers and coordinates) are downloaded (Figure A2c) and collected into a text file (Figure A2d).

Subsequently, the data file is adapted and implemented in a LAMMPS format file, which contains specific line codes and keywords (for more details and documentation, readers can refer to the LAMMPS documentation available at https://lammps.sandia.gov/doc/Manual.html). For instance, for this model, we used the keywords *meso* for the atom style and *bond* for the inter-atomic potential.

The unsymmetrical shape of the tricuspid valve system makes a 3D design necessary. This representation requires a fine mesh with a consequent number of nodes (418,743 particles) to ensure a realistic valve motion. Each node/particle is connected with at least three other nodes located nearby, increasing the data processing drastically. As a result, we developed an algorithm (C++ code) in order to set the bond definitions automatically. All particle positions are scanned by the code and for each particle, its neighbour (within a prescribed radius distance of interaction) is identified, numbered, and printed out in a bond file. Meanwhile, a second file is also created with the storage of the bond coefficients (potential force) and the distance (between two particles).

Finally, as the number of fluid particles is very important (342,358), the inclusion of the list into the data file (which is technically possible) makes the input file processing heavy. Instead, we preferred to generate, during the first step of the simulation, the fluid particles via the LAMMPS commands *'create_atoms'* and *'region'*. The command *'delete'* is also used to avoid overlapping.

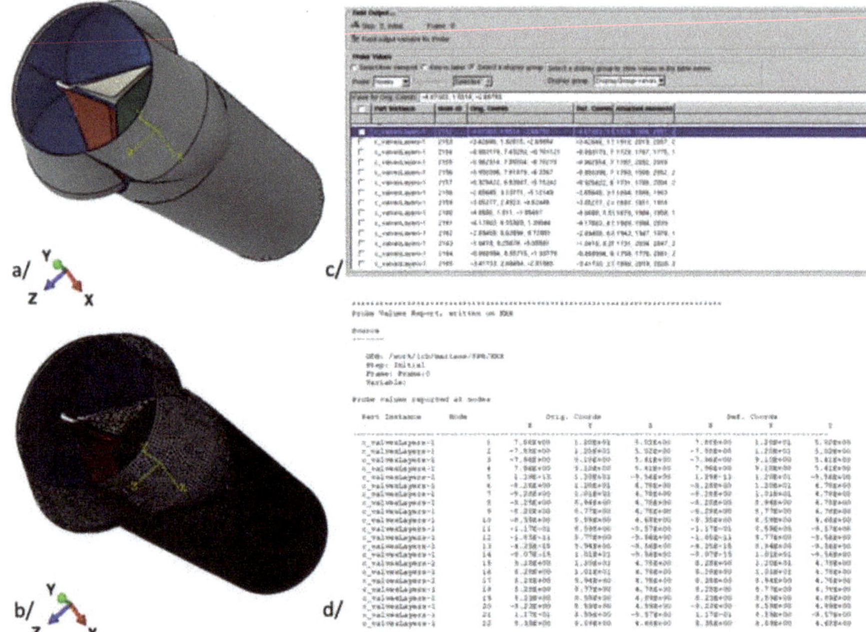

Figure A2. Tricuspid valve CAD part (**a**), tricuspid valve meshed CAD part (**b**), nodes generation using the CAD software (**c**), and external data file (**d**).

References

1. Fioretta, E.S.; Dijkman, P.E.; Emmert, M.Y.; Hoerstrup, S.P. The Future of Heart Valve Replacement: Recent Developments and Translational Challenges for Heart Valve Tissue Engineering: The Future of Heart Valve Replacement. *J. Tissue Eng. Regen. Med.* **2018**, *12*, e323–e335. [CrossRef] [PubMed]
2. Amindari, A.; Saltik, L.; Kirkkopru, K.; Yacoub, M.; Yalcin, H.C. Assessment of Calcified Aortic Valve Leaflet Deformations and Blood Flow Dynamics Using Fluid-Structure Interaction Modeling. *Inform. Med. Unlocked* **2017**, *9*, 191–199. [CrossRef]
3. Otto, C.M.; Prendergast, B. Aortic-Valve Stenosis—From Patients at Risk to Severe Valve Obstruction. *N. Engl. J. Med.* **2014**, *371*, 744–756. [CrossRef] [PubMed]
4. Fedele, M.; Faggiano, E.; Dedè, L.; Quarteroni, A. A Patient-Specific Aortic Valve Model Based on Moving Resistive Immersed Implicit Surfaces. *Biomech. Modeling Mechanobiol.* **2017**, *16*, 1779–1803. [CrossRef] [PubMed]
5. Youssefi, P.; Gomez, A.; He, T.; Anderson, L.; Bunce, N.; Sharma, R.; Figueroa, C.A.; Jahangiri, M. Patient-Specific Computational Fluid Dynamics—Assessment of Aortic Hemodynamics in a Spectrum of Aortic Valve Pathologies. *J. Thorac. Cardiovasc. Surg.* **2017**, *153*, 8–20. [CrossRef] [PubMed]
6. Bluestein, D.; Einav, S. The Effect of Varying Degrees of Stenosis on the Characteristics of Turbulent Pulsatile Flow through Heart Valves. *J. Biomech.* **1995**, *28*, 915–924. [CrossRef]
7. Hamid, M.S.; Sabbah, H.N.; Stein, P.D. Comparison of Finite Element Stress Analysis of Aortic Valve Leaflet Using Either Membrane Elements or Solid Elements. *Comput. Struct.* **1985**, *20*, 955–961. [CrossRef]
8. Arjunon, S.; Rathan, S.; Jo, H.; Yoganathan, A.P. Aortic Valve: Mechanical Environment and Mechanobiology. *Ann. Biomed. Eng.* **2013**, *41*, 1331–1346. [CrossRef]
9. Morganti, S.; Conti, M.; Aiello, M.; Valentini, A.; Mazzola, A.; Reali, A.; Auricchio, F. Simulation of Transcatheter Aortic Valve Implantation through Patient-Specific Finite Element Analysis: Two Clinical Cases. *J. Biomech.* **2014**, *47*, 2547–2555. [CrossRef]

10. Lazaros, G.; Drakopoulou, M.I.; Tousoulis, D. Transaortic Flow in Aortic Stenosis: Stroke Volume Index versus Flow Rate. *Cardiology* **2018**, *141*, 71–73. [CrossRef]

11. Ghasemi Bahraseman, H.; Mohseni Languri, E.; Yahyapourjalaly, N.; Espino, D.M. Fluid-Structure Interaction Modeling of Aortic Valve Stenosis at Different Heart Rates. *Acta Bioeng. Biomech.* **2016**, *18*, 11–20. [CrossRef]

12. Alexiadis, A. The Discrete Multi-Hybrid System for the Simulation of Solid-Liquid Flows. *PLoS ONE* **2015**, *10*, e0124678. [CrossRef] [PubMed]

13. Alexiadis, A. A Smoothed Particle Hydrodynamics and Coarse-Grained Molecular Dynamics Hybrid Technique for Modelling Elastic Particles and Breakable Capsules under Various Flow Conditions: SPH-CGMD HYBRID. *Int. J. Numer. Meth. Eng.* **2014**, *100*, 713–719. [CrossRef]

14. Alexiadis, A. A New Framework for Modelling the Dynamics and the Breakage of Capsules, Vesicles and Cells in Fluid Flow. *Procedia IUTAM* **2015**, *16*, 80–88. [CrossRef]

15. Ariane, M.; Allouche, M.H.; Bussone, M.; Giacosa, F.; Bernard, F.; Barigou, M.; Alexiadis, A. Discrete Multi-Physics: A Mesh-Free Model of Blood Flow in Flexible Biological Valve Including Solid Aggregate Formation. *PLoS ONE* **2017**, *12*, e0174795. [CrossRef]

16. Alexiadis, A.; Stamatopoulos, K.; Wen, W.; Batchelor, H.K.; Bakalis, S.; Barigou, M.; Simmons, M.J.H. Using Discrete Multi-Physics for Detailed Exploration of Hydrodynamics in an In Vitro Colon System. *Comput. Biol. Med.* **2017**, *81*, 188–198. [CrossRef]

17. Ariane, M.; Vigolo, D.; Brill, A.; Nash, F.G.B.; Barigou, M.; Alexiadis, A. Using Discrete Multi-Physics for Studying the Dynamics of Emboli in Flexible Venous Valves. *Comput. Fluids* **2018**, *166*, 57–63. [CrossRef]

18. Ariane, M.; Wen, W.; Vigolo, D.; Brill, A.; Nash, F.G.B.; Barigou, M.; Alexiadis, A. Modelling and Simulation of Flow and Agglomeration in Deep Veins Valves Using Discrete Multi Physics. *Comput. Biol. Med.* **2017**, *89*, 96–103. [CrossRef]

19. Ariane, M.; Kassinos, S.; Velaga, S.; Alexiadis, A. Discrete Multi-Physics Simulations of Diffusive and Convective Mass Transfer in Boundary Layers Containing Motile Cilia in Lungs. *Comput. Biol. Med.* **2018**, *95*, 34–42. [CrossRef]

20. Alexiadis, A. Deep Multiphysics: Coupling Discrete Multiphysics with Machine Learning to Attain Self-Learning in-Silico Models Replicating Human Physiology. *Artif. Intell. Med.* **2019**, *98*, 27–34. [CrossRef]

21. Liu, G.R.; Liu, M.B. *Smoothed Particle Hydrodynamics: A Meshfree Particle Method*; World Scientific: Hackensack, NJ, USA, 2003.

22. Kot, M.; Nagahashi, H.; Szymczak, P. Elastic Moduli of Simple Mass Spring Models. *Vis. Comput.* **2015**, *31*, 1339–1350. [CrossRef]

23. Gingold, R.A.; Monaghan, J.J. Smoothed Particle Hydrodynamics: Theory and Application to Non-Spherical Stars. *Mon. Not. R. Astron. Soc.* **1977**, *181*, 375–389. [CrossRef]

24. Lucy, L.B. A Numerical Approach to the Testing of the Fission Hypothesis. *Astron. J.* **1977**, *82*, 1013. [CrossRef]

25. Hopp-Hirschler, M.; Shadloo, M.S.; Nieken, U. A Smoothed Particle Hydrodynamics Approach for Thermo-Capillary Flows. *Comput. Fluids* **2018**, *176*, 1–19. [CrossRef]

26. Shadloo, M.S.; Zainali, A.; Sadek, S.H.; Yildiz, M. Improved Incompressible Smoothed Particle Hydrodynamics Method for Simulating Flow around Bluff Bodies. *Comput. Methods Appl. Mech. Eng.* **2011**, *200*, 1008–1020. [CrossRef]

27. Monaghan, J.J. Smoothed Particle Hydrodynamics. *Annu. Rev. Astron. Astrophys.* **1992**, *30*, 543–574. [CrossRef]

28. Rahmat, A.; Yildiz, M. A Multiphase ISPH Method for Simulation of Droplet Coalescence and Electro-Coalescence. *Int. J. Multiph. Flow* **2018**, *105*, 32–44. [CrossRef]

29. Schütt, M.; Stamatopoulos, K.; Simmons, M.J.H.; Batchelor, H.K.; Alexiadis, A. Modelling and Simulation of the Hydrodynamics and Mixing Profiles in the Human Proximal Colon Using Discrete Multiphysics. *Comput. Biol. Med.* **2020**, *121*, 103819. [CrossRef]

30. Ji, S.; Chen, X.; Liu, L. Coupled DEM-SPH Method for Interaction between Dilated Polyhedral Particles and Fluid. *Math. Probl. Eng.* **2019**, *2019*, 1–11. [CrossRef]

31. Li, D.; Zhen, Z.; Zhang, H.; Li, Y.; Tang, X. Numerical Model of Oil Film Diffusion in Water Based on SPH Method. *Math. Probl. Eng.* **2019**, *2019*, 1–14. [CrossRef]

32. Rahmat, A.; Barigou, M.; Alexiadis, A. Deformation and Rupture of Compound Cells under Shear: A Discrete Multiphysics Study. *Phys. Fluids* **2019**, *31*, 051903. [CrossRef]

33. Tran-Duc, T.; Phan-Thien, N.; Khoo, B.C. A Smoothed Particle Hydrodynamics (SPH) Study of Sediment Dispersion on the Seafloor. *Phys. Fluids* **2017**, *29*, 083302. [CrossRef]
34. Rahmat, A.; Tofighi, N.; Shadloo, M.S.; Yildiz, M. Numerical Simulation of Wall Bounded and Electrically Excited Rayleigh–Taylor Instability Using Incompressible Smoothed Particle Hydrodynamics. *Colloids Surf. A Physicochem. Eng. Asp.* **2014**, *460*, 60–70. [CrossRef]
35. Shadloo, M.S.; Yildiz, M. Numerical Modeling of Kelvin-Helmholtz Instability Using Smoothed Particle Hydrodynamics. *Int. J. Numer. Meth. Engng.* **2011**, *87*, 988–1006. [CrossRef]
36. Liu, M.B.; Liu, G.R. Restoring Particle Consistency in Smoothed Particle Hydrodynamics. *Appl. Numer. Math.* **2006**, *56*, 19–36. [CrossRef]
37. Shadloo, M.S.; Zainali, A.; Yildiz, M. Simulation of Single Mode Rayleigh–Taylor Instability by SPH Method. *Comput. Mech.* **2013**, *51*, 699–715. [CrossRef]
38. Pazdniakou, A.; Adler, P.M. Lattice Spring Models. *Transp. Porous. Med.* **2012**, *93*, 243–262. [CrossRef]
39. Lloyd, B.; Szekely, G.; Harders, M. Identification of Spring Parameters for Deformable Object Simulation. *IEEE Trans. Visual. Comput. Graph.* **2007**, *13*, 1081–1094. [CrossRef]
40. Stevens, S.A.; Lakin, W.D.; Goetz, W. A Differentiable, Periodic Function for Pulsatile Cardiac Output Based on Heart Rate and Stroke Volume. *Math. Biosci.* **2003**, *182*, 201–211. [CrossRef]
41. Plimpton, S. Fast Parallel Algorithms for Short-Range Molecular Dynamics. *J. Comput. Phys.* **1995**, *117*, 1–19. [CrossRef]
42. Bavo, A.M.; Rocatello, G.; Iannaccone, F.; Degroote, J.; Vierendeels, J.; Segers, P. Fluid-Structure Interaction Simulation of Prosthetic Aortic Valves: Comparison between Immersed Boundary and Arbitrary Lagrangian-Eulerian Techniques for the Mesh Representation. *PLoS ONE* **2016**, *11*, e0154517. [CrossRef] [PubMed]
43. Murgo, J.P.; Westerhof, N.; Giolma, J.P.; Altobelli, S.A. Aortic Input Impedance in Normal Man: Relationship to Pressure Wave Forms. *Circulation* **1980**, *62*, 105–116. [CrossRef] [PubMed]
44. Blais, C.; Burwash, I.G.; Mundigler, G.; Dumesnil, J.G.; Loho, N.; Rader, F.; Baumgartner, H.; Beanlands, R.S.; Chayer, B.; Kadem, L.; et al. Projected Valve Area at Normal Flow Rate Improves the Assessment of Stenosis Severity in Patients With Low-Flow, Low-Gradient Aortic Stenosis: The Multicenter TOPAS (Truly or Pseudo-Severe Aortic Stenosis) Study. *Circulation* **2006**, *113*, 711–721. [CrossRef]
45. Hammer, P.E.; Sacks, M.S.; del Nido, P.J.; Howe, R.D. Mass-Spring Model for Simulation of Heart Valve Tissue Mechanical Behavior. *Ann. Biomed. Eng.* **2011**, *39*, 1668–1679. [CrossRef] [PubMed]
46. Jermihov, P.N.; Jia, L.; Sacks, M.S.; Gorman, R.C.; Gorman, J.H.; Chandran, K.B. Effect of Geometry on the Leaflet Stresses in Simulated Models of Congenital Bicuspid Aortic Valves. *Cardiovasc. Eng. Tech.* **2011**, *2*, 48–56. [CrossRef] [PubMed]
47. Auricchio, F.; Conti, M.; Morganti, S.; Reali, A. Simulation of Transcatheter Aortic Valve Implantation: A Patient-Specific Finite Element Approach. *Comput. Methods Biomech. Biomed. Eng.* **2014**, *17*, 1347–1357. [CrossRef] [PubMed]
48. Kamimura, D.; Hans, S.; Suzuki, T.; Fox, E.R.; Hall, M.E.; Musani, S.K.; McMullan, M.R.; Little, W.C. Delayed Time to Peak Velocity Is Useful for Detecting Severe Aortic Stenosis. *JAHA* **2016**, *5*. [CrossRef]
49. Saeed, S.; Senior, R.; Chahal, N.S.; Lønnebakken, M.T.; Chambers, J.B.; Bahlmann, E.; Gerdts, E. Lower Transaortic Flow Rate Is Associated with Increased Mortality in Aortic Valve Stenosis. *JACC Cardiovasc. Imaging* **2017**, *10*, 912–920. [CrossRef]
50. Thubrikar, M.; Deck, J.; Aouad, J.; Nolan, S. Role of Mechanical Stress in Calcification of Aortic Bioprosthetic Valves. *J. Thorac. Cardiovasc. Surg.* **1983**, *86*, 115–125. [CrossRef]
51. Huntley, G.D.; Thaden, J.J.; Alsidawi, S.; Michelena, H.I.; Maleszewski, J.J.; Edwards, W.D.; Scott, C.G.; Pislaru, S.V.; Pellikka, P.A.; Greason, K.L.; et al. Comparative Study of Bicuspid vs. Tricuspid Aortic Valve Stenosis. *Eur. Heart J. Cardiovasc. Imaging* **2018**, *19*, 3–8. [CrossRef]

 *chemengineering*

Article

Modelling Complex Particle–Fluid Flow with a Discrete Element Method Coupled with Lattice Boltzmann Methods (DEM-LBM)

Wenwei Liu and Chuan-Yu Wu *

Department of Chemical and Process Engineering, University of Surrey, Guildford GU2 7XH, UK; wenwei.liu@surrey.ac.uk
* Correspondence: c.y.wu@surrey.ac.uk

Received: 1 July 2020; Accepted: 29 September 2020; Published: 7 October 2020

Abstract: Particle–fluid flows are ubiquitous in nature and industry. Understanding the dynamic behaviour of these complex flows becomes a rapidly developing interdisciplinary research focus. In this work, a numerical modelling approach for complex particle–fluid flows using the discrete element method coupled with the lattice Boltzmann method (DEM-LBM) is presented. The discrete element method and the lattice Boltzmann method, as well as the coupling techniques, are discussed in detail. The DEM-LBM is thoroughly validated for typical benchmark cases: the single-phase Poiseuille flow, the gravitational settling and the drag force on a fixed particle. In order to demonstrate the potential and applicability of DEM-LBM, three case studies are performed, which include the inertial migration of dense particle suspensions, the agglomeration of adhesive particle flows in channel flow and the sedimentation of particles in cavity flow. It is shown that DEM-LBM is a robust numerical approach for analysing complex particle–fluid flows.

Keywords: particle–fluid flow; discrete element method; lattice Boltzmann method

---

1. Introduction

Particulate flows are extensively encountered in nature and industrial processes, attracting tremendous engineering research interests in almost all areas of sciences [1–3], including astrophysics, chemical engineering, biology, life science and so on. Due to the complex particle–particle interactions and their interactions with the surrounding media, i.e., gas/liquid environment or electrostatic/magnetic field, the dynamic behaviour of particulate flow is very complicated. Therefore, it is of significant importance to understand the particle dynamics, in order to promote relevant manufacture and production processes.

With the rapid development of the modern modelling technique, abundant numerical studies on the particulate flows have sprung up in recent years. As the name suggests, the particulate flow can behave like a continuous fluid or a rigid solid. Therefore, studies of particulate flows can be classified into three categories based on the types of approach: (1) continuum modelling, which extends the continuum mechanics of single-phase fluid to describe the particle transportation, leading to a representative population balance method [4,5]; (2) developing the kinetic theory of particulate flow [6] based on the averaged equations for multi-particle systems, which generalise the dynamics of particle–particle collision processes; (3) discrete particle modelling, which solves the particle's motion individually based on certain interaction laws. The third approach is also classified as Lagrangian particle method and has many different variations, including the discrete element method (DEM), dissipative particle dynamics (DPD), molecular dynamics (MD) and Brownian dynamics (BD). These discrete particle methods differentiate themselves with different particle–particle interaction laws and their corresponding length and time scales [3]. DEM is widely used for modelling of granular materials and adhesive particulate flows [7–17], and was first introduced by Cundall and Strack [18] to study the rock mechanics. In DEM, particles are treated as individual entities, and interact with each other through contact and non-contact forces. The translational and rotational motions of the particles are described using the Newton's second law. The greatest advantage of DEM is that a large amount of dynamic information, which is almost unattainable experimentally, such as the transient forces and torques, can be obtained in great detail.

Apart from the particle–particle interaction, the hydrodynamic interaction associated with the surrounding fluid is also an important issue in the study of particulate flow. Various computational fluid dynamics (CFD) methods at different scales of time and length are developed to model the single-phase fluid flow, ranging from discrete models, e.g., the lattice Boltzmann method (LBM) [19–21], to continuum models, e.g., the direct numerical simulation (DNS) [22], the large eddy simulation (LES) [23] and the two-fluid model (TFM) [24]. Hence, a large number of hybrid models are developed for modelling particle–fluid flow [20,25–33]. A comprehensive comparison of various hybrid models was discussed in the literature [1,34].

Comparing with other conventional CFD approaches, LBM arises in the last two decades and has become a promising numerical method due to several advantages [19,20]. First, the primary computation procedure, i.e., the so-called collision operator, is performed locally with information only from where the computation takes place, facilitating straightforward implementation in a parallel way with high computational efficiency. Second, the LBM shows significant flexibility in handling complex boundary conditions, as they are treated with a local bounce-back fashion instead of the conventional boundary treatment. Third, the lattice grid adopted in LBM is traditionally orthogonal and smaller than the particle size, providing a high space resolution on the evaluation of the hydrodynamic interactions. As a result, the LBM was coupled with DEM and widely applied in analysing a large range of particle–fluid problems, such as particle suspensions [35–38], flow through porous media [39,40], multicomponent flows [20,41,42], particle-laden turbulent flows [31,43,44] and magnetohydrodynamics [45].

In this paper, a numerical approach to model complex particle–fluid flow using a coupled DEM-LBM is introduced. The fundamental equations of DEM and LBM, as well as various coupling techniques are presented in detail. The validity and accuracy of the numerical approach are demonstrated with some benchmark tests. Finally, case studies, including the migration, the agglomeration and the sedimentation of particle suspensions, are presented to illustrate the potential and applicability of the numerical approach. It should be noted that the interactions between particles and the surrounding environment, such as electrostatic interactions, liquid bridge effect and so on, are not within the scope of this work, but can be found in the literature [46–54].

2. Discrete Element Method (DEM)

In DEM, particle motion is solved individually using the Newton's equation of motion,

$$m\frac{d\mathbf{U}_p}{dt} = \mathbf{G} + \sum \mathbf{F},$$
$$I\frac{d\mathbf{\Omega}_p}{dt} = \sum \mathbf{M}, \tag{1}$$

where $\mathbf{U}_p$ and $\mathbf{\Omega}_p$ are the translational and rotational velocities of a particle, respectively. m and I are the mass and rotational inertia of the particle. $\mathbf{G}$ is the gravitational force. $\mathbf{F}$ and $\mathbf{M}$ represent the force and torque acting on the particle, which could come from various sources, such as interparticle collision, hydrodynamic interactions with the fluid, electrostatic interactions with charge and so on. As DEM was designed to deal with the numerous collision problems between the distinct particles, the collision forces and torques are the most important aspect. When two particles get in contact with each other, the collision force and torque can be decomposed as

$$\mathbf{F}_{collision} = F_n\mathbf{n} + F_s\mathbf{t}_s,$$
$$\mathbf{M}_{collision} = r_p F_s(\mathbf{n} \times \mathbf{t}_s) + M_r(\mathbf{t}_r \times \mathbf{n}) + M_t\mathbf{n}, \tag{2}$$

where F_n and F_s denote the contact forces in the normal and tangential direction, respectively. M_r and M_t are the rolling and twisting resistance torques, respectively. r_p represents the particle radius $\mathbf{n}$, $\mathbf{t}_s$ and $\mathbf{t}_r$ are the unit vectors in the normal, tangential and rolling direction at the contact point, respectively.

2.1. Normal Contact Force

The most common model for the normal contact force is the Hertz model [55], which can be expressed by a non-linear Hook's law. The Hertz model was validated and widely applied in the contact of spherical particles with size above millimeters. However, as the particle approaches the micrometer size range, the adhesion effect due to the van der Waals force must be taken into consideration. Extensions of the Hertz model, including the well-known Johnson–Kendall–Roberts (JKR) theory [56] and Derjaguin–Muller–Toporov (DMT) theory [57], were also proposed to describe the contact force in presence of adhesion.

Let us consider two contacting particles at centroid positions $\mathbf{x}_i$ and $\mathbf{x}_j$ with radii $r_{p,i}$ and $r_{p,j}$, respectively. The normal overlap δ_N between these two particles is derived as $\delta_N = r_{p,i} + r_{p,j} - |\mathbf{x}_j - \mathbf{x}_i|$. Then the normal forces between two particles are given as:

$$
\begin{aligned}
F_{ne-Hertz} &= -k_N \delta_N^{1.5}, \\
F_{ne-JKR} &= -4F_C(\hat{a}^3 - \hat{a}^{1.5}),
\end{aligned}
\tag{3}
$$

where k_N is normal stiffness coefficient in the Hertz model, $\hat{a} = a/a_0$ is the normalized contact area radius and F_C represents the critical pull-off force obtained from the JKR theory. In the Hertz model, the normal stiffness is analytically derived as $k_N = 4/3E\sqrt{R}$, where R is the effective particle radius and E is the effective elastic moduli, defined as:

$$
\begin{aligned}
\frac{1}{R} &= \frac{1}{r_{p,i}} + \frac{1}{r_{p,j}}, \\
\frac{1}{E} &= \frac{1-\sigma_i^2}{E_i} + \frac{1-\sigma_j^2}{E_j},
\end{aligned}
\tag{4}
$$

with E_i and E_j being elastic moduli and σ_i and σ_j being the Poisson's ratios, respectively. Note that the contact area radius a, the effective radius R and the normal overlap δ_N are related to each other via a simple geometrical relation, $a = \sqrt{R\delta_N}$. In the JKR theory, the particles will keep in contact due to the presence of van der Waals adhesion even if the external load is zero [56], which gives rise to an equilibrium contact radius a_0,

$$
a_0 = \left(\frac{9\pi\gamma R^2}{E}\right)^{1/3},
\tag{5}
$$

where γ is the particle's surface energy. The critical pull-off force in the JKR theory is then derived as $F_C = 3\pi\gamma R$ [56]. However, both the Hertz model and the JKR theory are energy conserved, i.e., there is no energy dissipation during the process of particle collision. To consider the energy dissipation during a collision, various dissipation mechanisms were proposed, including the viscoelastic response, plastic deformation as well as the fluid-phase dissipation [58]. For the low-speed impact regime, the viscoelastic dissipation is dominant. The dissipation force F_{nd} is approximated with a dashpot model, which is proportional to the approaching velocity of the particles, i.e.,

$$
F_{nd} = -\eta_N \mathbf{v}_R \cdot \mathbf{n},
\tag{6}
$$

where η_N is the normal dissipation coefficient, and $\mathbf{v}_R$ is the relative velocity at the contact point.

2.2. Sliding, Rolling and Twisting Resistance

The tangential contact force results from the relative sliding motion between two particles. Similar to the normal force, the standard sliding model is approximated by a linear spring-dashpot,

$$F_s = -k_T \xi_T - \eta_T \mathbf{v}_s \cdot \mathbf{t}_s, \tag{7}$$

where k_T is the spring stiffness coefficient in the tangential direction, $\mathbf{v}_s$ is relative sliding velocity, $\xi_T = \int_{t_0}^{t} \mathbf{v}_s(t) \cdot \mathbf{t}_s \, dt$ is the cumulative displacement in the tangential direction and η_T is the sliding dissipation coefficient. According to the Amonton's friction law, two contacting particles start to slide relative to each other when the tangential force reaches a critical value $F_{s,crit}$ and the dynamic friction force remains constant as $F_s = -F_{s,crit}$. For non-adhesive particles, the critical value is $F_{s,crit} = \mu_s |F_{ne}|$, where μ_s is sliding friction coefficient. Due to the presence of adhesion, the critical value for adhesive particles is larger, $F_{s,crit} = \mu_s |F_{ne} + 2F_C|$ [3,58].

The rolling and twisting resistances are similarly postulated in the form:

$$M_r = -k_R \xi_R - \eta_R \mathbf{v}_L \cdot \mathbf{t}_r,$$
$$M_t = -k_Q \xi_Q - \eta_Q \Omega_T, \tag{8}$$

where $\xi_R = \int_{t_0}^{t} \mathbf{v}_L(t) \cdot \mathbf{t}_r \, dt$ and $\xi_Q = \int_{t_0}^{t} \Omega_T(t) dt$ are the rolling and twisting displacements, $\mathbf{v}_L = -R(\Omega_{p,i} - \Omega_{p,j}) \times \mathbf{n}$ and $\Omega_T = (\Omega_{p,i} - \Omega_{p,j}) \cdot \mathbf{n}$ are the relative rolling and twisting velocity, respectively. The rolling direction is $\mathbf{t}_r = \mathbf{v}_L / |\mathbf{v}_L|$. k_R, k_Q are the rolling and twisting stiffness, respectively, and η_R, η_Q are the rolling and twisting dissipation coefficients, respectively. Correspondingly, the critical values for rolling and twisting resistances are given as:

$$M_{r,crit} = k_R \theta_{crit} R,$$
$$M_{t,crit} = 3\pi a F_{s,crit} / 16, \tag{9}$$

where θ_{crit} is the critical rolling angle in the relative rolling motion between two contacting particles.

3. Lattice Boltzmann Method (LBM)

3.1. Lattice Boltzmann Equation

Fundamentally, the LBM aims to establish a microscopic kinetic model involving essential physics, so as to make the average physical quantities in macroscopic scale follow the expected equations, which is completely different from the conventional numerical methods that directly discretise the macroscopic continuum equations. Hence an important premise lies in that the dynamic behaviour of a fluid in the macroscopic scale is the outcome from the collective behavior of numerous constitutions in the microscopic scale [19].

The LBM originated from lattice gas automata, which utilises a discrete lattice and discrete time. In each lattice, the fluid is described by a packet of microscopic particles, which only move in certain directions with certain discrete velocities. A number of lattice speed models are available, depending on the dimension of the problem. For example, the D2Q9 and D3Q19 are frequently used in the 2D and 3D problems, respectively, as illustrated in Figure 1. The individual particle motion is neglected in the LBM and a single-particle distribution function $f_i(\mathbf{x},t)$ is adopted instead.

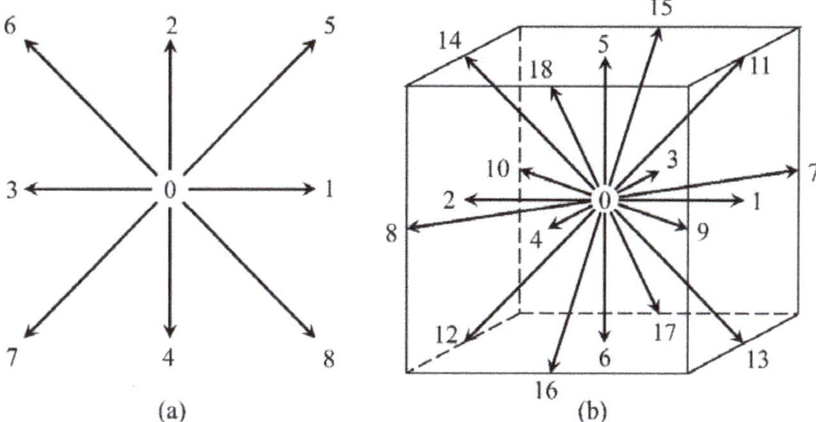

Figure 1. The D2Q9 (**a**) and D3Q19 (**b**) lattice speed model.

The evolution of the particle distribution functions $f_i(x,t)$ follows the lattice Boltzmann equation (LBE) [19] with an external force term,

$$f_i(\mathbf{x} + \mathbf{e}_i \Delta t, t + \Delta t) = f_i(\mathbf{x}, t) + \Omega_i[f_i(\mathbf{x}, t)] + F_i \Delta t, \tag{10}$$

where $\mathbf{x}$ is the position of the node, $\mathbf{e}_i$ is the unit vector of the lattice speed in the direction i as denoted in Figure 1, Δt is the explicit time step, F_i is the external body force and $\Omega_i[f_i(\mathbf{x},t)]$ is the collision operator. Equation (10) can be evaluated in two independent processes: collision and streaming, as shown below:

$$f_i(\mathbf{x}, t + \Delta t) = f_i(\mathbf{x}, t) + \Omega_i[f_i(\mathbf{x}, t)] + F_i \Delta t, \tag{11}$$

$$f_i(\mathbf{x} + \mathbf{e}_i \Delta t, t + \Delta t) = f_i(\mathbf{x}, t + \Delta t). \tag{12}$$

Equation (11) represents the collision process, which updates the distribution functions at position $\mathbf{x}$ from t to $t+\Delta t$ based on the relaxation operator $\Omega_i[f_i(\mathbf{x},t)]$. Then the streaming process (Equation (12)) takes place, which simply propagates the updated distribution functions from position $\mathbf{x}$ to their nearest neighbor nodes $\mathbf{x} + \mathbf{e}_i \Delta t$ according to the lattice speed model. The sequence of collision and streaming becomes irrelevant after a few time steps.

3.1.1. Single-Relaxation-Time Model

The single-relaxation-time (SRT) model refers to the Bhatnagar-Gross-Krook (BGK) approximation of the collision operator [59]. The collision operator is simplified to a linear form [59–62]:

$$\Omega_i = -\frac{\Delta t}{\tau}[f_i(\mathbf{x}, t) - f_i^{eq}(\mathbf{x}, t)], \tag{13}$$

where τ is the dimensionless relaxation parameter and $f_i^{eq}(\mathbf{x},t)$ is the equilibrium distribution function that is defined as:

$$f_i^{eq} = \rho \omega_i [1 + \frac{\mathbf{e}_i \cdot \mathbf{u}}{c_s^2} + \frac{(\mathbf{e}_i \cdot \mathbf{u})^2}{2 c_s^4} - \frac{u^2}{2 c_s^2}]. \tag{14}$$

In this equation, the weight coefficient ω_i depends on the lattice speed model. For example, $\omega_0 = 4/9$, $\omega_{1,2,3,4} = 1/9$ and $\omega_{5,6,7,8} = 1/36$ for D2Q9 model, while $\omega_0 = 1/3$, $\omega_{1,\ldots,6} = 1/18$ and $\omega_{7,\ldots,18} = 1/36$ for D3Q19 model. $c_s = 1/\sqrt{3}$ is the lattice sound speed.

The discrete body force term F_i in Equation (10) is given as [63]:

$$F_i = (1 - \frac{1}{2\tau})\omega_i [\frac{\mathbf{e}_i - \mathbf{u}}{c_s^2} + \frac{(\mathbf{e}_i \cdot \mathbf{u})}{c_s^4}\mathbf{e}_i] \cdot \mathbf{F}, \tag{15}$$

where $\mathbf{F}$ represents the macroscopic body force. The macroscopic fluid properties, including the density ρ, velocity $\mathbf{u}$ and pressure p are related to the microscopic particle distribution function, which are determined as:

$$\begin{aligned}
\rho &= \sum_i f_i, \\
\rho\mathbf{u} &= \sum_i f_i\mathbf{e}_i + \frac{\Delta t}{2}\mathbf{F}, \\
p &= c_s^2\rho.
\end{aligned} \tag{16}$$

The kinematic viscosity of the fluid v is related to the relaxation parameter as:

$$v = \frac{2\tau - 1}{6}. \tag{17}$$

3.1.2. Multi-Relaxation-Time Model

It was reported that the SRT LBE does not show good stability with small relaxation time τ or with high Reynolds number. To improve the performance of the LBM, the multi-relaxation-time (MRT) LBE was proposed, which can be expressed as [20]:

$$f_i(\mathbf{x} + \mathbf{e}_i\Delta t, t + \Delta t) = f_i(\mathbf{x}, t) - \sum_\alpha \Omega_{i\alpha}[f_\alpha(\mathbf{x}, t) - f_\alpha^{eq}(\mathbf{x}, t)] + [S_i(\mathbf{x}, t) - 0.5\sum_\alpha \Omega_{i\alpha}S_\alpha(\mathbf{x}, t)], \tag{18}$$

where S is the force term. From Equation (15), we have:

$$S_i = \omega_i [\frac{\mathbf{e}_i - \mathbf{u}}{c_s^2} + \frac{(\mathbf{e}_i \cdot \mathbf{u})}{c_s^4}\mathbf{e}_i] \cdot \mathbf{F}. \tag{19}$$

Equation (18) can be transformed into the matrix form as:

$$\mathbf{f}(\mathbf{x} + \mathbf{e}_i\Delta t, t + \Delta t) = \mathbf{f}(\mathbf{x}, t) - \mathbf{M}^{-1}\Lambda[\mathbf{m}(\mathbf{x}, t) - \mathbf{m}^{eq}(\mathbf{x}, t)] + \mathbf{M}^{-1}[\mathbf{I} - \frac{\Lambda}{2}]\overline{\mathbf{S}}(\mathbf{x}, t), \tag{20}$$

where $\mathbf{M}$ is the transformation matrix, $\mathbf{m}$ and $\mathbf{m}^{eq}$ are the moment spaces of the distribution function f_i and its equilibrium distribution f_i^{eq}, which can be derived as $\mathbf{m} = \mathbf{M}f_i$ and $\mathbf{m}^{eq} = \mathbf{M}f_i^{eq}$, respectively. The collision matrix Λ is a diagonal matrix in the moment space.

The matrixes $\mathbf{M}$ and Λ vary for different speed models. For D2Q9 model, the matrix $\mathbf{M}$ is:

$$\mathbf{M} = \begin{bmatrix}
1 & 1 & 1 & 1 & 1 & 1 & 1 & 1 & 1 \\
-4 & -1 & -1 & -1 & -1 & 2 & 2 & 2 & 2 \\
4 & -2 & -2 & -2 & -2 & 1 & 1 & 1 & 1 \\
0 & 1 & 0 & -1 & 0 & 1 & -1 & -1 & 1 \\
0 & -2 & 0 & 2 & 0 & 1 & -1 & -1 & 1 \\
0 & 0 & 1 & 0 & -1 & 1 & 1 & -1 & -1 \\
0 & 0 & -2 & 0 & 2 & 1 & 1 & -1 & -1 \\
0 & 1 & -1 & 1 & -1 & 0 & 0 & 0 & 0 \\
0 & 0 & 0 & 0 & 0 & 1 & -1 & 1 & -1
\end{bmatrix}. \tag{21}$$

The equilibrium distribution functions $\mathbf{m}^{eq}$ in the moment space are related to that in the velocity space as follows:

$$\mathbf{m}^{eq} = \mathbf{M}\mathbf{f}_i^{eq} = \begin{bmatrix} \rho \\ e^{eq} \\ \varepsilon^{eq} \\ j_x^{eq} \\ q_x^{eq} \\ j_y^{eq} \\ q_y^{eq} \\ p_{xx}^{eq} \\ p_{xy}^{eq} \end{bmatrix} = \begin{bmatrix} \rho \\ -2\rho + 3(u_x^2 + u_y^2) \\ \rho - 3(u_x^2 + u_y^2) \\ \rho u_x \\ -u_x \\ \rho u_y \\ -u_y \\ u_x^2 - u_y^2 \\ u_x u_y \end{bmatrix}. \tag{22}$$

The diagonal matrix Λ is $\Lambda = diag(s_1, s_2, s_3, s_4, s_5, s_6, s_7, s_8, s_9)$. In order to ensure the mass and momentum conservation after collision, it is easily derived that $s_1 = s_4 = s_6 = 0$. s_8 and s_9 are set as $s_8 = s_9 = 1/\tau$ in order to reproduce the same viscosity of SRT model [64]. The other relaxation parameters s_2, s_3, s_5 and s_7 can be decided with more flexibility according to the physical model.

For D3Q19 model, the matrix $\mathbf{M}$ is:

$$\mathbf{M} = \begin{bmatrix}
1 & 1 & 1 & 1 & 1 & 1 & 1 & 1 & 1 & 1 & 1 & 1 & 1 & 1 & 1 & 1 & 1 & 1 & 1 \\
-30 & -11 & -11 & -11 & -11 & -11 & -11 & 8 & 8 & 8 & 8 & 8 & 8 & 8 & 8 & 8 & 8 & 8 & 8 \\
12 & -4 & -4 & -4 & -4 & -4 & -4 & 1 & 1 & 1 & 1 & 1 & 1 & 1 & 1 & 1 & 1 & 1 & 1 \\
0 & 1 & 0 & 0 & -1 & 0 & 0 & 1 & 1 & 0 & -1 & -1 & 0 & 1 & 1 & 0 & -1 & -1 & 0 \\
0 & -4 & 0 & 0 & 4 & 0 & 0 & 1 & 1 & 0 & -1 & -1 & 0 & 1 & 1 & 0 & -1 & -1 & 0 \\
0 & 0 & 1 & 0 & 0 & -1 & 0 & 1 & 0 & 1 & -1 & 0 & -1 & -1 & 0 & 1 & 1 & 0 & -1 \\
0 & 0 & -4 & 0 & 0 & 4 & 0 & 1 & 0 & 1 & -1 & 0 & -1 & -1 & 0 & 1 & 1 & 0 & -1 \\
0 & 0 & 0 & 1 & 0 & 0 & -1 & 0 & 1 & 1 & 0 & -1 & -1 & 0 & -1 & -1 & 0 & 1 & 1 \\
0 & 0 & 0 & -4 & 0 & 0 & 4 & 0 & 1 & 1 & 0 & -1 & -1 & 0 & -1 & -1 & 0 & 1 & 1 \\
0 & 2 & -1 & -1 & 2 & -1 & -1 & 1 & 1 & -2 & 1 & 1 & -2 & 1 & 1 & -2 & 1 & 1 & -2 \\
0 & -4 & 2 & 2 & -4 & 2 & 2 & 1 & 1 & -2 & 1 & 1 & -2 & 1 & 1 & -2 & 1 & 1 & -2 \\
0 & 0 & 1 & -1 & 0 & 1 & -1 & 1 & -1 & 0 & 1 & -1 & 0 & 1 & -1 & 0 & 1 & -1 & 0 \\
0 & 0 & -2 & 2 & 0 & -2 & 2 & 1 & -1 & 0 & 1 & -1 & 0 & 1 & -1 & 0 & 1 & -1 & 0 \\
0 & 0 & 0 & 0 & 0 & 0 & 0 & 1 & 0 & 0 & 1 & 0 & 0 & -1 & 0 & 0 & -1 & 0 & 0 \\
0 & 0 & 0 & 0 & 0 & 0 & 0 & 0 & 0 & 1 & 0 & 0 & 1 & 0 & 0 & -1 & 0 & 0 & -1 \\
0 & 0 & 0 & 0 & 0 & 0 & 0 & 0 & 1 & 0 & 0 & 1 & 0 & 0 & -1 & 0 & 0 & -1 & 0 \\
0 & 0 & 0 & 0 & 0 & 0 & 0 & 1 & -1 & 0 & -1 & 1 & 0 & 1 & -1 & 0 & -1 & 1 & 0 \\
0 & 0 & 0 & 0 & 0 & 0 & 0 & -1 & 0 & 1 & 1 & 0 & -1 & 1 & 0 & 1 & -1 & 0 & -1 \\
0 & 0 & 0 & 0 & 0 & 0 & 0 & 0 & 1 & -1 & 0 & -1 & 1 & 0 & -1 & 1 & 0 & 1 & -1
\end{bmatrix}. \tag{23}$$

The equilibrium distribution functions $\mathbf{m}^{eq}$ in the moment space is:

$$\mathbf{m}^{eq} = \mathbf{M}f_i^{eq} = \begin{bmatrix} \rho \\ -11\rho + 19\rho(u_x^2 + u_y^2 + u_z^2) \\ 3\rho - \frac{11}{2}\rho(u_x^2 + u_y^2 + u_z^2) \\ \rho u_x \\ -\frac{2}{3}\rho u_x \\ \rho u_y \\ -\frac{2}{3}\rho u_y \\ \rho u_z \\ -\frac{2}{3}\rho u_z \\ \rho(2u_x^2 - u_y^2 - u_z^2) \\ -\frac{1}{2}\rho(2u_x^2 - u_y^2 - u_z^2) \\ \rho(u_y^2 - u_z^2) \\ -\frac{1}{2}\rho(u_y^2 - u_z^2) \\ \rho u_x u_y \\ \rho u_y u_z \\ \rho u_x u_z \\ 0 \\ 0 \\ 0 \end{bmatrix}. \tag{24}$$

The diagonal matrix Λ is $\Lambda = diag(s_1, s_2, \ldots \ldots, s_{18}, s_{19})$, where $s_1 = s_4 = s_6 = s_8 = 1.0$, $s_2 = s_5 = s_7 = s_9 = 1.2$, $s_3 = s_{11} = s_{13} = 1.4$, $s_{10} = s_{12} = s_{14} = s_{15} = s_{16} = 1/\tau$, $s_{17} = s_{18} = s_{19} = 1.98$.

3.2. Boundary Conditions

In the LBM, the conventional velocity and pressure boundary conditions cannot be applied directly, since these quantities are not primary variables in the LBE. Instead, the boundary conditions are imposed with the distribution functions. In this section, only the 'no-slip' wall and the periodic boundary conditions are introduced. It should be noted that particles are treated as moving boundaries, which play important roles in the coupling between LBM and DEM. The interactions between the moving particles and the fluid will be discussed separately later.

The 'no-slip' boundary condition is implemented with the so-called bounce-back rule [65,66]. As shown in Figure 2, if f_i is a distribution function entering into the solid stationary wall, the undetermined distribution function that comes out of the wall is simply a reflection of f_i, i.e.,

$$f_{-i}(\mathbf{x}, t + \Delta t) = f_i^+(\mathbf{x}, t), \tag{25}$$

where f_{-i} and f_i^+ denote the distribution functions in the opposite direction of i and after collision, respectively. This simple bounce-back rule ensures a strict 'no-slip' condition at the wall position, as the tangential velocity is zero. It should be noted that the accuracy of the simple bounce-back rule is only first order, except that the wall is just in the middle of the link between a solid node and a fluid node, where the accuracy is second order. However, the bounce-back rule is easy to implement and can be extended to obstacles with arbitrary shape, which makes the LBM quite suitable for problems with complex geometries.

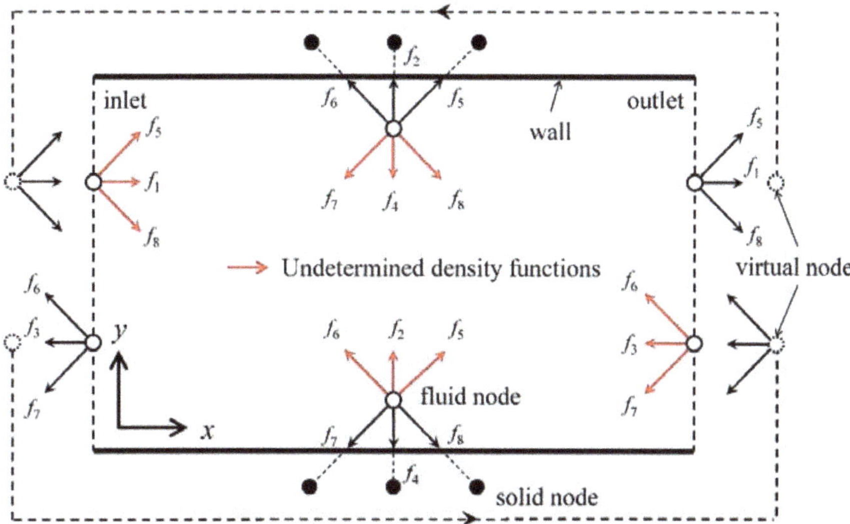

Figure 2. Schematic of the bounce-back rules and periodic boundary conditions for D2Q9 model.

Periodic boundary conditions are also straightforward to impose in the LBM. As shown in Figure 2, distribution functions propagating out of the domain stream into the corresponding boundary node at the opposite boundary in the same direction. Take Figure 2 as an example, the inlet and outlet planes are periodic. As a result, the distribution functions f_1, f_5 and f_8 at the outlet boundary will stream into the corresponding node at the inlet, while f_3, f_6 and f_7 at the inlet propagate into the corresponding node at the outlet.

3.3. Unit Conversion in LBM

In the LBM, the physical parameters in the computation are commonly dimensionless, which necessitates a unit conversion. In principle, one only needs to determine the conversion coefficients of three fundamental units, i.e., length, time and mass, and then the conversion coefficients of all the other physical parameters can be sequentially decided.

Normally, the lattice space Δx, the time step Δt, and the fluid density ρ are set to be one in the LBM computation. How to perform the unit conversion will be explained using an example of pipe flow in 3D. Suppose that the pipe size is $L \times D = 8$ mm $\times$ 4 mm, where L is the pipe length and D is the pipe diameter, and water flows through this pipe with an average velocity of $V = 0.0125$ m/s. First, set the number of lattice as, for instance, $n_x \times n_y \times n_z = 100 \times 50 \times 50$, which results in a lattice length of $dx = L/n_x = D/n_y = D/n_z = 0.08$ mm. Note that the lattice should be square in 2D or cube in 3D, indicating that the lattice length is identical in every dimension. Then the conversion coefficient of length is decided by $C_l = dx/\Delta x = 8 \times 10^{-5}$ m. Second, given that the density of water is $\rho_f = 1000$ kg/m^3, the conversion coefficient of density is $C_\rho = \rho_f/\rho = 1000$ kg/m^3. Then the conversion coefficient of mass is determined by $C_m = C_\rho C_l^3 = 5.12 \times 10^{-10}$ kg. For better numerical stability, the relaxation time τ shall be at least 0.55, implying that the kinematic viscosity ν shall be at least 0.017. Here let us set $\nu = 0.05$. Given that the kinematic viscosity of water is $\nu_{water} = 1.0 \times 10^{-6}$ m^2/s, the conversion coefficient of viscosity is $C_\nu = \nu_{water}/\nu = 2.0 \times 10^{-5}$ m^2/s. Then the conversion coefficient of time is determined as $C_t = C_l^2/C_\nu = 3.2 \times 10^{-4}$ s. So far, the conversion coefficients of the three fundamental units are decided. Other parameters that need to be nondimensionalized can be converted accordingly. As a summary, the unit conversion of this pipe flow problem is listed in Table 1.

Table 1. Unit conversion of 3D pipe flow.

Physical Parameters	Unit	Real Value	Lattice Value	Conversion Coefficient
Pipe length (L)	m	0.008	100	8.0×10^{-5}
Pipe diameter (D)	m	0.004	50	8.0×10^{-5}
Fluid density (ρ_f)	kg/m^3	1000	1	1000
Fluid kinematic viscosity (v_f)	m^2/s	1.0×10^{-6}	0.05	2.0×10^{-5}
Average fluid velocity (V)	m/s	0.0125	0.05	0.025
Pressure drop (ΔP)	Pa	0.2	3.2×10^{-3}	62.5
Time step (Δt)	s	3.2×10^{-4}	1	3.2×10^{-4}

4. LBM-DEM Coupling

For coupling LBM with DEM, it is of significant importance to properly describe the fluid-solid interactions. The first step is to establish a lattice discretisation of the solid particle on the lattice grid. As illustrated in Figure 3, four types of lattice nodes are obtained as: (1) the pure fluid node without connection to any pure solid node; (2) the pure solid node that is not adjacent to any pure fluid node; (3) the fluid boundary node that is linked with both pure fluid node and solid boundary node; and (4) the solid boundary node that is connected with both fluid boundary node and pure solid node. Obviously, it becomes very easy and straightforward to handle any shaped particles with the discrete lattice representation. Hence, there was a variety of techniques with different levels of complexity and accuracy to compute the fluid-solid interaction, including the modified bounce-back (MBB) scheme [65,66], interpolated bounce-back (IBB) scheme [67–72] and immersed boundary methods (IBM) [73–81].

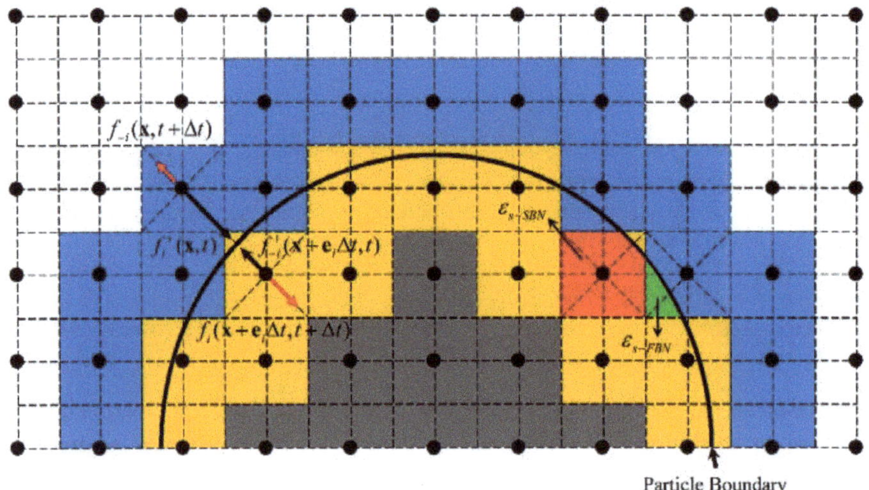

Figure 3. The mapping of a circular solid particle on the lattice grid, showing solid boundary nodes (**orange**), fluid boundary nodes (**blue**), internal solid nodes (**grey**) and pure fluid node (**white**).

4.1. Modified Bounce-Back Scheme

The MBB scheme was proposed by Ladd [65,66] in order to improve the simple bounce-back rules to incorporate the influence of moving fluid-solid interfaces. As shown in Figure 3, this method assumes that the exact solid surface lies halfway between fluid boundary node and solid boundary node, where the bounce-back condition is enforced. The particle is treated as a shell full of fluid. Therefore, the bounce-back takes place on both side of the fluid-solid interface.

The first step is to estimate the velocity at the fluid-solid interface based on the solid particle's velocity, which is given as:

$$\mathbf{v}_b = \mathbf{U}_p + \boldsymbol{\Omega}_p \times (\mathbf{x} + \frac{1}{2}\mathbf{e}_i\Delta t - \mathbf{X}_p), \tag{26}$$

where $\mathbf{X}_p$ is centre of the solid particle. Then the undetermined distribution functions at the fluid boundary node and solid boundary node are computed respectively, according to the modified bounce-back rule,

$$\begin{aligned} f_{-i}(\mathbf{x}, t + \Delta t) &= f_i^+(\mathbf{x}, t) - 2\omega_i\rho\mathbf{v}_b \cdot \mathbf{e}_i, \\ f_i(\mathbf{x} + \mathbf{e}_i\Delta t, t + \Delta t) &= f_{-i}^+(\mathbf{x} + \mathbf{e}_i\Delta t, t) + 2\omega_i\rho\mathbf{v}_b \cdot \mathbf{e}_i. \end{aligned} \tag{27}$$

It can be seen that Equation (27) reduces to the simple bounce-back scheme (Equation (25)) when $\mathbf{v}_b = 0$. Once the bounce-back procedure is accomplished, the hydrodynamic force on the solid particle at this particular boundary node is calculated with the momentum exchange method,

$$\mathbf{F}_{f,-i}(\mathbf{x} + \frac{1}{2}\mathbf{e}_i\Delta t, t + \frac{1}{2}\Delta t) = 2[f_i^+(\mathbf{x}, t) - f_{-i}^+(\mathbf{x} + \mathbf{e}_i\Delta t, t) - 2\omega_i\rho\mathbf{v}_b \cdot \mathbf{e}_i] \cdot \mathbf{e}_i. \tag{28}$$

The total hydrodynamic force and torque are computed by summing up the contributions from every lattice speed direction and boundary node,

$$\begin{aligned} \mathbf{F}_f &= \sum_{link} \sum_i \mathbf{F}_{f,-i}, \\ \mathbf{M}_f &= \sum_{link} \sum_i (\mathbf{x} + \frac{1}{2}\mathbf{e}_i\Delta t - \mathbf{X}_p) \times \mathbf{F}_{f,-i}. \end{aligned} \tag{29}$$

Apparently, the discrete representation of the solid particle surface on the lattice grid is in a stepwise fashion. It is reported that the accuracy of MBB is of first order only due to the zig-zag staircases of a curved surface. Large fluctuations of the hydrodynamic interactions are also observed. Using a sufficiently high lattice resolution could provide much more smooth and accurate force evaluations, which, however, will lead to a huge increase of the computational cost. To overcome these drawbacks of MBB, several improved techniques were proposed, which will be introduced below.

4.2. Interpolated Bounce-Back Scheme

To precisely describe the actual boundary of the solid particle, several bounce-back schemes using the spatial interpolation were also proposed [67–72]. In all these techniques, the relative location of the exact fluid-solid interface is interpolated with a location parameter q. Hence more accurate and smoother evaluations of the hydrodynamic force and torque are obtained, which is demonstrated to maintain the second-order accuracy. Here a double interpolation scheme proposed by Yu et al. [71,72] is introduced to show how these schemes work. As shown in Figure 4, the undetermined distribution function is the one coming out of the solid boundary $f_{-i}(\mathbf{x}_f, t + \Delta t)$. In most of time, the solid boundary is not exactly located on a lattice node. Then the relative location of the solid boundary can be described by a weighting parameter q by means of spatial interpolation,

$$q = |\mathbf{x}_f - \mathbf{x}_b| / |\mathbf{x}_f - \mathbf{x}_s|, \tag{30}$$

where $\mathbf{x}_s$, $\mathbf{x}_b$ and $\mathbf{x}_f$ represent the positions of the solid node, the solid boundary and the first fluid node next to the solid boundary, respectively. The weighting parameter q varies between 0 and 1. When $q = 0$, the solid boundary is exactly located on the nearest fluid node, which is turned into a solid node. When $q = 1$, the solid boundary lies just on the nearest solid node. Then the density function at the temporary location $\mathbf{x}_b$ that propagates exactly to the solid boundary during the streaming process can be interpolated with the existing density functions $f_i(\mathbf{x}_f, t + \Delta t)$ and $f_i(\mathbf{x}_{ff}, t + \Delta t)$, through a first order form,

$$f_i(\mathbf{x}_b, t + \Delta t) = q f_i(\mathbf{x}_f, t) + (1 - q) f_i(\mathbf{x}_{ff}, t). \tag{31}$$

Next, a bounce-back operation takes place instantaneously at the solid boundary as:

$$f_{-i}(\mathbf{x}_b, t + \Delta t) = f_i(\mathbf{x}_b, t + \Delta t) - 2\omega_i \rho \mathbf{u}_b \cdot \mathbf{e}_i, \tag{32}$$

which is the same as the MBB described in Equation (27). In the last step, the unknown density function $f_{-i}(\mathbf{x}_f, t + \Delta t)$ is determined from $f_{-i}(\mathbf{x}_b, t + \Delta t)$ and $f_{-i}(\mathbf{x}_{ff}, t + \Delta t)$ through a first order interpolation,

$$f_{-i}(\mathbf{x}_f, t + \Delta t) = \frac{1}{1+q} f_{-i}(\mathbf{x}_b, t + \Delta t) + \frac{q}{1+q} f_{-i}(\mathbf{x}_{ff}, t + \Delta t). \tag{33}$$

It should be noted that the interpolation can also be performed with second order form, where the density function from further fluid node $\mathbf{x}_{fff}$ is needed. The force and torque on the solid particle can be evaluated similarly with a momentum exchange method,

$$\begin{aligned}
\mathbf{F}_f &= \sum_{\text{all } \mathbf{x}_f} \sum_i [f_i^+(\mathbf{x}_f, t) + f_{-i}(\mathbf{x}_f, t + \Delta t)] \mathbf{e}_i, \\
\mathbf{M}_f &= \sum_{\text{all } \mathbf{x}_f} \sum_i (\mathbf{x}_b - \mathbf{X}_p) \times [f_i^+(\mathbf{x}_f, t) + f_{-i}(\mathbf{x}_f, t + \Delta t)] \mathbf{e}_i.
\end{aligned} \tag{34}$$

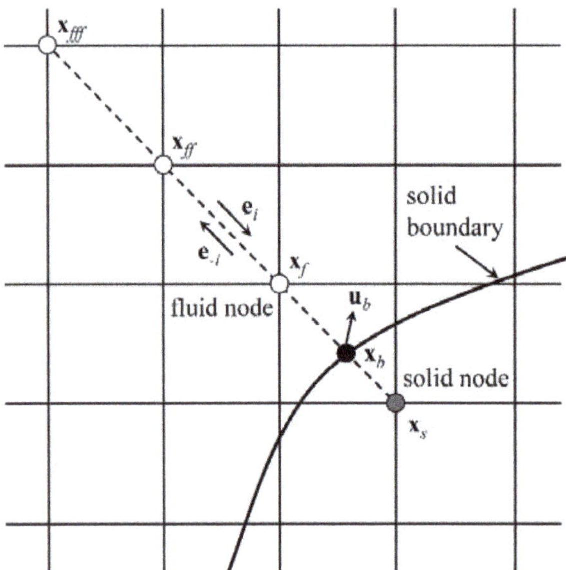

Figure 4. Schematic of the double interpolation scheme.

4.3. Immersed Moving Boundary Method

The immersed boundary method (IBM) was proposed by Peskin [73,74] and further implemented in LBM by other researchers [75–80]. In this method, extra moving Lagrangian nodes are introduced to represent the solid particle shape, which are assumed to be deformable with a large stiffness. The effects of the immersed particle boundary on the fluid are modelled by restoring forces based on the 'no-slip' boundary condition, which tend to keep the particle to its original shape. Then the restoring forces are distributed to their surrounding fluid nodes and considered as the external force terms in the governing equations. The drawbacks of the IBM include the introduction of additional free parameters, the problem-sensitive determination of the spring stiffness and the damping constant, as well as the severely restricted computational time step.

As an alternative, Noble and Torczynski [81] proposed an immersed moving boundary technique based on the local solid fraction in each lattice cell, which is also known as the partially-solid scheme. This method not only overcomes the momentum discontinuity of MBB-based techniques and provides an adequate representation of complex boundaries at relative lower lattice resolutions, but also retains the prominent advantages of the LBM, i.e., the simple linear collision operator and its locality in computation. In this method, the lattice Boltzmann equation is modified to include an additional collision term, which depends on the local volume fraction of solid (see Figure 3 right part),

$$f_i(\mathbf{x} + \mathbf{e}_i \Delta t, t + \Delta t) = f_i(\mathbf{x}, t) - (1 - B_n)[\frac{\Delta t}{\tau}(f_i(\mathbf{x}, t) - f_i^{eq}(\mathbf{x}, t))] + \sum_s B_s \Omega_i^s + (1 - B_n)F_i \Delta t, \tag{35}$$

where B_n is a total weighting function in each lattice cell, B_s is the weighting function from each solid particle in the same lattice cell, and $\Omega_i{}^s$ is the additional collision term. The total weighting function is calculated by summing up the weighting function of each solid particle that intersect with the same lattice cell such that $B_n = \sum_s B_s$. Each particle's weighting function is determined by its own solid fraction and the relaxation parameter as:

$$B_s(\varepsilon_s, \tau) = \frac{\varepsilon_s(\tau/\Delta t - 0.5)}{(1 - \varepsilon_n) + (\tau/\Delta t - 0.5)}, \tag{36}$$

where ε_s is the corresponding particle's solid fraction and ε_n is total solid fraction in the same lattice cell that is the sum of each particle's contribution, i.e., $\varepsilon_n = \sum_s \varepsilon_s$. It can be seen that when the solid fraction ε_s varies from 0 (a completely fluid cell) to 1 (a completely solid cell), B_s varies from 0 to 1. Equation (35) returns to the original SRT-LBE for pure fluid when $B_s = 0$, and returns the new collision operator Ω_i^s plus the distribution from the previous time step when $B_s = 1$. The new collision operator is given by

$$\Omega_i^s = f_{-i}(\mathbf{x}, t) - f_{-i}^{eq}(\rho, \mathbf{u}) + f_i^{eq}(\rho, \mathbf{U}_p) - f_i(\mathbf{x}, t). \tag{37}$$

The total hydrodynamic force and torque acting on the solid particle can be evaluated using the momentum exchange method with the additional collision operator over all lattice directions at each node and then over all fluid boundary, solid boundary and internal solid nodes, which are expressed as:

$$\begin{aligned}
\mathbf{F}_f &= -\sum_n B_s(\sum_i \Omega_i^s \mathbf{e}_i), \\
\mathbf{T}_f &= -\sum_n [(\mathbf{x}_n - \mathbf{X}_p) \times B_s(\sum_i \Omega_i^s \mathbf{e}_i)].
\end{aligned} \tag{38}$$

Note that the implementation of the immersed moving boundary method is straightforward. Only quantities already available in the lattice cell or easily derived are used. No additional data storage or organization is needed, which is a crucial advantage over other moving boundary techniques. However, it is also noticed that the critical step in this method is the calculation of the solid fraction in each lattice cell, which is primarily a geometry problem. Many methods are proposed to estimate the solid coverage ratio of a geometry with a square or a cube, including analytically solution, Monte Carlo sampling technique, cell decomposition method, polygonal approximation, edge-intersection averaging method, linear approximation and so on [82,83]. In the current LBM-DEM numerical framework, the analytical solution is used for 2D problems, while for 3D a linear approximation method is adopted, in order to save computational cost but still retain relatively accurate results.

4.4. Time Steps in the LBM-DEM Coupling

The time step coupling is also a crucial issue in the LBM-DEM coupling [31,82]. As we discussed previously, the time step of LBM is generally set as a dimensionless value of one in the computation. The real physical time is implicitly determined by computational parameters as well as the unit conversion scheme. Based on Li and Marshall's work [3,58], the DEM time step depends on the interparticle collision time scale,

$$t_c \approx R \left(\frac{\rho_p^2}{E^2 v_R} \right)^{1/5} ,$$
(39)

which typically varies around $10^{-6} \sim 10^{-9}$ s. Therefore, it is believed that the LBM time step is generally larger than the DEM time step. A time step ratio λ is then introduced as $\Delta t_{LBM} = \lambda \Delta t_{DEM}$, which can be decided after all the parameters are settled. It should be noted that the time step ratio λ could be either greater or lower than one. If $\lambda < 1$, the critical time step can be simply set as Δt_{LBM}, and the DEM time step can be equal to Δt_{LBM}. When $\lambda > 1$, a number of DEM computation steps are performed within one LBM time step, during which the fluid forces and torques are kept unchanged.

5. Validation

In this section, a few validation studies are performed to demonstrate the accuracy of our LBM-DEM numerical approach, including single-phase Poiseuille flow, gravitational settling of a particle, and the drag force on a stationary particle.

5.1. Poiseuille Flow

Single-phase Poiseuille flows in both 2D and 3D are first analysed using the LBM-DEM. The validation is performed with both SRT and MRT models. As shown in Figure 5, the boundaries in are periodic, and y-direction are bounded by two 'no-slip' walls. A constant pressure gradient is imposed in the channel in x-direction to drive the flow. The channel size is $H \times H$ for the 2D case and $10 \times H \times H$ for the 3D case. A few lattice resolutions are selected in the range $H = 11 \sim 101$ to examine the convergence of the numerical model. The relaxation parameter is fixed at 0.65. Different channel Reynolds numbers are achieved by changing the magnitude of the body force. Figures 6 and 7 show the normalised velocity profiles for 2D and 3D Poiseuille flow, respectively. The L-2 norm calculation is used to compute the relative error:

$$relative\ error = \sum_n \sqrt{\frac{(U - U_{theory})^2}{U_{theory}^2}},$$
(40)

where U is the flow velocity obtained from the simulation and U_{theory} is the analytical solution of the Poiseuille flow [84]. It can be inferred that the numerical results agree very well with the theoretical prediction. The accuracy is second order for the 2D flow, while it is slightly lower than second order for the 3D duct flow.

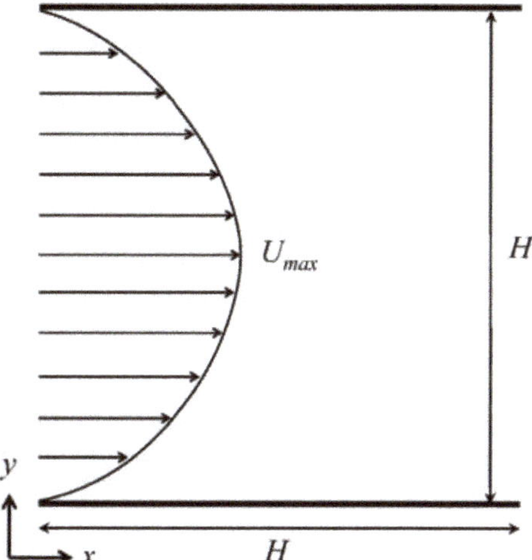

Figure 5. Schematic of single-phase Poiseuille flow in 2D.

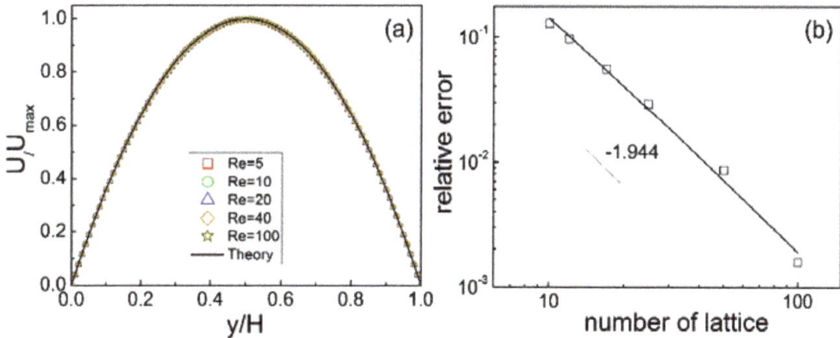

Figure 6. (**a**) Normalised velocity profile for 2D Poiseuille flow with channel size 101×101. (**b**) The relative error of the fluid velocity as a function of the lattice resolution.

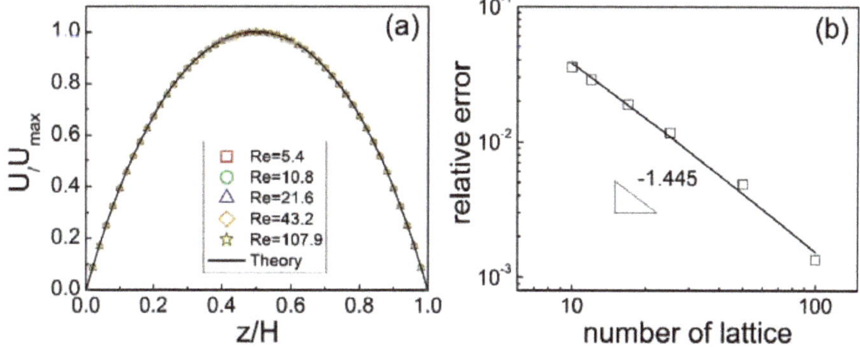

Figure 7. (**a**) Normalised velocity profile for 3D duct flow with channel size $10 \times 51 \times 51$. The velocity profile is along the vertical direction in the mid-plane in y-direction. (**b**) The relative error of the fluid velocity as a function of the lattice resolution.

5.2. Gravitational Settling of a Particle

For gravitational settling of a particle, both 2D and 3D cases are simulated to evaluate the validity of our numerical approach in a dynamic system. This validation is performed with SRT-LBM along with the Hertz model in DEM. For the 2D case, the computational setup is identical to that in Wen et al.'s work [85]. As shown in Figure 8, a cylinder with diameter d_p = 0.1 cm is initially located 0.076 cm away from the left wall of a vertical channel with width 0.4 cm. The fluid density and kinematic viscosity are 1000 kg/m^3 and 1×10^{-6}m^2/s, respectively. The mass density of the cylinder is 1030 kg/m^3. Therefore, the cylinder will settle under the gravitational force and finally reaches a steady state that moves along the centreline at a constant velocity. The computational setup in the lattice unit scheme is as follows. The size of the channel is $n_x \times n_y$ = 121 × 1201 and 'no-slip' wall boundaries are imposed on all the four faces. The cylinder's diameter and mass density are d_p = 30 and ρ_p = 1.03, respectively. The dimensionless relaxation parameter is set as 0.6. Note that the IBM is applied in the solid–fluid coupling in this particular problem.

A detailed comparison between the present numerical results and that obtained by the arbitrary Lagrangian–Eulerian technique (ALE) [85] is presented in Figure 9, where very good agreement is reached. Furthermore, the force profiles are quite smooth without any large fluctuations.

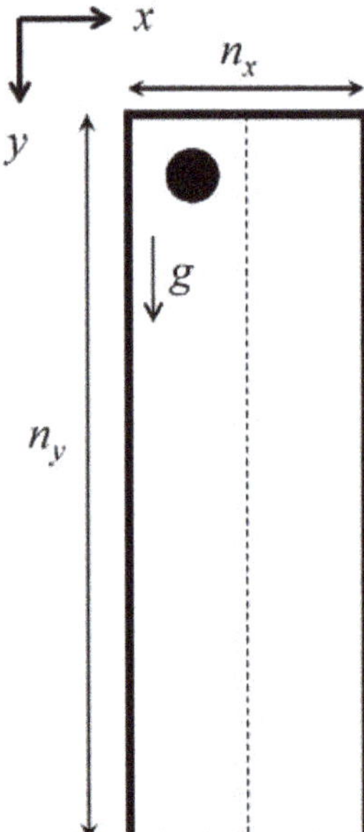

Figure 8. Schematic of a single particle settling under gravity.

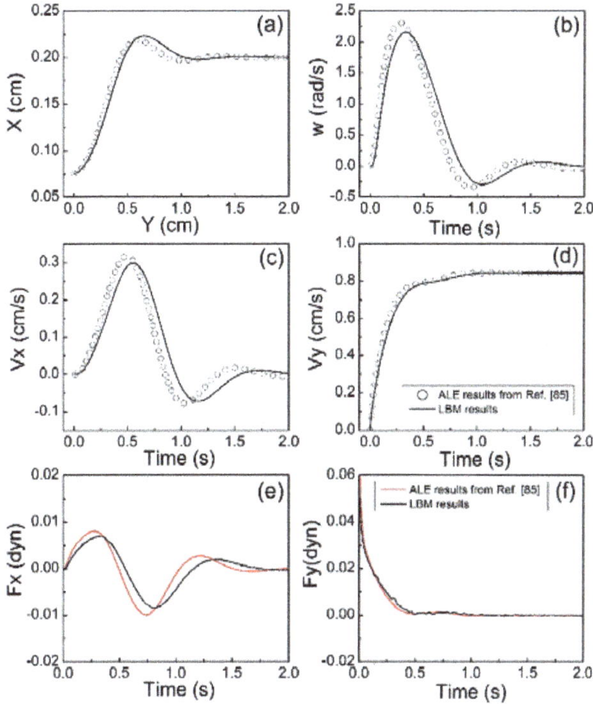

Figure 9. Time-dependent (**a**) particle trajectory, (**b**) angular velocity, (**c**) horizontal velocity, (**d**) vertical velocity, (**e**) horizontal force and (**f**) vertical force.

For the 3D case, the single particle settling experiment reported by Ten Cate et al. [86] is numerically reproduced. A spherical particle with diameter d_p = 15 mm and density ρ_p = 1,120 kg/m^3 is initially released from a height h = 120 mm in a cuboid box with a size of 100 × 100 × 160 mm^3. The box is full of silicon oil, and four different types of oil are used in the experiment. The densities and dynamic viscosities of the silicon oil are ρ_f = 970, 965, 962, 960 kg/m^3 and v_f = 373, 212, 113, 58 Pa·s, respectively. The terminal settling velocity of the particle in the four oils are u_{ter} = 0.038, 0.060, 0.091, 0.128 m/s, respectively, which results in four different particle Reynolds number Re_p = 1.5, 4.1, 11.6, 31.9, respectively. The computational parameters in the dimensionless lattice unit are given below. The domain size is 51 × 51 × 81 and the particle diameter is d_p − 7.5. The density of the fluid is fixed at ρ_f = 1.0, so that the corresponding particle density is ρ_p = 1.155, 1.161, 1.164, and 1.167 for different fluids considered, respectively. The relaxation parameter is set as 0.65. Figure 10 shows the time evolution of particle trajectory and velocity profiles for different Reynolds numbers. It is apparent that the numerical results are in excellent agreement with the experimental results, which further demonstrates the validity and accuracy of the present numerical approach.

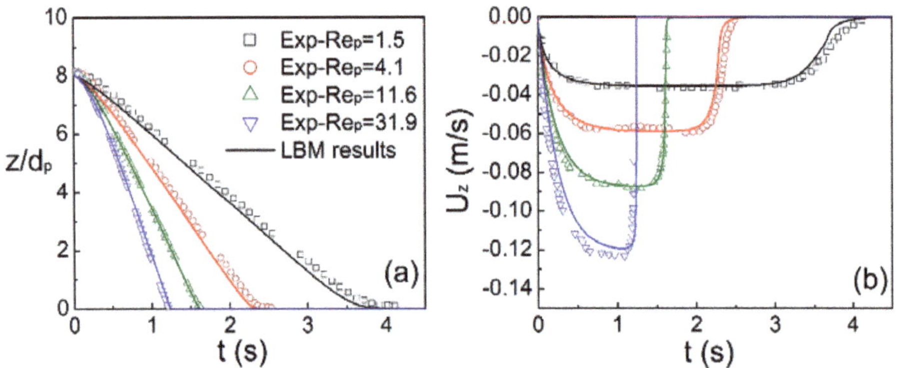

Figure 10. (a) Particle trajectory and (b) velocity profiles as a function of time.

5.3. Drag Force on a Stationary Particle

The classical drag force problem is also simulated to verify the accuracy of the force evaluation in our LBM-DEM coupling, which is carried out with the SRT-LBM. As depicted in Figure 11, a circular (spherical) particle is placed in the centre of a rectangular (cuboid) domain and kept stationary. The domain size is $L \times H = 50d_p \times 50d_p$ in 2D and $L \times H \times H = 20d_p \times 10d_p \times 10d_p$ in 3D. The inlet boundary is constant flow with velocity U_0 and the outlet is set as constant pressure boundary. All the other boundaries are set as open boundary (zero gradient). The particle Reynolds number and the drag coefficient are calculated as

$$\mathrm{Re}_p = \frac{U_0 d_p}{\nu},$$
$$C_d = \frac{8F_d}{\rho_f U_0^2 \pi d_p^2},$$

(41)

where F_d is the fluid drag force on the particle. Note that both the IBB and IBM are used in the drag force validation.

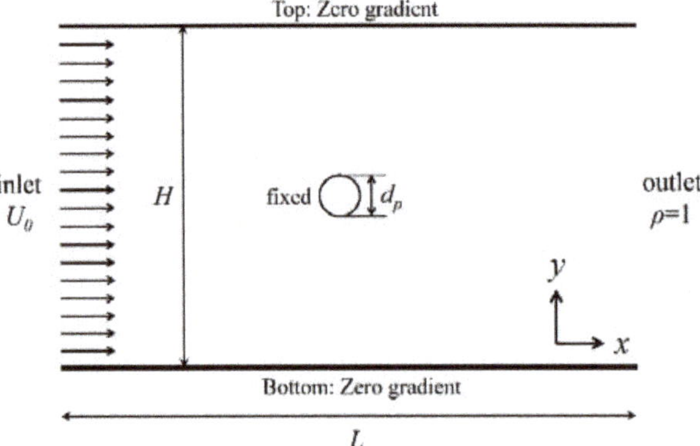

Figure 11. Schematic of flow past a stationary particle.

Figure 12 shows the drag coefficient C_d as a function of the particle Reynolds number Re_p for both 2D and 3D cases. For the 2D case shown in Figure 12a, the experimental results reported by Tritton are also superimposed for comparison [87]. It is clear that both the results obtained by IBB and IBM agree well with the experiments. Second order accuracy of the force evaluation is achieved for IBM, while the accuracy for IBB is lower than second order but higher than first order. For the 3D case shown in Figure 12a,b widely accepted empirical law of the drag coefficient is introduced to serve as the theoretical prediction [88]

$$C_d = \frac{24}{\mathrm{Re}_p}(1 + 0.15\mathrm{Re}_p^{0.687}). \tag{42}$$

We can see that the simulation results agree well with the above theoretical prediction, which demonstrates the validity and accuracy of our coupled LBM-DEM. Furthermore, it is noticed that accurate drag coefficients are obtained using both IBB and IBM even with a relatively low size resolution $d_p = 6$, implying that acceptable accuracy can still be retained when the computational cost is reduced.

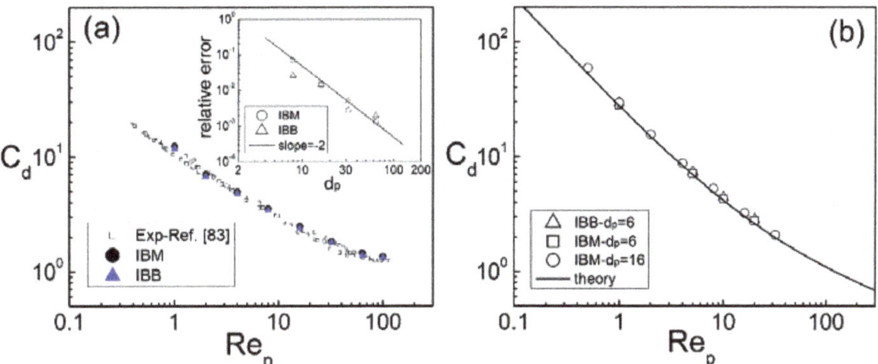

Figure 12. Drag coefficient as a function of particle Reynolds number for (**a**) 2D and (**b**) 3D cases.

6. LBM-DEM Applications

In this section, a few numerical examples are presented to demonstrate the capability of LBM-DEM, which include inertial migration of dense particle suspensions, agglomeration of adhesive particles in channel flow, and sedimentation of particle suspensions in a cavity flow.

6.1. Inertial Migration of Dense Particle Suspensions

In an experimental study of pipe flow of a dilute suspension consisting of neutrally buoyant particles, Segré and Silberberg [89,90] first observed that the particle suspensions tend to migrate laterally and focus in an annulus at radial position $r \approx 0.6R$, with R being the pipe radius. This phenomenon is named as the tubular pinch effect (or the Segré–Silberberg effect). Due to the lack of theoretical explanation at the time when it was discovered, this interesting phenomenon prompted a great deal of interest to further explore the underlying mechanism. Since then, many studies were carried out on this fascinating phenomenon experimentally [91–95], theoretically [96–99] and computationally [100–106]. The lateral focusing of the particle suspensions is found to be closely related to the fluid inertia. It is well recognised that the lateral migration exists in both 2D and 3D flows, and the equilibrium position moves closer to the wall as the channel Reynolds number increases, which is successfully described with the theoretical solution based on the perturbation theory and the asymptotic expansion method. However, most of the previous works are limited to very dilute suspensions with a particle concentration lower than 1%. When the particle concentration is increased, whether the lateral focusing phenomenon still exists remains less explored. Therefore, the coupled LBM-DEM is employed in the current study to explore this problem.

As shown in Figure 13, let us consider a pressure-driven flow of non-Brownian particle suspensions in a 2D channel. Periodic boundary conditions are applied in x-direction and the top and bottom planes in y-direction are set as 'no-slip' walls. The particles are neutrally buoyant, i.e., the fluid and the particle have the same mass density. The particles' initial positions are randomly generated inside the channel at the beginning of the simulation. Driven by the pressure gradient, the particle suspensions will transport and migrate to their equilibrium positions. A steady particle–fluid flow is expected after a sufficient long time of computation. The parameters used in the simulation are as follows. The size of the channel is $L \times H = 501 \times 101$. The particle diameter is $d_p = 6$ and 12. The particle concentration ϕ ranges between 1% and 50%. The channel Reynolds number Re_0 is tuned in the range 4 ~ 100 by varying the pressure gradient. Note that the channel Reynolds number refers to the one under a single phase flow with no particles. The SRT model and the Hertz contact model are used in the LBM and DEM for this particular problem, respectively.

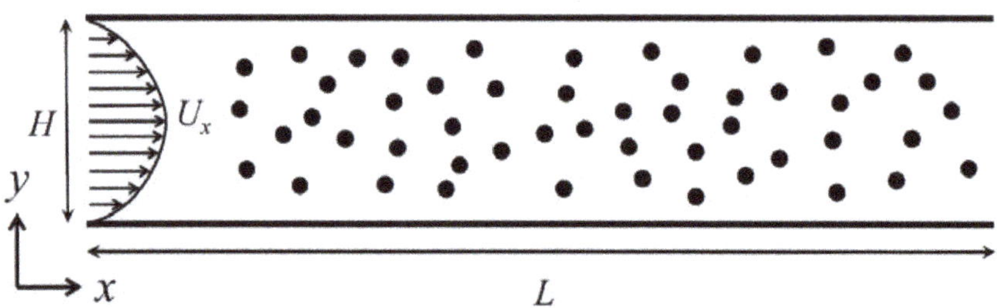

Figure 13. Schematic of pressure-driven flow of dense particle suspensions.

Figure 14 displays the snapshots of the particle suspension with different channel Reynolds numbers, where the particle concentration is fixed at $\phi = 10\%$. It is observed that only a small part of the particles undergoes the lateral migration at a relatively small Reynolds number $Re_0 = 4$. As Re_0 increases, a complete migration takes place at both $Re_0 = 40$ and $Re_0 = 100$, where all the particles are focused at a certain lateral position. Furthermore, it is found out that the larger Re_0 is, the shorter time it takes to develop into a full migration. For example, it occurs around $t = 5.5 \times 10^4$ for $Re_0 = 40$, while it is much earlier around $t = 4 \times 10^4$ for $Re_0 = 100$. Figure 15 illustrates the effects of particle concentration on the particle migration. It is clear that with relatively low particle concentration ($\phi = 1\%$ and 10%), the lateral migration is still notable. However, as ϕ increases to 40%, the migration is dramatically suppressed and the particle suspensions appear to jam near the wall.

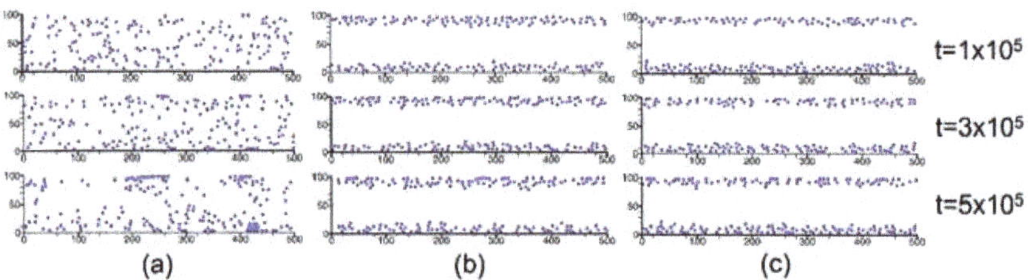

Figure 14. Snapshots of the migration process of particle suspensions with different channel Reynolds numbers, (**a**) $Re_0 = 4$, (**b**) $Re_0 = 40$, and (**c**) $Re_0 = 100$. The particle diameter is $d_p = 6$ and the concentration is $\phi = 10\%$.

To quantify the influence of the particle concentration, the degree of inertial migration can be defined as [94],

$$P_f = \sum PDF\left(|y| \geq \frac{1}{2}H\right), \tag{43}$$

which estimates the total probability distribution function (PDF) of the particles in the upper and lower quarter of the channel. Since the lateral equilibrium position is around 0.6 according to the Segré and Silberberg effect, the degree of the migration must be $P_f = 1$ if all the particles are laterally focused. On the other hand, if the particle suspension remains randomly distributed in the channel, P_f is expected to be 0.5. Figure 16 shows the variation of the degree of inertial migration with particle concentration for different d_p and Re_0. It is observed that P_f equals one for all the Re_0 at a low particle concentration ($\phi \leq 5\%$), while P_f remains around 0.5 ~ 0.6 at $\phi = 50\%$ for all the Re_0, implying a huge impact of the particle concentration on the lateral migration. A full migration is always accessible when the particle suspension is not dense, but it is completely suppressed with a very large particle concentration for the range of Re_0 in the present study. However, when ϕ is between 5% and 50%, the variation of P_f strongly depends on Re_0. With a fixed ϕ, a larger Re_0 leads to a larger P_f, i.e., a higher degree of lateral migration.

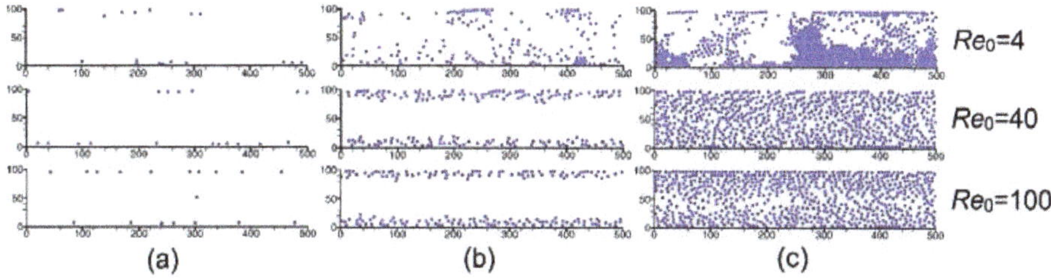

Figure 15. Snapshots of the particle positions with different particle concentrations, (a) $\phi = 1\%$, (b) $\phi = 10\%$, (c) $\phi = 40\%$. The particle diameter is $d_p = 6$ and all the snapshots are captured at the steady state.

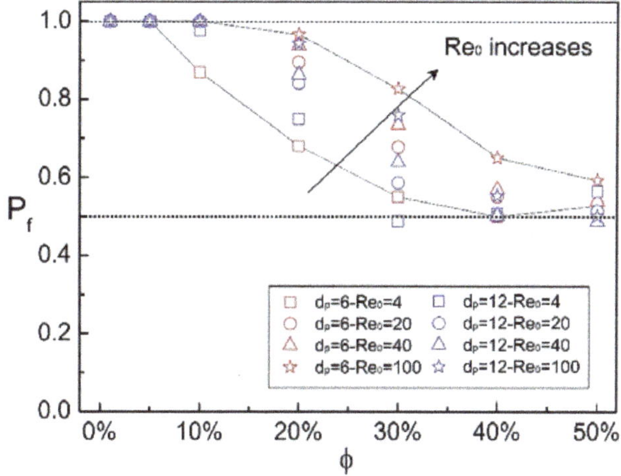

Figure 16. The degree of inertial migration P_f as a function of particle concentration for different particle sizes d_p and channel Reynolds numbers Re_0. The data was originally reported in [37].

The above discussion reveals that the degree of the migration is determined by the particle concentration and channel Reynolds number. A dimensionless focusing number was proposed to characterise the degree of migration [37],

$$F_c = \frac{Re_0^m}{\phi^n},$$

(44)

where $m = 0.36$ and $n = 2.33$ are obtained from the fitting of simulation data. Figure 17 shows the degree of migration P_f as a function of the focusing number F_c, where an empirical fitting is given by an exponential equation $P_f = 1 - 0.5 \times 10^{-0.0043 \times (\log F_c)^{6.51}}$. It can be seen that the numerical results coalesce onto a single master curve and three distinct regimes can be identified based on the value of F_c: a laterally unfocused regime ($F_c < 30$), a partially migration regime ($30 < F_c < 300$) and a fully migration regime ($F_c > 300$).

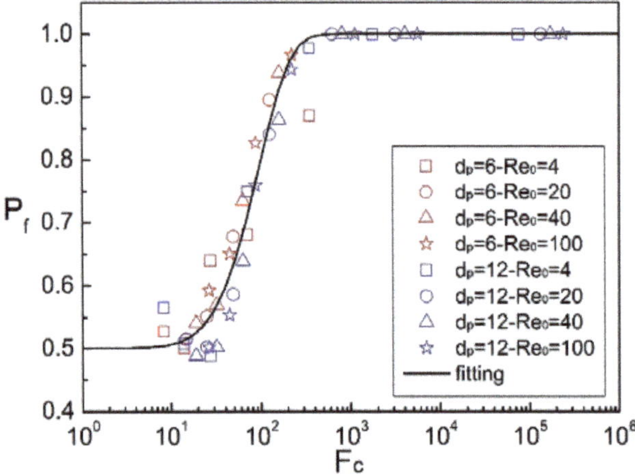

Figure 17. The degree of inertial migration as a function of the focusing number. The data was originally reported in [37].

6.2. Agglomeration of Adhesive Particles in Channel Flow

The inertial focusing introduced in the previous section has prompted a variety of novel applications in the microfabrication, such as particle filtration, segregation and flow cytometry in the micro channels [107–109]. An inevitable problem of particle agglomeration due to the cohesive nature when the size reduces to the micron scale must be taken into consideration. The agglomeration is ubiquitous in nature and industry, and sometimes will cause unfavourable effects in industrial facility. For example, the agglomeration of asphaltenes or gas hydrates results in the pipeline blockage in the oil and gas engineering [110,111]. Therefore, it is worth investigating the fundamental mechanism of micro-particle agglomeration.

As displayed in Figure 18, a stream of dilute particle suspension flowing through a 3D square channel is considered. Periodic boundary conditions are imposed at the inlet and outlet, while other four faces of the channel are set as 'no-slip' walls. The flow is driven by a constant pressure gradient. The particles are neutrally buoyant and adhesive, with initial positions randomly distributed in the channel. The computational parameters are set as follow. The dimension of the channel is $L \times H \times H = 201 \times 61 \times 61$. By varying the pressure gradient, the channel Reynolds number ranges between 5.4 and 108. The particle size is $d_p = 5$ and the particle concentration is fixed at 1%, corresponding to a particle number of 110. The strength of the particle adhesion is represented by the surface energy, which is chosen as 0.075 and 0.0075. The SRT model and the JKR contact theory are used in the LBM and DEM for this problem, respectively.

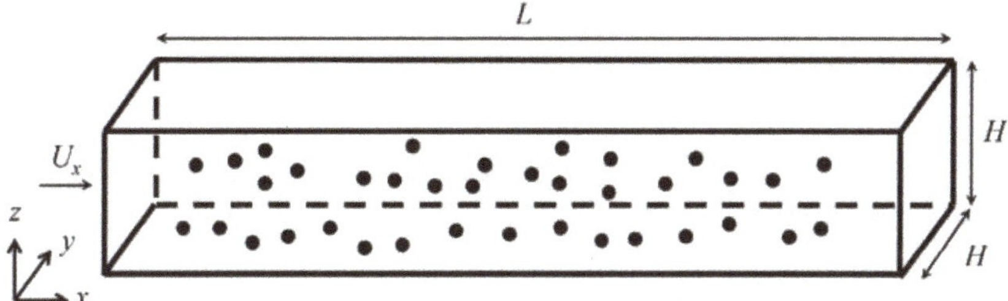

Figure 18. Schematic of agglomeration of adhesive particles in channel flow.

Figures 19 and 20 display the snapshots of adhesive particle suspensions with different surface energies $\gamma = 0.075$ and $\gamma = 0.0075$, respectively. It can be found that large sized agglomerates are formed with relatively higher surface energy ($\gamma = 0.075$). Some particles even stick on the wall due to strong adhesion. However, when the surface energy is reduced by one order of magnitude ($\gamma = 0.0075$), the size of the agglomerates becomes smaller and the particles appear to focus around a certain lateral position with $Re = 54$ and 108 (see Figure 20c,d), which resembles the Segré-Silberberg effect as discussed in the last section.

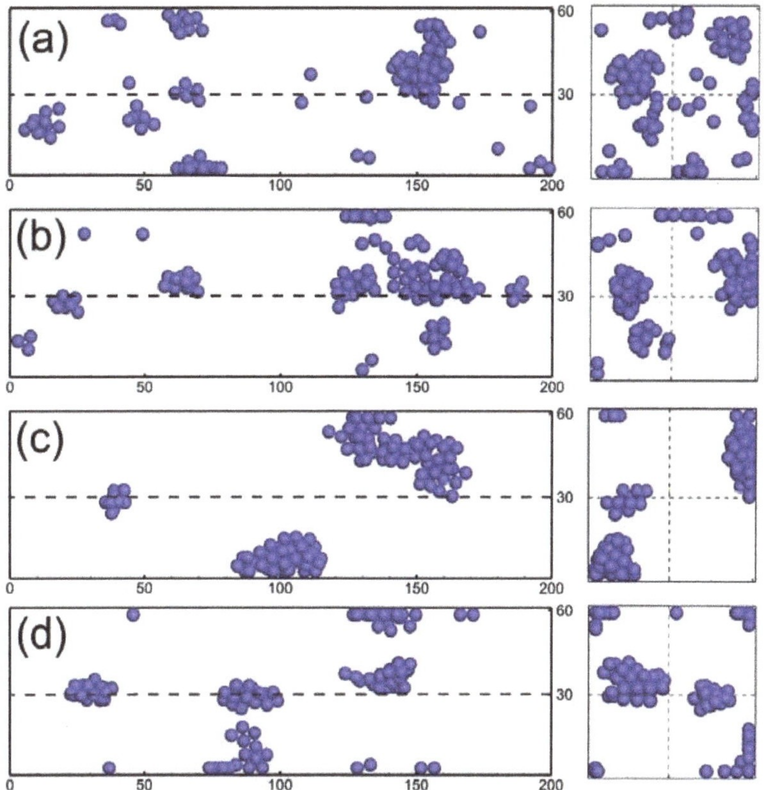

Figure 19. Snapshots of adhesive particle suspensions with $\gamma = 0.075$ and different Reynolds numbers (**a**) $Re = 5.4$, (**b**) $Re = 21.6$, (**c**) $Re = 54$ and (**d**) $Re = 108$. The left columns are side views and the right columns are cross-sectional view.

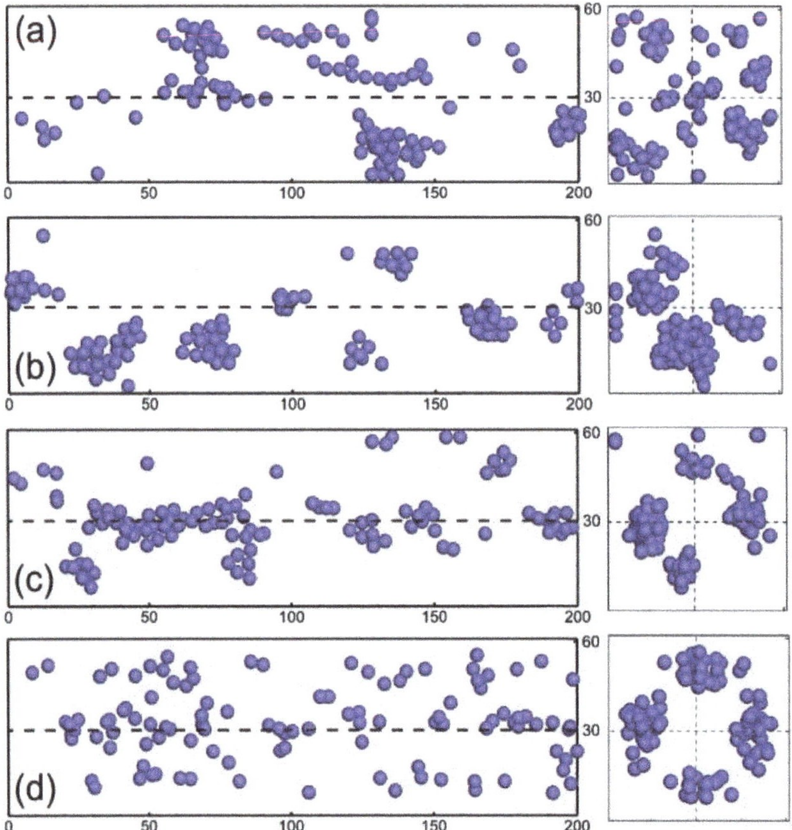

Figure 20. Snapshots of adhesive particle suspensions with $\gamma = 0.0075$ and different Reynolds numbers (**a**) $Re = 5.4$, (**b**) $Re = 21.6$, (**c**) $Re = 54$ and (**d**) $Re = 108$. The left columns are side views and the right columns are cross-sectional view.

Figure 21 shows the lateral PDF of the particle suspensions based on the following definition,

$$f(l) = \frac{N(l, l + \Delta l)}{N},\qquad(45)$$

where N is the total number of particles in the channel and $N(l, l+\Delta l)$ is the number of particles in the square annulus between l and $l + \Delta l$. For comparison, the channel flow with non-adhesive particles is also modelled and the results are included. It can be seen from Figure 21a that a single peak at the lateral position around 0.6 is observed in the PDF for non-adhesive particles, which moves closer to the wall as Re increases and is in consistent with the Segré and Silberberg effect. However, the PDFs are quite distinct for adhesive particles. The single peak distribution is not clearly observed for particles with $\gamma = 0.075$ (see Figure 21b). Moreover, there appears to be a peak close to the wall position, where a number of particles are found to stick to wall because of the strong adhesion, as indicated by Figure 19. On the other hand, for much less adhesive particles ($\gamma = 0.0075$), the PDFs become similar to those of non-adhesive particles (see Figure 21c). The average lateral position as a function of Re is further shown in Figure 22. Note that the relatively larger value of average lateral position for $Re = 5.4$ is caused by the statistical effect [38], because the fluid inertia is relatively small to drive the lateral migration and the particles almost remain randomly distributed in the cross-sectional area. Except for the data for $Re = 5.4$, it is observed that the average lateral position increases monotonically with the increase of Re for both non-adhesive and less adhesive particles ($\gamma = 0.0075$), which demonstrates a similar dynamics between non-adhesive and weak adhesive particles. However, for strongly adhesive particles ($\gamma = 0.075$), the monotonic increase of the average lateral position disappears because of the agglomeration.

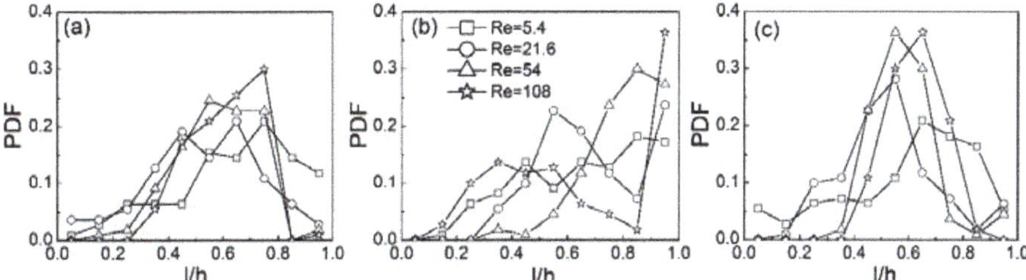

Figure 21. Lateral probability distribution functions for different cases, (**a**) non-adhesive particles, (**b**) adhesive particles with $\gamma = 0.075$, and (**c**) adhesive particles with $\gamma = 0.0075$.

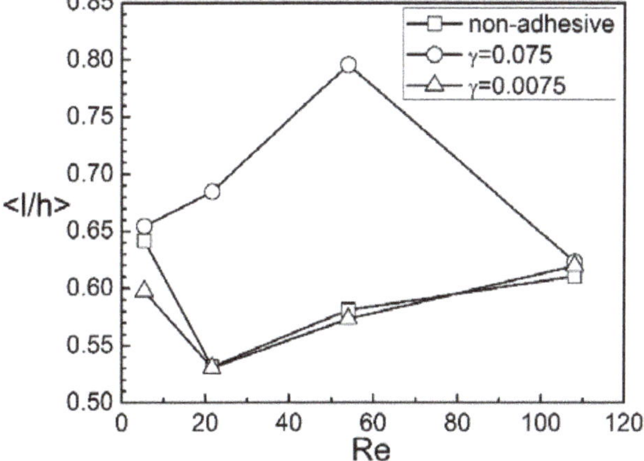

Figure 22. The average lateral position as a function of Reynolds number.

The agglomeration mechanism for such channel flow can be attributed to the competition between the interparticle adhesive contact force and the hydrodynamic force. The adhesive force sticks particles together, which imposes a positive effect on the agglomeration. The hydrodynamic force exerts two opposite effects on the particle suspensions. First, it induces the particle lateral migration, which motivates the interparticle collision and facilitates the agglomeration. Secondly, the fluid inertia also tears particles apart from each other, resulting in the breakup of agglomerates. As a consequence, a new dimensionless adhesion number Ad is proposed based on the balance of the critical pull-off force and the drag force [11,12,38],

$$Ad = \frac{\gamma}{\mu_f U},$$ (46)

where μ_f is the fluid dynamic viscosity and U is the mean fluid velocity. Using the adhesion number, the degree of the agglomeration can be characterised and represented by the agglomerate ratio that is defined as the ratio of the number of particles involved in agglomerate to the total particle number. Figure 23 shows the relationship between the agglomerate ratio and the adhesion number. A monotonic increase of the agglomerate ratio as a function of *Ad* is observed. After *Ad* reaches a critical value, the agglomerate ratio seems to converge to one. Therefore, two different regimes can be identified, i.e., a partial agglomeration regime where the hydrodynamic force dominates over the adhesive force, and a complete agglomeration regime where all the particles are involved in agglomerates.

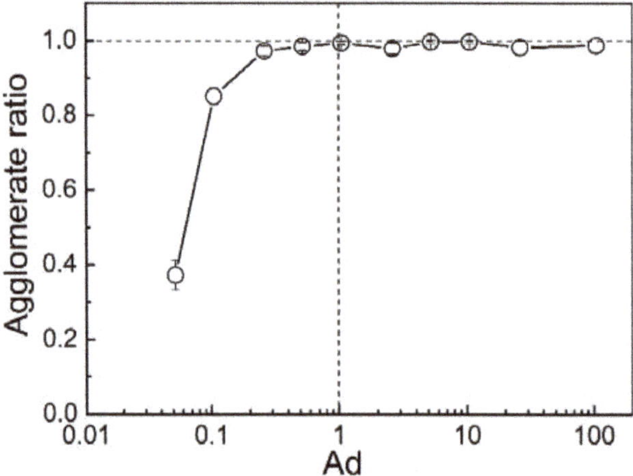

Figure 23. The agglomerate ratio as a function of the adhesion number. The vertical dashed line denotes the critical value of adhesion number in the present study. The data was originally reported in [38].

6.3. Sedimentation of Particle Suspensions in Cavity Flow

Sedimentation is extensively used in industries to separate particles from fluid or other particles with different sizes or velocities, such as separating particles and cells in microfluids [112], dewatering coal slurries [113], and post-treating wastewater [114–116]. Furthermore, sediments are also physical pollutants in the river, which could have great influence on the ecosystem and human health [117–120]. Therefore, it is of great importance to study the sedimentation, which is a very complicated process, due to the complex fluid mechanics involved with the sediment particles. With LBM-DEM, a numerical investigation is then performed to gain some insights to the sedimentation problem.

As shown in Figure 24, we consider a sedimentation system composed of a cuboid channel and a cavity, which is placed in the middle of the channel. The boundaries in x and y directions are periodic, and all the other faces are set as no-slip walls. The particles are initially generated at random positions in the main channel (no particles in the cavity). The particle density is larger than the fluid, so that they can sedimentate into the cavity under the gravity. The size of the main channel is $L \times W \times H = 321 \times 81 \times 49$. The cavity has the same width as the main channel. The length of the cavity is $l = 40$ and 80, while the height is $h = 16$ and 32, which gives rise to four configurations of the cavity. The channel Reynolds number, defined as $Re = U_x H / v$ varies between 28.44 and 88.32. The particle diameter is $d_p = 6$ and the density is in the range 1.1 ~ 2.0. The total number of particles is fixed at 110, which is equal to an overall concentration less than 1%. The SRT model and the Hertz model are used in the LBM and DEM for this particular problem, respectively.

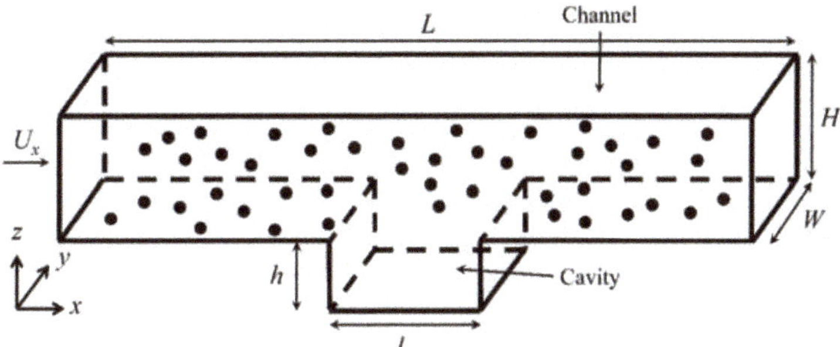

Figure 24. Schematic of the sedimentation system.

Figure 25 shows the snapshots of the particle suspensions at different time. The particles are initially located at random positions in the main channel with no particles trapped inside the cavity. Then the particles settle to the bottom of the channel and flow over the cavity. During the transportation of the suspensions, some particles fall into the cavity and stay stationary inside, while some other keep circulating around an orbit in the central vortex inside the cavity. The total number of particles trapped inside the cavity is evaluated after a sufficiently long time, when the transportation of the particles reaches a steady state. From Figure 25 the typical trajectories of particle suspensions in the cavity are analysed, where three distinct dynamic behaviours can be identified: resuspension, deposition and circulation. Resuspension means that the particles fall into the cavity and then come out due to the lift force from the fluid. Deposition happens mostly at the rear edge of the cavity, where the particles deposit to the corner of the cavity and stay motionless. Circulation indicates that the particles are captured by the central vortex inside the cavity and keeps moving along an orbit in a periodic way. Both deposition and circulation are treated as the entrapment of a particle.

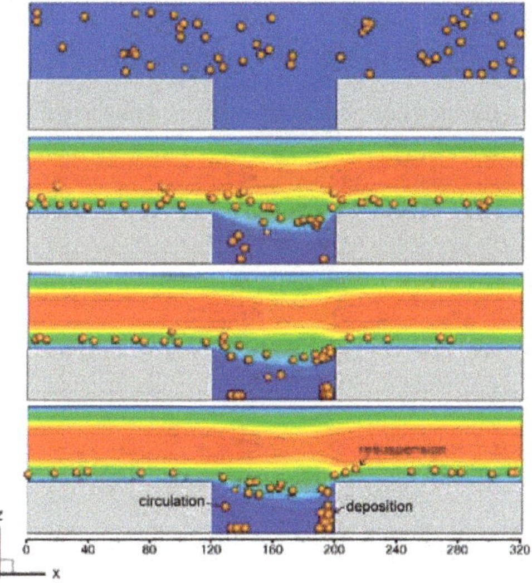

Figure 25. Snapshots of the particle suspensions for $Re = 56.88$, $h = 32$, $l = 80$ and $\rho_p = 1.5$. The time point for each subplot is $t = 0$, $t = 50{,}000$, $t = 200{,}000$ and $t = 500{,}000$ from top to bottom, respectively. The flow field only contains a slice at the mid-plane in y direction. Some particles may be sheltered and look smaller.

Figure 26 shows the trap efficiency η, defined as the ratio of the trapped particle number to the total particle number, as a function of the particle density for different cavity configurations and Reynolds numbers. It is clear that as the particle density increases, the trap efficiency increases continuously from 0 to 1. For a fixed Reynolds number, the largest cavity ($h = 32$, $l = 80$) has the highest trap efficiency, while the smallest cavity ($h = 16$, $l = 40$) has the lowest trap efficiency. The results of the other two cavity configurations are in between, which indicates that increasing either length or depth leads to a higher trap efficiency. Furthermore, for a fixed cavity configuration, an increase in the Reynolds number results in a decrease in the trap efficiency. With a larger Reynolds number, the particle's translational velocity in x direction becomes larger, so that it has less time to enter the cavity in the vertical direction. Meanwhile, a larger Reynolds number produces a larger lift force. As a consequence, the particles are less easily trapped in the cavity.

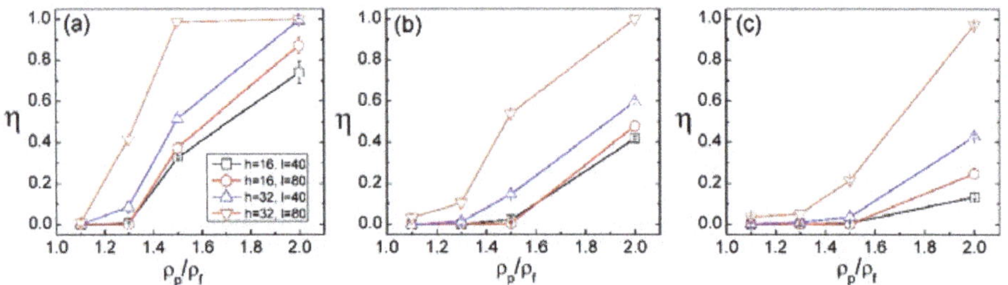

Figure 26. The trap efficiency as a function of particle density for different Reynolds numbers: (**a**) $Re = 28.44$, (**b**) $Re = 56.88$, (**c**) $Re = 85.32$.

7. Summary

An overview of the coupled discrete element method with the lattice Boltzmann method is presented together with its applications in solid–liquid flows. The fundamentals of DEM and LBM, including the contact mechanics based on Hertz model and JKR theory, the lattice Boltzmann equation with both single-relaxation-time model and multi-relaxation-time model are introduced. Several solid–fluid interaction models, i.e., the coupling between DEM and LBM, are discussed and compared in detail. The validity and accuracy of the numerical method are validated for three classical solid–liquid flow problems, and the results are in good quantitative agreement with those from experiments and other numerical approaches. Three case studies, including the inertial migration of particle suspensions, the agglomeration of adhesive particles, and the sedimentation problem, are also performed to demonstrate the capability of the coupled DEM-LBM to solid–liquid flow problems and potential engineering applications.

Author Contributions: Conceptualization, C.-Y.W.; formal analysis, W.L.; funding acquisition, C.-Y.W.; methodology, W.L. and C.-Y.W.; project administration, C.-Y.W.; supervision, C.-Y.W.; validation, W.L.; writing—original draft, W.L.; writing—review and editing, C.-Y.W. All authors have read and agreed to the published version of the manuscript.

Funding: This research was funded by the Engineering and Physical Sciences Research Council (EPSRC), grant number No: EP/N033876/1.

Acknowledgments: The authors acknowledge Duo Zhang and Nicolin Govender in the University of Surrey for their helpful suggestions and fruitful discussion in developing the numerical approach.

Conflicts of Interest: The authors declare no conflict of interest.

References

1. Zhu, H.; Zhou, Z.; Yang, R.; Yu, A. Discrete particle simulation of particulate systems: Theoretical developments. *Chem. Eng. Sci.* **2007**, *62*, 3378–3396. [CrossRef]
2. Zhu, H.; Zhou, Z.; Yang, R.; Yu, A. Discrete particle simulation of particulate systems: A review of major applications and findings. *Chem. Eng. Sci.* **2008**, *63*, 5728–5770. [CrossRef]
3. Li, S.; Marshall, J.S.; Liu, G.; Yao, Q. Adhesive particulate flow: The discrete-element method and its application in energy and environmental engineering. *Prog. Energy Combust. Sci.* **2011**, *37*, 633–668. [CrossRef]
4. Hounslow, M.; Ryall, R.L.; Marshall, V.R. A discretized population balance for nucleation, growth, and aggregation. *AIChE J.* **1988**, *34*, 1821–1832. [CrossRef]
5. Lister, J.D.; Smit, D.J.; Hounslow, M. Adjustable discretized population balance for growth and aggregation. *AIChE J.* **1995**, *41*, 591–603. [CrossRef]
6. Gidaspow, D. *Multiphase Flow and Fluidization*; Elsevier: Amsterdam, The Netherlands, 1994.
7. Zhu, R.; Zhu, W.; Xing, L.; Sun, Q. DEM simulation on particle mixing in dry and wet particles spouted bed. *Powder Technol.* **2011**, *210*, 73–81. [CrossRef]
8. Liu, G.; Li, S.; Yao, Q. A JKR-based dynamic model for the impact of micro-particle with a flat surface. *Powder Technol.* **2011**, *207*, 215–223. [CrossRef]
9. Yang, M.; Li, S.; Yao, Q. Mechanistic studies of initial deposition of fine adhesive particles on a fiber using discrete-element methods. *Powder Technol.* **2013**, *248*, 44–53. [CrossRef]
10. Chen, S.; Li, S.; Yang, M. Sticking/rebound criterion for collisions of small adhesive particles: Effects of impact parameter and particle size. *Powder Technol.* **2015**, *274*, 431–440. [CrossRef]
11. Liu, W.; Li, S.; Baule, A.; Makse, H.A. Adhesive loose packings of small dry particles. *Soft Matter* **2015**, *11*, 6492–6498. [CrossRef]
12. Liu, W.; Li, S.; Chen, S. Computer simulation of random loose packings of micro-particles in presence of adhesion and friction. *Powder Technol.* **2016**, *302*, 414–422. [CrossRef]
13. Liu, W.; Jin, Y.; Li, S.; Chen, S.; Makse, H.A. Equation of state for random sphere packing with arbitrary adhesion and friction. *Soft Matter* **2017**, *13*, 421–427. [CrossRef] [PubMed]
14. Liu, W.; Chen, S.; Li, S. Random adhesive loose packings of micron-sized particles under a uniform flow field. *Powder Technol.* **2018**, *335*, 70–76. [CrossRef]
15. Chen, S.; Liu, W.; Li, S. A fast adhesive discrete element method for random packings of fine particles. *Chem. Eng. Sci.* **2019**, *193*, 336–345. [CrossRef]
16. Zhang, H.; Sharma, G.; Wang, Y.; Li, S.; Biswas, P. Numerical modeling of the performance of high flow DMAs to classify sub-2 nm particles. *Aerosol Sci. Technol.* **2018**, *53*, 106–118. [CrossRef]
17. Dong, M.; Li, J.; Shang, Y.; Li, S. Numerical investigation on deposition process of submicron particles in collision with a single cylindrical fiber. *J. Aerosol Sci.* **2019**, *129*, 1–15. [CrossRef]
18. Cundall, P.A.; Strack, O.D.L. A discrete numerical model for granular assemblies. *Géotechnique* **1979**, *29*, 47–65. [CrossRef]
19. Chen, S.; Doolen, G.D. Lattice boltzmann method for fluid flows. *Annu. Rev. Fluid Mech.* **1998**, *30*, 329–364. [CrossRef]
20. Aidun, C.K.; Clausen, J.R. Lattice-Boltzmann Method for Complex Flows. *Annu. Rev. Fluid Mech.* **2010**, *42*, 439–472. [CrossRef]
21. Krüger, T.; Kusumaatmaja, H.; Kuzmin, A.; Shardt, O.; Silva, G.; Viggen, E.M. *The Lattice Boltzmann Method*; Springer Science and Business Media: Berlin, Germany, 2017; Volume 10, pp. 4–15.
22. Moin, P.; Mahesh, K. Direct Numerical Simulation: A Tool in Turbulence Research. *Annu. Rev. Fluid Mech.* **1998**, *30*, 539–578. [CrossRef]
23. Sagaut, P. *Large Eddy Simulation for Incompressible Flows*; Springer Science and Business Media: Berlin, Germany, 2006.
24. Ishii, M.; Mishima, K. Two-fluid model and hydrodynamic constitutive relations. *Nucl. Eng. Des.* **1984**, *82*, 107–126. [CrossRef]
25. Tsuji, Y.; Kawaguchi, T.; Tanaka, T. Discrete particle simulation of two-dimensional fluidized bed. *Powder Technol.* **1993**, *77*, 79–87. [CrossRef]

26. Kuipers, J.; Van Duin, K.; Van Beckum, F.; Van Swaaij, W. A numerical model of gas-fluidized beds. *Chem. Eng. Sci.* **1992**, *47*, 1913–1924. [CrossRef]

27. Kafui, K.; Thornton, C.; Adams, M. Discrete particle-continuum fluid modelling of gas-solid fluidised beds. *Chem. Eng. Sci.* **2002**, *57*, 2395–2410. [CrossRef]

28. Guo, Y.; Kafui, K.D.; Wu, C.-Y.; Thornton, C.; Seville, J.P.K. A coupled DEM/CFD analysis of the effect of air on powder flow during die filling. *AIChE J.* **2009**, *55*, 49–62. [CrossRef]

29. Guo, Y.; Wu, C.-Y.; Kafui, K.; Thornton, C. 3D DEM/CFD analysis of size-induced segregation during die filling. *Powder Technol.* **2011**, *206*, 177–188. [CrossRef]

30. Guo, Y.; Wu, C.-Y.; Thornton, C. Modeling gas-particle two-phase flows with complex and moving boundaries using DEM-CFD with an immersed boundary method. *AIChE J.* **2012**, *59*, 1075–1087. [CrossRef]

31. Feng, Y.T.; Han, K.; Owen, D.R.J. Coupled lattice Boltzmann method and discrete element modelling of particle transport in turbulent fluid flows: Computational issues. *Int. J. Numer. Methods Eng.* **2007**, *72*, 1111–1134. [CrossRef]

32. Strack, O.E.; Cook, B.K. Three-dimensional immersed boundary conditions for moving solids in the lattice-Boltzmann method. *Int. J. Numer. Methods Fluids* **2007**, *55*, 103–125. [CrossRef]

33. van der Hoef, M.A.; Annaland, M.V.S.; Deen, N.; Kuipers, H. Numerical Simulation of Dense Gas-Solid Fluidized Beds: A Multiscale Modeling Strategy. *Annu. Rev. Fluid Mech.* **2008**, *40*, 47–70. [CrossRef]

34. Van Wachem, B.; Schouten, J.C.; Bleek, C.M.V.D.; Krishna, R.; Sinclair, J.L. Comparative analysis of CFD models of dense gas–solid systems. *AIChE J.* **2001**, *47*, 1035–1051. [CrossRef]

35. Ladd, A.J.C.; Verberg, R. Lattice-Boltzmann Simulations of Particle-Fluid Suspensions. *J. Stat. Phys.* **2001**, *104*, 1191–1251. [CrossRef]

36. Nguyen, N.-Q.; Ladd, A.J.C. Lubrication corrections for lattice-Boltzmann simulations of particle suspensions. *Phys. Rev. E* **2002**, *66*, 046708. [CrossRef] [PubMed]

37. Liu, W. Analysis of inertial migration of neutrally buoyant particle suspensions in a planar Poiseuille flow with a coupled lattice Boltzmann method-discrete element method. *Phys. Fluids* **2019**, *31*, 063301. [CrossRef]

38. Liu, W.; Wu, C.-Y. Migration and agglomeration of adhesive microparticle suspensions in a pressure-driven duct flow. *AIChE J.* **2020**, *66*, 16974. [CrossRef]

39. Maier, R.S.; Kroll, D.M.; Kutsovsky, Y.E.; Davis, H.T.; Bernard, R.S. Simulation of flow through bead packs using the lattice Boltzmann method. *Phys. Fluids* **1998**, *10*, 60–74. [CrossRef]

40. Guo, Z.; Zhao, T. Lattice Boltzmann model for incompressible flows through porous media. *Phys. Rev. E* **2002**, *66*, 036304. [CrossRef] [PubMed]

41. Inamuro, T.; Ogata, T.; Tajima, S.; Konishi, N. A lattice Boltzmann method for incompressible two-phase flows with large density differences. *J. Comput. Phys.* **2004**, *198*, 628–644. [CrossRef]

42. Li, Q.; Luo, K.H.; Kang, Q.; He, Y.; Chen, Q.; Liu, Q. Lattice Boltzmann methods for multiphase flow and phase-change heat transfer. *Prog. Energy Combust. Sci.* **2016**, *52*, 62–105. [CrossRef]

43. Han, Y.; Cundall, P.A. LBM-DEM modeling of fluid-solid interaction in porous media. *Int. J. Numer. Anal. Methods Geomech.* **2012**, *37*, 1391–1407. [CrossRef]

44. Peng, C.; Ayala, O.M.; Wang, L.-P. A direct numerical investigation of two-way interactions in a particle-laden turbulent channel flow. *J. Fluid Mech.* **2019**, *875*, 1096–1144. [CrossRef]

45. Chen, S.; Chen, H.; Martnez, D.; Matthaeus, W. Lattice Boltzmann model for simulation of magnetohydrodynamics. *Phys. Rev. Lett.* **1991**, *67*, 3776–3779. [CrossRef] [PubMed]

46. Pei, C.; Wu, C.-Y.; England, D.; Byard, S.; Berchtold, H.; Adams, M. DEM-CFD modeling of particle systems with long-range electrostatic interactions. *AIChE J.* **2015**, *61*, 1792–1803. [CrossRef]

47. Pei, C.; Wu, C.-Y.; Adams, M.; England, D.; Byard, S.; Berchtold, H. Contact electrification and charge distribution on elongated particles in a vibrating container. *Chem. Eng. Sci.* **2015**, *125*, 238–247. [CrossRef]

48. Pei, C.; Wu, C.-Y.; Adams, M. Numerical analysis of contact electrification of non-spherical particles in a rotating drum. *Powder Technol.* **2015**, *285*, 110–122. [CrossRef]

49. Chen, S.; Li, S.; Liu, W.; Makse, H.A. Effect of long-range repulsive Coulomb interactions on packing structure of adhesive particles. *Soft Matter* **2016**, *12*, 1836–1846. [CrossRef]

50. Chen, S.; Liu, W.; Li, S. Effect of long-range electrostatic repulsion on pore clogging during microfiltration. *Phys. Rev. E* **2016**, *94*, 063108. [CrossRef]

51. Chen, S.; Liu, W.; Li, S. Scaling laws for migrating cloud of low-Reynolds-number particles with Coulomb repulsion. *J. Fluid Mech.* **2017**, *835*, 880–897. [CrossRef]

52. Zhu, R.; Li, S.; Yao, Q. Effects of cohesion on the flow patterns of granular materials in spouted beds. *Phys. Rev. E* **2013**, *87*, 022206. [CrossRef]

53. Zhang, H.; Li, S. DEM simulation of wet granular-fluid flows in spouted beds: Numerical studies and experimental verifications. *Powder Technol.* **2017**, *318*, 337–349. [CrossRef]

54. Chen, H.; Liu, W.; Li, S. Deposition of wet microparticles on a fiber: Effects of impact velocity and initial spin. *Powder Technol.* **2019**, *357*, 83–96. [CrossRef]

55. Hertz, H. Ueber die Berührung fester elastischer Körper. *J. für die reine und angewandte Mathematik (Crelles J.)* **1882**, *1882*, 156–171. [CrossRef]

56. Johnson, K.L.; Kendall, K.; Roberts, A.D. Surface energy and the contact of elastic solids. *Proc. R. Soc. London. Ser. A Math. Phys. Sci.* **1971**, *324*, 301–313. [CrossRef]

57. Derjaguin, B.; Muller, V.; Toporov, Y. Effect of contact deformations on the adhesion of particles. *J. Colloid Interface Sci.* **1975**, *53*, 314–326. [CrossRef]

58. Marshall, J.S.; Li, S. *Adhesive Particle Flows*; Cambridge University Press (CUP): Cambridge, UK, 2014.

59. Bhatnagar, P.L.; Gross, E.P.; Krook, M. A Model for Collision Processes in Gases. I. Small Amplitude Processes in Charged and Neutral One-Component Systems. *Phys. Rev.* **1954**, *94*, 511–525. [CrossRef]

60. Chen, H.; Chen, S.; Matthaeus, W.H. Recovery of the Navier-Stokes equations using a lattice-gas Boltzmann method. *Phys. Rev. A* **1992**, *45*, R5339–R5342. [CrossRef] [PubMed]

61. Qian, Y.H.; D'Humières, D.; Lallemand, P. Lattice BGK Models for Navier-Stokes Equation. *EPL (Europhysics Lett.)* **1992**, *17*, 479–484. [CrossRef]

62. Wolf-Gladrow, D.A. *Lattice-Gas Cellular Automata and Lattice Boltzmann Models: An Introduction*; Springer: New York, NY, USA, 2004.

63. Guo, Z.; Zheng, C.; Shi, B. Discrete lattice effects on the forcing term in the lattice Boltzmann method. *Phys. Rev. E* **2002**, *65*, 046308. [CrossRef]

64. Lallemand, P.; Luo, L.-S. Theory of the lattice Boltzmann method: Dispersion, dissipation, isotropy, Galilean invariance, and stability. *Phys. Rev. E* **2000**, *61*, 6546–6562. [CrossRef]

65. Ladd, A.J.C. Numerical simulations of particulate suspensions via a discretized Boltzmann equation. Part 1. Theoretical foundation. *J. Fluid Mech.* **1994**, *271*, 285. [CrossRef]

66. Ladd, A.J.C. Numerical simulations of particulate suspensions via a discretized Boltzmann equation. Part 2. Numerical results. *J. Fluid Mech.* **1994**, *271*, 311. [CrossRef]

67. Bouzidi, M.; Firdaouss, M.; Lallemand, P. Momentum transfer of a Boltzmann-lattice fluid with boundaries. *Phys. Fluids* **2001**, *13*, 3452–3459. [CrossRef]

68. Filippova, O.; Hänel, D. Grid Refinement for Lattice-BGK Models. *J. Comput. Phys.* **1998**, *147*, 219–228. [CrossRef]

69. Mei, R.; Luo, L.-S.; Shyy, W. An Accurate Curved Boundary Treatment in the Lattice Boltzmann Method. *J. Comput. Phys.* **1999**, *155*, 307–330. [CrossRef]

70. Mei, R.; Shyy, W.; Yu, D.; Luo, L.-S. Lattice Boltzmann Method for 3-D Flows with Curved Boundary. *J. Comput. Phys.* **2000**, *161*, 680–699. [CrossRef]

71. Yu, D.; Mei, R.; Luo, L.-S.; Shyy, W. Viscous flow computations with the method of lattice Boltzmann equation. *Prog. Aerosp. Sci.* **2003**, *39*, 329–367. [CrossRef]

72. Peng, C.; Teng, Y.; Hwang, B.; Guo, Z.; Wang, L.-P. Implementation issues and benchmarking of lattice Boltzmann method for moving rigid particle simulations in a viscous flow. *Comput. Math. Appl.* **2016**, *72*, 349–374. [CrossRef]

73. Peskin, C.S. Numerical analysis of blood flow in the heart. *J. Comput. Phys.* **1977**, *25*, 220–252. [CrossRef]

74. Peskin, C.S. The immersed boundary method. *Acta Numer.* **2002**, *11*, 479–517. [CrossRef]

75. Feng, Z.-G.; Michaelides, E.E. The immersed boundary-lattice Boltzmann method for solving fluid–particles interaction problems. *J. Comput. Phys.* **2004**, *195*, 602–628. [CrossRef]

76. Uhlmann, M. An immersed boundary method with direct forcing for the simulation of particulate flows. *J. Comput. Phys.* **2005**, *209*, 448–476. [CrossRef]

77. Breugem, W.-P. A second-order accurate immersed boundary method for fully resolved simulations of particle-laden flows. *J. Comput. Phys.* **2012**, *231*, 4469–4498. [CrossRef]

78. Favier, J.; Revell, A.; Pinelli, A. A Lattice Boltzmann–Immersed Boundary method to simulate the fluid interaction with moving and slender flexible objects. *J. Comput. Phys.* **2014**, *261*, 145–161. [CrossRef]

79. Valero-Lara, P.; Igual, F.D.; Prieto, M.; Pinelli, A.; Favier, J. Accelerating fluid–solid simulations (Lattice-Boltzmann & Immersed-Boundary) on heterogeneous architectures. *J. Comput. Sci.* **2015**, *10*, 249–261. [CrossRef]

80. Valero-Lara, P.; Pinelli, A.; Prieto, M. Accelerating Solid-fluid Interaction using Lattice-boltzmann and Immersed Boundary Coupled Simulations on Heterogeneous Platforms. *Procedia Comput. Sci.* **2014**, *29*, 50–61. [CrossRef]

81. Noble, D.R.; Torczynski, J.R. A Lattice-Boltzmann Method for Partially Saturated Computational Cells. *Int. J. Mod. Phys. C* **1998**, *9*, 1189–1201. [CrossRef]

82. Owen, D.R.J.; Leonardi, C.R.; Feng, Y. An efficient framework for fluid-structure interaction using the lattice Boltzmann method and immersed moving boundaries. *Int. J. Numer. Methods Eng.* **2010**, *87*, 66–95. [CrossRef]

83. Jones, B.D.; Williams, J.R. Fast computation of accurate sphere-cube intersection volume. *Eng. Comput.* **2017**, *34*, 1204–1216. [CrossRef]

84. Sutera, S.P.; Skalak, R. The history of Poiseuille's law. *Annu. Rev. Fluid Mechan.* **1993**, *25*, 1–20. [CrossRef]

85. Wen, B.; Zhang, C.; Tu, Y.; Wang, C.; Fang, H. Galilean invariant fluid–solid interfacial dynamics in lattice Boltzmann simulations. *J. Comput. Phys.* **2014**, *266*, 161–170. [CrossRef]

86. Cate, A.T.; Nieuwstad, C.H.; Derksen, J.J.; Akker, H.V.D. Particle imaging velocimetry experiments and lattice-Boltzmann simulations on a single sphere settling under gravity. *Phys. Fluids* **2002**, *14*, 4012–4025. [CrossRef]

87. Tritton, D.J. Experiments on the flow past a circular cylinder at low Reynolds numbers. *J. Fluid Mech.* **1959**, *6*, 547. [CrossRef]

88. Schiller, L.; Naumann, A. A drag coefficient correlation. *Z. des Ver. Deutsch. Ing.* **1935**, *77*, 318–320.

89. Segre, G.; Silberberg, A. Behaviour of macroscopic rigid spheres in Poiseuille flow Part 1. Determination of local concentration by statistical analysis of particle passages through crossed light beams. *J. Fluid Mech.* **1962**, *14*, 115–135. [CrossRef]

90. Segre, G.; Silberberg, A. Behaviour of macroscopic rigid spheres in Poiseuille flow Part 2. Experimental results and interpretation. *J. Fluid Mech.* **1962**, *14*, 136–157. [CrossRef]

91. Han, M.; Kim, C.; Kim, M.; Lee, S. Particle migration in tube flow of suspensions. *J. Rheol.* **1999**, *43*, 1157–1174. [CrossRef]

92. Matas, J.-P.; Morris, J.F.; Guazzelli, É. Inertial migration of rigid spherical particles in Poiseuille flow. *J. Fluid Mech.* **2004**, *515*, 171–195. [CrossRef]

93. Di Carlo, D.; Irimia, D.; Tompkins, R.G.; Toner, M. Continuous inertial focusing, ordering, and separation of particles in microchannels. *Proc. Natl. Acad. Sci. USA* **2007**, *104*, 18892–18897. [CrossRef]

94. Choi, Y.-S.; Seo, K.W.; Lee, S.J. Lateral and cross-lateral focusing of spherical particles in a square microchannel. *Lab. Chip* **2011**, *11*, 460–465. [CrossRef]

95. Seo, K.W.; Kang, Y.J.; Lee, S.J. Lateral migration and focusing of microspheres in a microchannel flow of viscoelastic fluids. *Phys. Fluids* **2014**, *26*, 063301. [CrossRef]

96. Ho, B.P.; Leal, L.G. Inertial migration of rigid spheres in two-dimensional unidirectional flows. *J. Fluid Mech.* **1974**, *65*, 365–400. [CrossRef]

97. Schonberg, J.A.; Hinch, E.J. Inertial migration of a sphere in Poiseuille flow. *J. Fluid Mech.* **1989**, *203*, 517. [CrossRef]

98. Asmolov, E.S. The inertial lift on a spherical particle in a plane Poiseuille flow at large channel Reynolds number. *J. Fluid Mech.* **1999**, *381*, 63–87. [CrossRef]

99. Matas, J.-P.; Morris, J.F.; Guazzelli, É. Lateral force on a rigid sphere in large-inertia laminar pipe flow. *J. Fluid Mech.* **2009**, *621*, 59. [CrossRef]

100. Feng, J.J.; Hu, H.H.; Joseph, D.D. Direct simulation of initial value problems for the motion of solid bodies in a Newtonian fluid. Part 2. Couette and Poiseuille flows. *J. Fluid Mech.* **1994**, *277*, 271. [CrossRef]

101. Yang, B.H.; Wang, J.; Joseph, D.D.; Hu, H.H.; Pan, T.-W.; Glowinski, R. Migration of a sphere in tube flow. *J. Fluid Mech.* **2005**, *540*, 109. [CrossRef]

102. Shao, X.; Yu, Z.; Sun, B. Inertial migration of spherical particles in circular Poiseuille flow at moderately high Reynolds numbers. *Phys. Fluids* **2008**, *20*, 103307. [CrossRef]

103. Inamuro, T.; Maeba, K.; Ogino, F. Flow between parallel walls containing the lines of neutrally buoyant circular cylinders. *Int. J. Multiph. Flow* **2000**, *26*, 1981–2004. [CrossRef]

104. Chun, B.; Ladd, A.J.C. Inertial migration of neutrally buoyant particles in a square duct: An investigation of multiple equilibrium positions. *Phys. Fluids* **2006**, *18*, 31704. [CrossRef]

105. Sun, D.-K.; Bo, Z. Numerical simulation of hydrodynamic focusing of particles in straight channel flows with the immersed boundary-lattice Boltzmann method. *Int. J. Heat Mass Transf.* **2015**, *80*, 139–149. [CrossRef]

106. Hu, J.; Guo, Z. A numerical study on the migration of a neutrally buoyant particle in a Poiseuille flow with thermal convection. *Int. J. Heat Mass Transf.* **2017**, *108*, 2158–2168. [CrossRef]

107. Ookawara, S.; Agrawal, M.; Street, D.; Ogawa, K. Quasi-direct numerical simulation of lift force-induced particle separation in a curved microchannel by use of a macroscopic particle model. *Chem. Eng. Sci.* **2007**, *62*, 2454–2465. [CrossRef]

108. Di Carlo, D. Inertial microfluidics. *Lab Chip* **2009**, *9*, 3038. [CrossRef] [PubMed]

109. Ahn, S.W.; Lee, S.S.; Lee, S.J.; Kim, J.M. Microfluidic particle separator utilizing sheathless elasto-inertial focusing. *Chem. Eng. Sci.* **2015**, *126*, 237–243. [CrossRef]

110. Eskin, D.; Ratulowski, J.; Akbarzadeh, K.; Pan, S. Modelling asphaltene deposition in turbulent pipeline flows. *Can. J. Chem. Eng.* **2011**, *89*, 421–441. [CrossRef]

111. Balakin, B.V.; Hoffmann, A.; Kosinski, P. Experimental study and computational fluid dynamics modeling of deposition of hydrate particles in a pipeline with turbulent water flow. *Chem. Eng. Sci.* **2011**, *66*, 755–765. [CrossRef]

112. Bernate, J.A.; Liu, C.; Lagae, L.; Konstantopoulos, K.; Drazer, G. Vector separation of particles and cells using an array of slanted open cavities. *Lab Chip* **2013**, *13*, 1086–1092. [CrossRef] [PubMed]

113. Davis, R.H.; Acrivos, A. Sedimentation of noncolloidal particles at low Reynolds numbers. *Annu. Rev. Fluid Mechan.* **1985**, *17*, 91–118. [CrossRef]

114. Mahmoud, M.; Tawfik, A.; El-Gohary, F. Use of down-flow hanging sponge (DHS) reactor as a promising post-treatment system for municipal wastewater. *Chem. Eng. J.* **2011**, *168*, 535–543. [CrossRef]

115. Al-Sammarraee, M.; Chan, A.; Salim, S.M.; Mahabaleswar, U.S. Large-eddy simulations of particle sedimentation in a longitudinal sedimentation basin of a water treatment plant. Part I: Particle settling performance. *Chem. Eng. J.* **2009**, *152*, 307–314. [CrossRef]

116. Samaras, K.; Zouboulis, A.; Karapantsios, T.; Kostoglou, M. A CFD-based simulation study of a large scale flocculation tank for potable water treatment. *Chem. Eng. J.* **2010**, *162*, 208–216. [CrossRef]

117. Black, K.S.; Tolhurst, T.J.; Paterson, D.M.; Hagerthey, S.E. Working with Natural Cohesive Sediments. *J. Hydraul. Eng.* **2002**, *128*, 2–8. [CrossRef]

118. Vignati, D.A.L.; Pardos, M.; Diserens, J.; Ugazio, G.; Thomas, R.; Dominik, J. Characterisation of bed sediments and suspension of the river Po (Italy) during normal and high flow conditions. *Water Res.* **2003**, *37*, 2847–2864. [CrossRef]

119. Walker, S.; Narbaitz, R.M. Hollow fiber ultrafiltration of Ottawa River water: Floatation versus sedimentation pre-treatment. *Chem. Eng. J.* **2016**, *288*, 228–237. [CrossRef]

120. Su, S.; Xiao, R.; Mi, X.; Xu, X.; Zhang, Z.; Wu, J. Spatial determinants of hazardous chemicals in surface water of Qiantang River, China. *Ecol. Indic.* **2013**, *24*, 375–381. [CrossRef]

chemengineering

MDPI

Article

Mass Spring Models of Amorphous Solids

Maciej Kot

Old Technologies sp. z o.o., Kombatantów 11, 23-400 Biłgoraj, Poland; eustachy@gmail.com

Abstract: In this paper we analyse static properties of mass spring models (MSMs) with the focus of modelling non crystalline materials, and explore basic improvements, which can be made to MSMs with disordered point placement. Presented techniques address the problem of high variance of MSM properties which occur due to randomised nature of point distribution. The focus is placed on tuning spring parameters in a way which would compensate for local non-uniformity of point and spring density. We demonstrate that a simple force balancing algorithm can improve properties of the MSM on a global scale, while a more detailed stress distribution analysis is needed to achieve local scale improvements. Considered MSMs are three dimensional.

Keywords: mass spring model; soft body deformation; physically based modelling

Citation: Kot, M. Mass Spring Models of Amorphous Solids. *ChemEng* **2021**, 5, 3. https://doi.org/10.3390/chemengineering5010003

Received: 4 August 2020
Accepted: 4 January 2021
Published: 11 January 2021

1. Introduction

Mass spring models are used for the purpose of simulating elastic objects in various fields. Their potential applications include computer games, animations, virtual reality environments as well as engineering of structures or materials in which deformation under stress is considered, crack propagation studies or human tissue simulations. Related fields range from ultramicroscopy to astrophysics [1–6]. Specific needs of specific application may vary, but the general theory of how mass spring models (MSM) deform concerns all of them.

We distinguish two broad classes of mass-spring networks—crystal (lattice) based and disordered. In crystal based networks, mass points are placed on a periodic lattice and mechanical properties of such networks can be expressed with analytical formulas. This allows for more precise analysis and description of their behaviour. In disordered networks, on the other hand, the mass placement rules are relaxed and do not have to follow any regular order (Figure 1). The downside of such models is reduced accuracy and increased difficulty of estimating their mechanical properties [7]. Consequently, lattice based networks are usually preferred over disordered ones and are even used to model materials, which are known not to be crystals.

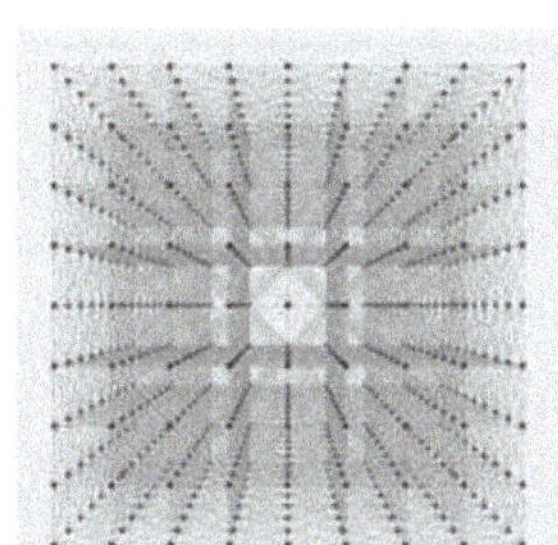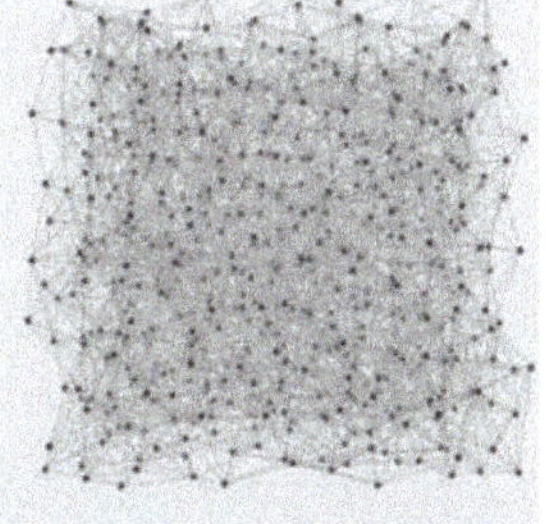

Figure 1. Cubic lattice and disordered mass placement.

For numerous applications this is not a problem, however, there is a danger that in certain situations the material may inherit some unexpected properties of the periodic network and exhibit unintended behaviours. An example is the crack propagation problem, where geometry of the network may influence the observed crack patterns considerably. If the numerical model has a regular, lattice-like topology which determines the potential crack placement, it creates "easy propagation planes" for cracks (whether it is based on MSM or FEM, or some other modelling technique). In such models cracks tend to form and propagate along lattice dependent directions and resulting patterns may reflect lattice properties instead of material ones (Figure 2). In such cases it is difficult to asses, if the obtained results follow from the mathematical model, or if they are only artefacts of the numerical representation. For example, in a model with cubic lattice topology a curved crack may appear as two straight segments approaching each other at 45 degrees, or worse, the crack may not bend at all, if breaking of subsequent segments influences tensions within the system.

Some attempts have been made to address these problems, such as Chen et al. [8], where introduction of non-local interactions into the model is shown to reduce negative effects of a lattice topology. Such solutions may mitigate most apparent artefacts in certain situations, however they do not eliminate the problem completely. A class of conditions for which such methods are unlikely to work involves situations where cracking is induced by a shrinkage front progressing through a material, such as in the case of drying or cooling materials [9–13]. For example if a cooling front is advancing through solidifying lava in presence of water, the problem has a translational symmetry. Cracks which appear at the top, allow for the water to be in contact with the top of the front, keeping the temperature at 100 °C, while the bottom of the front is in contact with non-solidified lava at an approximately constant temperature T_L. The front itself is expected to have a linear gradient of temperature between these two values [14]. Modelling such process with lattice-based networks will result with exactly the same conditions at each cracking step, and since the tip of the crack can follow only discretised paths, it may never turn if the incentive to turn is too weak.

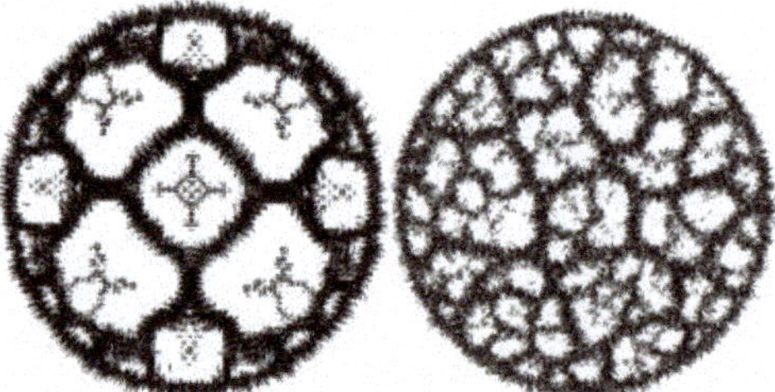

Figure 2. Example crack patterns observed on cubic lattice and disordered networks.

In such situations it may be beneficial to use disordered MSMs instead of lattice-based ones. As mentioned, the downside of such models is the reduced accuracy when compared to lattice-based MSMs, which in turn may cause discrepancies between modelled and theoretical properties of the material e.g., such as reported in [15]. In this work we place focus on disordered MSMs.

Mass spring models studied in this work are based on [7,16] and the reader is encouraged to refer to these papers for a more detailed introduction. We achieve varying values of Poisson's ratio ν in our models, by introducing a force dissipation mechanism. This technique offers a considerable simplicity in terms of implementation as well as low conceptual and computational complexity, when compared to alternative methods capable of achieving full spectrum of ν. A comprehensive review of other MSM-like techniques can be found in [17].

In the following sections we show how to formalise the force dissipation mechanism proposed in [16], with a focus on ease of implementation. We introduce a new pair of elastic constants for the material, which translate into simple properties of springs and nodes of the network. We also demonstrate how to mitigate the accuracy problems of disordered MSMs by adjusting stiffness coefficients of the springs. Presented techniques allow to improve both global and local behaviour of disordered mass-spring networks.

It should be noted, that we investigate only the static properties of MSMs. We do not address here the question of how these systems evolve over time or what are the most efficient numerical schemes to track their dynamics. Efficient ways of simulating the dynamics can be found e.g., in [18,19].

2. Elastic Moduli

In linear elasticity, the relation between tensions and deformations of a continuous medium is linear. This can be expressed in general form by:

$$\sigma_{ij} = \Lambda_{ijkl}\epsilon_{kl},\tag{1}$$

where $\hat{\sigma}$ denotes the stress and $\hat{e}$ the strain tensor; $\hat{\Lambda}$ is a tensor of elastic constants. Components of $\hat{\Lambda}$ are tied together by various relations which follow from symmetry of $\hat{\sigma}$ and $\hat{e}$. In this paper we discuss homogeneous isotropic media, which further restricts the values of Λ_{ijkl}.

In a medium which behaves like fluid or gas, in an equilibrium state, all the "tensions" are distributed equally in all directions, that is $\sigma_{ij} = -\delta_{ij}p$, where p denotes pressure (for fluids or gasses we should not be using quantities like stress or strain, however the purpose here is only to highlight analogies in the description of various media, in situations, where it is applicable to do so; same comment applies to E and v below). More precisely $\sigma_{ij} = -\delta_{ij}dp$ if we consider a small deformation from a reference configuration, so that:

$$K = -V\frac{dp}{dV} = -\frac{dp}{\epsilon_{kk}},$$
$$\epsilon_{ij} = -\delta_{ij}\frac{dp}{KN},\tag{2}$$
$$\sigma_{ij} = \delta_{ij}\epsilon_{kk}K,$$
$$\Lambda_{ijkl} = K\delta_{ij}\delta_{kl}$$

where K is a bulk modulus, V volume and N is the number of dimensions of the space. In this case Lamé parameters are equal $\lambda = K$, $\mu = 0$, Young's modulus $E = 0$ and Poisson's ratio $v = \frac{1}{N-1}$. One realisation of such a medium is a disordered collection of particles, which collide with each other, bouncing off randomly, so that there is no way to induce a momentum flow/surface pressure higher in one direction than in others. We will say, that the interactions which lead to such relations, are of a completely dispersive nature.

Let us now take our ensemble of particles and impose fixed relative positions on them, so that deformations are properly expressed by the strain tensor. We will assume here, that a change in the distance r between a pair of particles gives rise to a force (acting between them) which can be represented by a central potential $V(r)$. In such setup there is no way to displace a chosen particle and induce tension in only one given direction. If we move a particle by dl along x axis, y components of the distances to its neighbours will be affected as well. This is a geometrical property of euclidian space and it gives rise to the elastic moduli tensor of the from [7,17,20]:

$$\Lambda_{ijkl} = \mu(\delta_{ij}\delta_{kl} + \delta_{ik}\delta_{jl} + \delta_{il}\delta_{jk})\tag{3}$$

There is, similarly as in the case of gas/fluid, only one elastic constant here, however, this time bulk modulus depends on the connectivity, which follows from the dimensionality (the higher the dimension, the more particles are present in the immediate neighbourhood). It is given by $K = \mu(1 + \frac{2}{N})$. Similarly $E = 2\mu\frac{N+2}{N+1}$ and $v = \frac{1}{N+1}$. We also have $\lambda = \mu$. This one elastic constant corresponds to direct interactions.

There exist materials with these properties, an example of which is a diamond, however this model is too simplistic to properly describe elastic solids in general, as it results in only one degree of freedom, corresponding to a single elastic constant, while real materials are characterised by two.

Possible solutions, which allow to achieve two degrees of freedom, include introduction of angular springs or beams into the model, usage of non central forces or introduction of a volume change component into the potential energy [17,21–25] (some of them limited to very specific lattice topologies). However, a particularly simple way of extending this model, which works both with lattice based networks as well as disordered ones, is to allow for the force to be partially dispersed, so that all the interactions become superpositions between direct and dispersive forces [16].

$$\bar{F} = \bar{F}^\mu + \bar{F}^*, \tag{4}$$

where $\bar{F}^\mu$ denotes the direct, and $\bar{F}^*$ dispersive component of the force. The resulting elastic body then becomes a superposition between fluid and diamond like materials [16,17,26].

The reasoning behind this is, that direct interactions are in essence interactions by means of idealised classical Newtonian force, which acts across time and space, instantly, without accounting for relative velocities or other possible characteristics of the bodies involved in the problem, such as their shape, which could potentially influence the net effect of one body acting on another from a distance (There is plenty of evidence suggesting that the real character of the interaction on a distance depends on factors other than just relative position of the bodies in question, the prime example of which is electrodynamics, where the presence of relative velocity gives rise to a magnetic force).

If we take:

$$\sigma_{ij} = \mu\delta_{ij}\epsilon_{kk} + 2\mu\epsilon_{ij} + B\delta_{ij}\epsilon_{kk}, \qquad B = \lambda - \mu$$
$$Q = \frac{B}{\mu(1 + \frac{2}{N})}, \tag{5}$$

then Q denotes the ratio between dissipative forces and the direct ones, with its values spanning from -1 to infinity. $\bar{F}^\mu \sim \mu$, and $\bar{F}^* \sim \mu Q$. Other elastic constants become:

$$K = B(1 + \frac{1}{Q}),$$
$$E = \frac{B}{Q}\frac{2N^2(Q+1)}{(N-1)(N+2)(Q+1)+2} \tag{6}$$
$$\nu = \frac{(1 + \frac{2}{N})Q + 1}{(N-1)(1 + \frac{2}{N})Q + N + 1}.$$

In practice, when implementing a mass spring model for the representation of an elastic material, it may be more convenient to introduce yet another pair of constants.

$$C = \mu + \mu|Q|, \tag{7}$$

$$D = \frac{Q}{1 + |Q|}. \tag{8}$$

C now can be interpreted as the total amount of interaction potential (e.g., a density of interaction carriers between two particles). The dissipative part is given by $C \cdot D$, and the direct one by $C \cdot (1 - |D|)$. D is the dispersion fraction with values between -0.5 and 1. With this description we avoid the singularity in Q around the $\nu = 0.5$ point and allow for expressing possible dynamic differences between media with $K = 0$, but varying C (the aspects of MSMs not explicitly addressed in this work). Other elastic constants become:

$$\mu = C(1 - |D|) \tag{9}$$

$$\lambda = C\left(\left(1+\frac{2}{N}\right)D - |D| + 1\right) \tag{10}$$

$$K = C\left(1+\frac{2}{N}\right)(D - |D| + 1) \tag{11}$$

$$\nu = \frac{\left(1+\frac{2}{N}\right)D + 1 - |D|}{(N-1)\left(1+\frac{2}{N}\right)D + (N+1)(1-|D|)} \tag{12}$$

$$E = 2\mu\frac{N\lambda + 2\mu}{(N-1)\lambda + 2\mu} \tag{13}$$

A few examples of the decomposition of the interactions into dissipative and direct parts are illustrated in the Figure 3 (with $E = const$). The first extreme, $\nu = 0.49$, corresponds to $D \approx 1$, where almost all the forces are dispersed, making the material fluid-like ($\nu = 0.5$ in 3D means no change in volume under unidirectional compression; the limit value of $\nu = 0.5$ is unstable). The second extreme, $\nu = -0.99$ has $D \approx -0.5$ ($\nu = -1$ once again is an unstable limit), which means that half of the compressing force acts in a dispersive manner, but with a "reversed sign". In such regime $K = 0$ and the material can be uniformly compressed without changing its elastic energy. As we can see, the unidirectional compression gives rise to considerably high forces, both direct and dispersive, which cancel out each other almost perfectly in the resulting concave shape. The $\nu = 0.25$ has no dispersive forces at all and for $\nu = 0$, we have $D = -3/8$.

Figure 3. A cube compressed in x direction by 10% (x-borders frozen). First row $\nu = 0.49$, second $\nu = 0.25$, third $\nu = 0$, fourth $\nu = -0.99$. First column: total force on springs, second: direct component, third: dispersive component. Red colour indicates expansive force, blue compressive.

3. Mass Spring Models

Mass spring models represent elastic solids with discrete points connected by springs. As mentioned in Section 1, we distinguish two broad classes here—a regular crystal-like models and disordered ones [7]. In both cases, elastic constants of the material follow from spring coefficients. As elucidated in the previous section, if classical, direct force springs are used, the value of Poisson's ratio for three dimensional isotropic and homogeneous models is fixed and equal to $\frac{1}{4}$. The relation between bulk modulus K and spring parameters is given by:

$$K = \frac{5}{3}\mu = \frac{1}{9V}\sum_i k_i L_i^2, \tag{14}$$

where L_i and k_i denote the length of i-th spring and its stiffness coefficient. V is the volume of the object and the sum is taken over all springs inside this volume.

In order to model materials with other values of Poisson's ratio, we can introduce the force dissipation mechanism discussed in previous sections with the C modulus obeying:

$$\frac{5}{3}C = \frac{1}{9V}\sum_i k_i L_i^2, \tag{15}$$

where the classical stiffness coefficient of a spring is now $\kappa_i = k_i(1 - |D|)$. And the force propagation follows the algorithm [16]:

For each node:

- Compute forces from springs $F_i = -k_i\Delta L_i$.
- Apply $F_i^\mu = (1 - |D|)F_i$ as a regular force
- Additionally accumulate $0.5DF_iL_i$ as J_{acc}.

In the second pass:

- Redistribute the accumulated force as
$F^* = J_{acc}/(L_i b)$
to all the springs connected with the node (by applying it on both nodes the spring connects), where b denotes the number of these springs

Estimates given by Equations (14) and (15) can be applied to any mass spring network, however the more homogeneous and isotropic the network is, the better it will approximate elastic properties of a given material.

4. Accuracy of Random MSM Models

In this study we consider MSM networks with randomly generated points connected by springs with their nearest neighbours. The network topology is characterised by two parameters—first, *minL*, gives a minimal allowed distance between any pair of MSM nodes; second, *maxL*, is the range within which spring connections are formed (any two nodes which are less than *maxL* apart get connected by a spring) [7]. Please note, that the random point generation mechanism described in [7] is flawed. The article suggests a performance improvement for the implementation, which advocates a usage of a small spherical brush-like window traveling through the material in order to fill it with random points. The performance improvement for point collision calculation is real, however the "window" should be moved after each point addition, in contrast to a procedure which fills the window region fully first, before advancing to the next region of the material. The latter may introduce a non-homogeneity at the window borders. This was not stated clearly in the original article.

For such MSMs, our previous experiments show, that as long as the average number of springs attached to one node is sufficiently high, Equations (12) and (14) give the elastic properties of the network in bulk with a reasonable accuracy. The values of Poisson's ratio for such models are very close to theoretical predictions (observed discrepancies were in many cases of magnitude of measurement errors) and the values of E diverge from theoretical by no more than 10% [7,16]. Such accuracy may be satisfactory for many applications, but there is certainly a room for improvement.

Currently we will focus not only on bulk properties, but also investigate how MSMs behave locally, on a scale of internode distances. We perform a simple test, in which we consider a model of a cube $10a_0 \times 10a_0 \times 10a_0$, that undergoes compression (expansion) uniformly in all directions and observe how the MSM reacts. The following setup was used for this test. The elastic constants of the material were set to $K = 100\frac{k_0}{a_0}$ and $\nu = 0.25$ (this means $D = 0$, which makes the theoretical analysis simpler. Materials with $D \neq 0$ are expected to inherit the same properties here, which was confirmed by additional tests afterwards). The material was modelled with a random MSM with $minL = 0.92 \cdot 0.5a_0$ and $maxL = 1.77 \cdot 0.5a_0$, which gives around 6700 nodes, 55,000 springs and approximately $\langle S \rangle = 18$ springs connected to one node inside of the material. All springs were assigned the same spring constant $k[k_0]$ calculated with Equation (14). The cube was expanded uniformly in all directions by 1% (We will use expansion and compression interchangeably in the rest of the article when discussing the phenomenon analysed in this experiment; for technical reasons expansion is easier to perform). In such setup the values of diagonal components of the strain tensor should be equal to 0.01 and corresponding diagonal elements of the stress tensor should be equal to $3\frac{k_0}{a_0}$ (with a_0 being the unit of length and k_0 the unit of force).

In our numerical experiments we measure the stress inside of MSM according to the procedure proposed by Hardy [27,28]. It estimates the stress by measuring the tension inside a probe sphere placed in the point of interest. The tension in each spring in the surroundings of the point is weighted by a bond function. If only a fraction of the spring is contained within the probe sphere, the corresponding fraction of the tension is taken into account:

$$\hat{\sigma} = \frac{1}{2} \sum_{\alpha=1}^{N} \sum_{\beta \neq \alpha}^{N} \bar{x}_{\alpha\beta} \otimes \bar{F}_{\alpha\beta} B_{\alpha\beta}(\bar{x}), \tag{16}$$

where:
$x_{\alpha\beta}$—distance between nodes α and β
$\bar{F}_{ab}$—force acting through the spring ab
$B_{\alpha\beta}$—bond value

In order to calculate the bond value, we associate an influence sphere with each spring—the center of the sphere is placed in the middle of the spring, and its radius is equal to half of the length of the spring. Then the bond value is calculated as a fraction of the volume of that sphere, intersecting with a probing region. This is a slight modification to the original method, which measured intersections between raw springs and a probing region. Our experiments shown that calculating intersections with three dimensional spheres surrounding the springs reduces noice and allows for more localised measurements (smaller probe regions).

Experiments presented in this article are using spherical probing region with the radius $1.5\langle L \rangle$, where $\langle L \rangle$ denotes the average length of a spring in MSM.

The test consists of two steps. First we scale the cube uniformly, mimicking the perfect uniform deformation, then we let the MSM relax (holding only borders in place). In both stages we measure stress and strain present inside of the cube. In a perfect model the relaxation phase should not lead to any additional displacements.

As an example the ϵ_{22} strain tensor component and the σ_{22} stress tensor component measured in this experiment are presented on Figures 4 and 5. The character of discrepancies in other tensor components is similar. The measurements were taken along the line in X direction in the middle of the cube, with regions in close proximity to the border omitted (to avoid various problems with stress measurements near the border). Standard deviation from theoretical value of the final (relaxed) stress is denoted by Δ_t, and the standard deviation from the average value by Δ. Standard deviation between the stress in compressed (scaled) state σ^s and the stress in relaxed state σ^r is denoted by Δ_r.

$$\Delta = \sqrt{\frac{1}{M} \sum (A - \langle A \rangle)^2}, \tag{17}$$

$$\Delta_t = \sqrt{\frac{1}{M} \sum (A - A_0)^2}, \tag{18}$$

$$\Delta_r = \sqrt{\frac{1}{M} \sum (\sigma^r - \sigma^s)^2},\tag{19}$$

where A denotes measured, $\langle A \rangle$ average and A_0 theoretical value of a strain or stress component; M is a number of measurements.

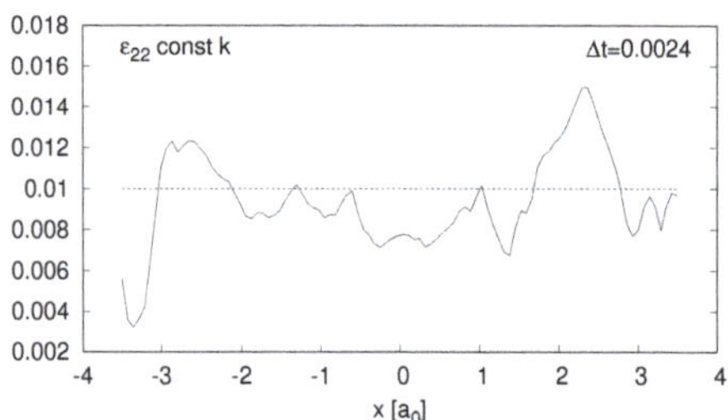

Figure 4. The strain component ϵ_{22} in a uniformly compressed solid modeled with mass spring models (MSM) with constant k, same for all springs. Dotted lines represent theoretical values. Solid lines indicate the strain after material has relaxed to the equilibrium compressed state.

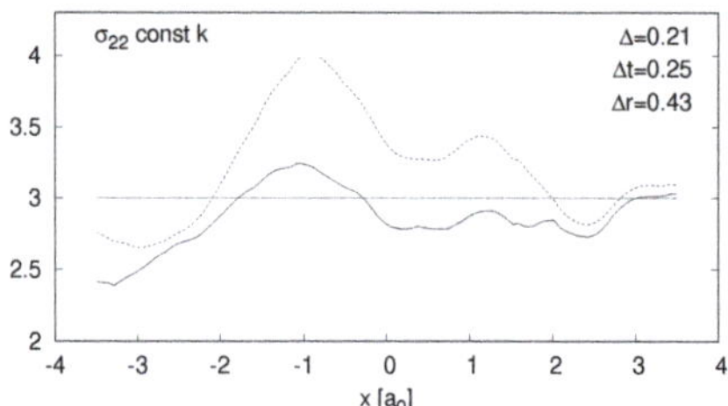

Figure 5. Same as Figure 4, but for $\sigma_{22}[\frac{k_0}{a_0}]$. Dotted lines represent stress in uniformly compressed material. Solid lines indicate stress after the material has relaxed to its new equilibrium state.

As we can see, the plotted tensor components average out to match theoretical predictions reasonably well. Local divergences, however, are high. The root cause of this behaviour is that the random MSM achieves isotropy statistically in bulk, but is not isotropic on singular nodes. To understand what exactly we mean by that, let us look at a simplified, one dimensional example from Figure 6. In disordered MSMs lengths of the springs are not equal. In uniform compression $\Delta L / L_0 = \epsilon$ and if all the springs have the same k, the force acting on a spring is $F_A = -kL_A\epsilon$, and as a result springs do not compress by the same degree under a constant pressure. If they did, the red coloured node from Figure 6 would experience non-zero net force and would move. In 3D networks, additionally, the number of springs facing each direction may differ, which further influences local isotropy.

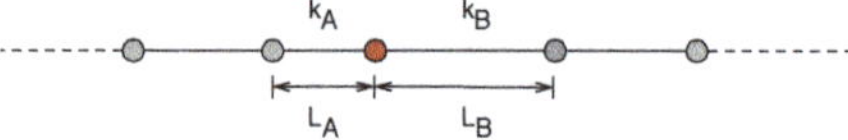

Figure 6. An MSM node with springs of different lengths connected to it.

These are the reasons for local discrepancies in deformation as well as Δ_r for stress in the material from Figures 4 and 5. Forces on the nodes do not balance to zero when the springs get compressed by the same degree. This also affects the estimation of C (and consequently K and E) from Equations (14) and (15). These equations measure the energy needed to compress material uniformly, but in this case the energy is lower, and the corresponding K is lower as well. The values of σ_{22} average to the desired $3\frac{k_0}{a_0}$ when the

MSM is scaled uniformly (indicated by dotted line on the Figure 5), however after the relaxation this value drops to around $2.8\frac{k_0}{a_0}$.

5. MSM Tuning

Our goal is to adjust stiffness coefficients for springs in such way, that in an MSM subjected to uniform compression (realized by simple scaling):

(a) forces acting on inner MSM nodes sum to zero,
(b) stress is isotropic.

Achieving (a) would reduce both, the variations of strain, and the change in stress which occurs when a scaled model relaxes (Δ_r), and in consequence would allow for a better estimation of the actual K in bulk. Achieving (b) would reduce the variations of stress. It should be noted, that condition (a) does not apply to the nodes which lie on the border of the MSM, where forces will sum to a value that counteracts the outside pressure. To distinguish which nodes are "inner", and which lie on the border, for each node, we measure the intersection between modelled object and a sphere centered on the node, with a radius equal to $maxL$. If the whole sphere intersects we treat the node as inner. Otherwise we assume it lies on the border.

To give some notion about how well (a) holds for a chosen MSM, we introduce the following measure, which we call kL-score:

$$S_{kL} = \frac{1}{N} \sum_n \left| \frac{\sum_b k_b \bar{L}_b}{\sum_b k_b |\bar{L}_b|} \right|, \tag{20}$$

where the inner sums indexed by b are taken over all springs connected to a node n. Outer sum is taken over all the nodes, the total number of nodes is N, and $\bar{L}_b$ is the vector between nodes connected by a spring b; $|\bar{x}|$ is the length of a vector $\bar{x}$. The lower the value of S_{kL} is, the closer to zero is the sum of forces acting on an MSM node (the lower the better). Such inner kl-score of the model from Figures 4 and 5 is equal $S_{kL} = 0.11$.

5.1. Constant kL

Since we have identified that in uniform compression the force exerted by a spring equals $F = -kL\epsilon$, our first attempt at improving accuracy of random MSMs could be by setting the stiffness coefficients of the springs in such way that $kL = const$. This would certainly help in 1-d case from Figure 6, however for three dimensional MSMs effects of such modification should be rather limited. In 3D the number of springs facing each direction is of bigger importance than differences in their lengths.

Setting $kL = const$ does however improve the MSM slightly with S_{kL} dropping to 0.091 and divergences from Figures 4 and 5 becoming: $\Delta_t \epsilon_{22} = 0.0023$, $\Delta_t \sigma_{22} = 0.17$ $\Delta_r \sigma_{22} = 0.4$.

This actually results, right away, in improvement for the estimation of the bulk value of K by a few percent (more results in following sections).

5.2. kL-Tuning

As described in previous sections, the source of divergences in strain within MSM are unbalanced forces which appear during uniform compression. They cause the nodes to move and springs relax to new equilibrium lengths.

Our first goal is to construct an MSM in which the unbalanced forces do not appear. It can be achieved in a number of ways, one of which is simply by solving as a set of linear equations for stiffness coefficients k_i. As the number of springs is an order of magnitude larger than the number of nodes, this may turn out to be problematic (but not impossible) for large systems, unless we have a whole computational cluster at our disposal. An alternative way, is to use the original MSM (with either $k = const$ or $kL = const$) and follow the same relaxation path which our MSM from Figure 5 traveled when reaching the equilibrium state. If we consider a simple time integrator (e.g., Euler or Verlet scheme) its basic steps are:

- compute forces
- update velocities
- update positions

the "update positions" step will result in changes in lengths of springs and consequently in changes of forces exerted by these springs. We can achieve the same change in forces by changing stiffness coefficients of the springs k_i instead of moving the nodes. Such algorithm (with damping) will find an MSM with $S_{kL} \approx 0$.

This is the approach we use, however in our case we skip the "update velocities" step and modify spring coefficients directly, increasing or decreasing their value depending on the magnitude of the projection of the force vector onto the direction of the spring. This is analogous to a simulation of an overdamped movement and basically is a variation of steepest descent minimisation procedure. The tuning procedure starts by artificially changing natural lengths of all springs by the same degree and then following with a number of minimisation steps. In our implementation we use $\frac{L - L_0}{L_0} = \epsilon = 1\%$. Additionally appropriate border conditions need to be set, similarly as it was done in compression/expansion experiment, where borders were frozen in place. This basically means that the relaxation should only be applied to inner nodes. Nodes that lie near the border have by definition non-isotropic spring connections and achieving $S_{kL} = 0$ is not even possible for them.

Each iteration of the algorithm reduces S_{kL} and increases the MSM quality. We stop when a certain value of S_{kL} is reached (0.0003), or the progress becomes too slow (max relative change of k smaller than 0.002).

Additionally, we place a restriction on the maximal allowed change of k for each spring to 50% of its original value. This way we explore only the set of solutions close to our original choice of k, and avoid degenerate solutions with negative k. In the last step we restore natural lengths of springs to their original values.

The values of ϵ_{22} and σ_{22} for the resulting MSM of the cube test are presented on Figures 7 and 8. The divergence in strain is now practically zero and Δ_r is two orders of magnitude smaller than for $k = const$. The profile of stress in this case is exactly the same as the relaxed profile from k-const model (Figure 5), but shifted towards desired value of $3\frac{k_0}{a_0}$. This means that we have managed to improve the estimate of bulk K, however local divergences did not change.

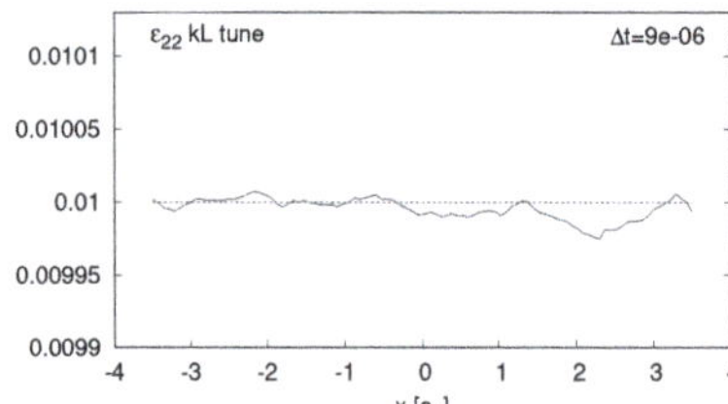

Figure 7. Same as Figure 4, but for kl–tune MSM.

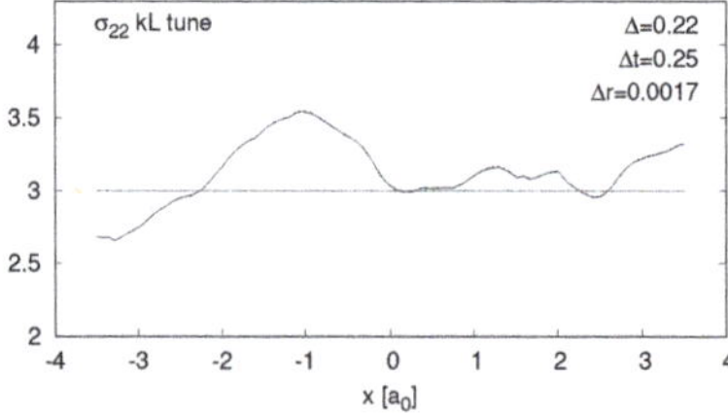

Figure 8. Same as Figure 5, but for kl–tune MSM.

5.3. ikL-Tuning

Figure 8 shows, that $S_{kL} \approx 0$ does not necessary translate into isotropy on a node basis. To further improve MSM representations we need to make sure, that not only forces on singular nodes sum up to zero, but also that their magnitude is approximately the same in all directions. Figure 9 illustrates an MSM node for which this is not the case; in this example forces do sum to zero, but their magnitude is different in X and Y directions.

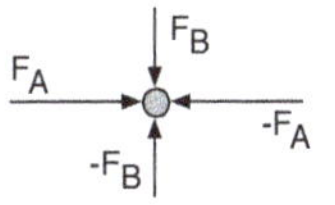

Figure 9. An MSM node for which forces balance to zero but are not isotropic.

In kL-tuning at each step of the algorithm we modified k_i based on the current value of the force acting on each node. In the improved procedure, instead of calculating the force, we will calculate the stress tensor at the point where the node is placed and see how much it diverges from the expected tensor:

$$\hat{\sigma}_{ij}^{err} = \hat{\sigma}_{ij} - 3K\epsilon\delta_{ij}, \tag{21}$$

where δ_{ij} is the Kronecker's delta.

The contribution of each spring to $\hat{\sigma}$ can be expressed as:

$$\delta\hat{\sigma} = \frac{Bk\epsilon}{V}\bar{L} \otimes \bar{L}_0, \tag{22}$$

where ϵ is the relative length change of the spring, L the current length and L_0 the neutral length. V is the volume of the region in which we are measuring the stress and B is a bond between this region and the spring. The bond can be calculated in a number of ways [27,28]; in our case we use the percentage of the overlap between probing region, which is a sphere and another sphere placed in the middle of the spring with radius equal $0.5L$. As we remember L_0 is artificially changed in the first step of the tuning procedure.

We project the spring influence onto the $\hat{\sigma}^{err}$ and compute the Δk for each spring as:

$$\Delta k = -t\frac{V}{B\epsilon LL_0}\frac{\bar{L} \otimes \bar{L}_0 : \hat{\sigma}^{err}}{LL_0}, \tag{23}$$

where $\hat{A} : \hat{B} = \sum_{ij} A_{ij}B_{ij}$, and t is step size for the steepest descent algorithm ($t \sim \frac{1}{b}$, where b is the number of springs connected to a node). The stress tensor tuning is run in addition to kL-tuning, and it also has a limit for the relative change of k on each spring to prevent degenerate solutions in which k becomes negative. After this procedure the strain tensor is as close to theoretical as it was in Figure 7. This time however we observe an improvement in stress tensor. The σ_{22} component is presented on Figure 10. The divergence $\Delta\sigma_{22}$ dropped to 0.067, and is about three times smaller than it was for kL-tune.

Comparison of the four presented algorithms is terms of divergences $\Delta_t\epsilon$, $\Delta_r\sigma$ and $\Delta_t\sigma$ observed in the cube compression test is presented on Figures 11–13. As we can see kL-tune procedure eliminates divergencies $\Delta_t\epsilon$, $\Delta_r\sigma$, which are caused by internal relaxation of unbalanced forces. The ikL-tune additionally reduces local variations in stress and reduces divergencies between stress present in the deformed body and its theoretical value.

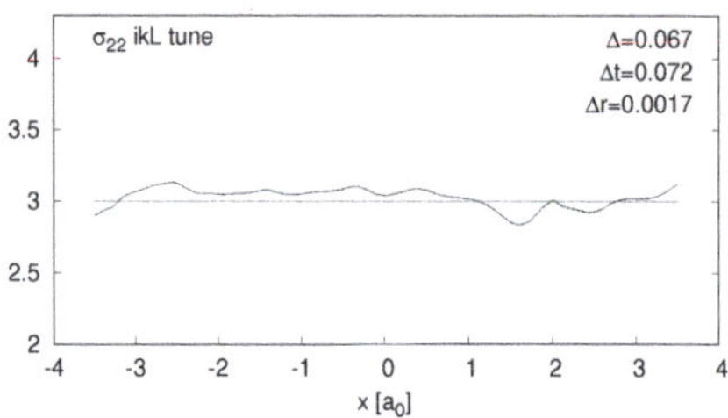

Figure 10. Same as Figure 5, but for ikl–tune MSM.

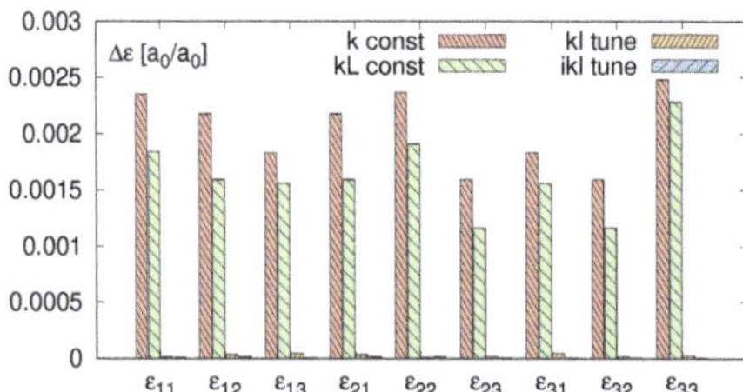

Figure 11. Comparison of $\Delta_t \epsilon$ for four variants of random MSM.

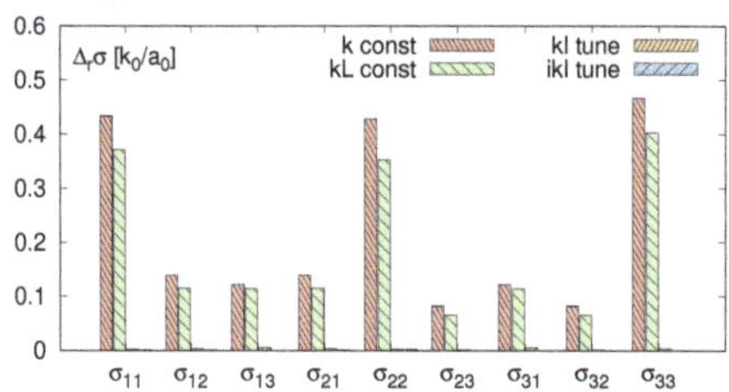

Figure 12. Comparison of $\Delta_r \sigma$ for four variants of random MSM.

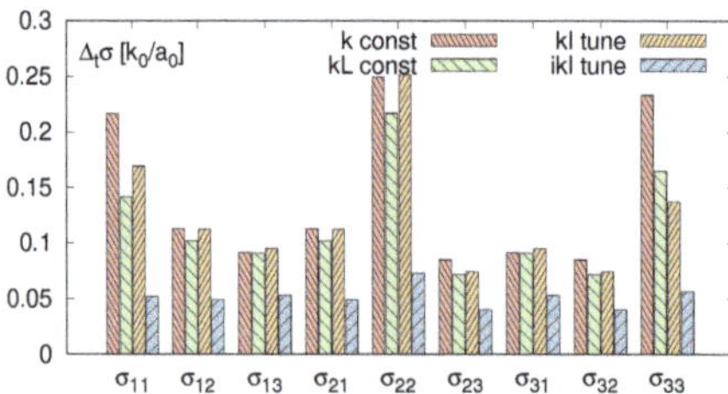

Figure 13. Comparison of $\Delta_t \sigma$ for four variants of random MSM.

6. Effects on Young's Modulus

Next we investigate what are the effects of tuning, on global material properties and how reduced variations of local stress and strain translate into the value of Young's modulus E.

We measure the value of E by performing a numerical experiment similar to the one described in [7]. We apply a static displacement to a block of material with dimensions $35a_0 \times 7.5a_0 \times 7.5a_0$, so that it gets compressed along X axis and we estimate E by measuring forces present in the material. In [7] we have measured the force passing through YZ plane in the middle of the system, following the equation:

$$E = \frac{F/A}{\Delta x / L_x},$$

where F is the reaction force, A the cross-sectional area of the block (in YZ plane), and Δx is the deformation of the block along x direction.

This equation however does not account for the possibility of non-zero stress in directions other than X in the system. As we have seen in previous section such stress may be present even for stress components which theoretically should be equal zero. A more accurate equation gives E as:

$$E = \frac{F_x/A_{yz} - v \cdot (F_y/A_{xz} + F_z/A_{xy})}{\Delta x / L_x}. \tag{24}$$

This time we measure not only a force through yz plane, but also xy and xz planes. After applying this correction to the measurement procedure, the resulting estimates of E become smaller by 2–5%, showing that the E calculated this way may diverge from theoretical value by a higher degree than our original estimates.

The results for the four variants of random MSM are presented in the Figure 14. First we notice that simple setting $kL = const$ instead of $k = const$ does improve the value of E noticeably. Introduction of kL-tune gives further improvement, however ikL-tune does not. For high $\langle S \rangle$ it is as good as kL-tune, however for lower values it is worse. The curve seems however more stable with lower variations around its trend value.

In all the cases the lower the number of springs, the higher divergence from theoretical value we observe. The average number of springs connected to one node for the cubic lattice MSM is $\langle S \rangle = 18$, and it seems that in our experiment all the curves reach plateau around this point. While the effects of tuning are positive for MSM with mid to high $\langle S \rangle$, for values lower than 14 other issues related to low network connectivity seem to dominate and the tuning can no longer compensate for them. For $\langle S \rangle < 12$ we observe a rapid decrease of rigidity of the network. This result is consistent with [29], where the loss of rigidity caused by reduced network connectivity is studied. Authors show, that weakened FCC networks with $\langle S \rangle$ around 8–10 suffer from the loss of rigidity and become floppy for lower values of $\langle S \rangle$ (exact numbers depend on applied strain).

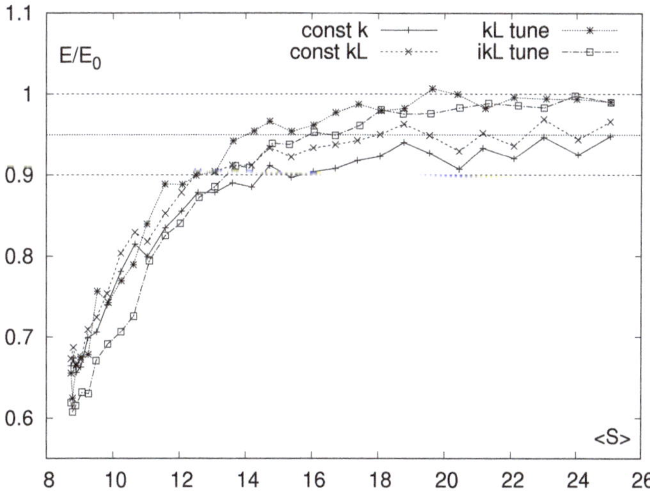

Figure 14. Values of Young's modulus relative to theoretical value for four variants of random MSM.

7. Conclusions

In this article, we have demonstrated possible accuracy problems one may face when using disordered MSMs. We have proposed two tuning procedures which aim to improve the accuracy of such MSMs. The first one, kL-tuning, eliminates unbalanced forces on MSM nodes which allows for better estimates of K (and E) of the whole network. The second, ikL-tuning aims to additionally reduce local stress variations. In both cases, the implementation details are of a lesser importance, than the effects each tuning procedure has on the MSM quality (as mentioned one might simply use linear solvers).

The presented analysis gives an overview of what should be expected of randomised models and what are their limitations. The proposed techniques can be used to reduce both global and local discrepancies and inhomogeneities present in the material, however one should keep in mind that such tuning may not always be necessary or even desirable. As mentioned in the introduction, real materials are not perfect, and the presence of inhomogeneities may be advantageous for certain applications and purposes. Overaggressive tuning may simply destroy the desired properties of our models (For the same reason, we did not even consider annealing-like approaches which modify position of nodes—such procedures may very well lead to node reorganisation into crystalline structures).

In the view of this, one may decide to abandon the tuning altogether and use $k = const$ for the whole network, with k scaled to match the desired K, not based on Equation (14), but rather on experiments. Typically a given network characteristics ($minL$ and $maxL$) lead to discrepancies of the same order, independent on the specific specimen of the MSM, therefore once we establish, that e.g., k should be increased by 10% relative to the original estimate, we can simply apply the 10% increase for all models generated with these parameters and get plausible results. The stress and strain figures in Section 5 should give a good intuition as to what to expect.

In our reference single threaded implementations, for the model with 55,000 springs, kL-tuning took 1s to complete, while ikL-tuning about 10s. The achieved reduction of stress variations is about 60% for the stress measured with resolution comparable to internode distances. However if such localised measurements are not needed, the kL-tune alone may be sufficient as it is considerably faster and much easier to implement.

Finally, we should remember, that problems discussed here concern disordered MSMs. If for a given application, a crystal-like structure is appropriate, then cubic lattice MSMs offer very high accuracy for typical deformation scenarios [7,30], and if we consider simple compression or shearing, the errors in local stresses and deformations are of an order of numerical precision of floating point numbers.

Funding: This research received no external funding.

Institutional Review Board Statement: Not applicable.

Informed Consent Statement: Not applicable.

Conflicts of Interest: The author declare no conflict of interest.

References

1. Klapetek, P.; Charvátová Campbell, A.; Buršíková, V. Fast mechanical model for probe–sample elastic deformation estimation in scanning probe microscopy. *Ultramicroscopy* **2019**, *201*, 18–27. [CrossRef] [PubMed]
2. Cristoforetti, A.; Masè, M.; Bonmassari, R.; Dallago, M.; Nollo, G.; Ravelli, F. A patient-specific mass-spring model for biomechanical simulation of aortic root tissue during transcatheter aortic valve implantation. *Phys. Med. Biol.* **2019**, *64*, 085014. [CrossRef] [PubMed]
3. Quillen, A.C.; Martini, L.; Nakajima, M. Near/far side asymmetry in the tidally heated Moon. *Icarus* **2019**, *329*, 182–196. [CrossRef] [PubMed]
4. Quillen, A.C.; Zhao, Y.; Chen, Y.; Sánchez, P.; Nelson, R.C.; Schwartz, S.R. Impact excitation of a seismic pulse and vibrational normal modes on asteroid Bennu and associated slumping of regolith. *Icarus* **2019**, *319*, 312–333. [CrossRef] [PubMed]
5. Sahputra, I.; Alexiadis, A.; Adams, M. A Coarse Grained Model for Viscoelastic Solids in Discrete Multiphysics Simulations. *ChemEngineering* **2020**, *4*, 30. [CrossRef]

6. Vicente, G.S.; Buchart, C.; Borro, D.; Celigüeta, J.T. Maxillofacial surgery simulation using a mass-spring model derived from continuum and the scaled displacement method. *Int. J. Comput. Assist. Radiol. Surg.* **2009**, *4*, 89–98. [CrossRef]
7. Kot, M.; Nagahashi, H.; Szymczak, P. Elastic moduli of simple mass spring models. *Vis. Comput.* **2015**, *31*, 1339–1350. [CrossRef]
8. Chen, H.; Lin, E.; Jiao, Y.; Liu, Y. A generalized 2D non-local lattice spring model for fracture simulation. *Comput. Mech.* **2014**, *54*, 1541–1558. [CrossRef]
9. Goehring, L.; Mahadevan, L.; Morris, S.W. Nonequilibrium scale selection mechanism for columnar jointing. *Proc. Natl. Acad. Sci. USA* **2009**, *106*, 387–392. [CrossRef]
10. Aydin, A.; Degraff, J. Evoluton of Polygonal Fracture Patterns in Lava Flows. *Science* **1988**, *239*, 471–476. [CrossRef]
11. Hofmann, M.; Anderssohn, R.; Bahr, H.A.; Weiß, H.J.; Nellesen, J. Why Hexagonal Basalt Columns? *Phys. Rev. Lett.* **2015**, *115*, 154301. [CrossRef] [PubMed]
12. Maurini, C.; Bourdin, B.; Gauthier, G.; Lazarus, V. Crack patterns obtained by unidirectional drying of a colloidal suspension in a capillary tube: Experiments and numerical simulations using a two-dimensional variational approach. *Int. J. Fract.* **2013**, *184*. [CrossRef]
13. Gauthier, G.; Lazarus, V.; Pauchard, L. Shrinkage star-shaped cracks: Explaining the transition from 90 degrees to 120 degrees. *EPL* **2010**, *89*, 26002. [CrossRef]
14. Goehring, L. Evolving fracture patterns: Columnar joints, mud cracks and polygonal terrain. *Philos. Trans. Ser. A Math. Phys. Eng. Sci.* **2013**, *371*, 20120353. [CrossRef] [PubMed]
15. Frouard, J.; Quillen, A.; Efroimsky, M.; Giannella, D. Numerical Simulation of Tidal Evolution of a Viscoelastic Body Modelled with a Mass-Spring Network. *Mon. Not. R. Astron. Soc.* **2016**, *458*, stw491. [CrossRef]
16. Kot, M.; Nagahashi, H. Mass Spring Models with Adjustable Poisson's Ratio. *Vis. Comput.* **2017**, *33*, 283–291. [CrossRef]
17. Golec, K.; Palierne, J.F.; Zara, F.; Nicolle, S.; Damiand, G. Hybrid 3D mass-spring system for simulation of isotropic materials with any Poisson's ratio. *Vis. Comput.* **2019**. [CrossRef]
18. Press, W.H.; Teukolsky, S.A.; Vetterling, W.T.; Flannery, B.P. *Numerical Recipes 3rd Edition: The Art of Scientific Computing*, 3rd ed.; Cambridge University Press: New York, NY, USA, 2007.
19. Liu, T.; Bargteil, A.W.; O'Brien, J.F.; Kavan, L. Fast Simulation of Mass-Spring Systems. *ACM Trans. Graph.* **2013**, *32*. [CrossRef]
20. Love, A.E.H. *A Treatise on the Mathematical Theory of Elasticity*; Cambridge University Press: Cambridge, UK, 1906.
21. Baudet, V.; Beuve, M.; Jaillet, F.; Shariat, B.; Zara, F. *Integrating Tensile Parameters in 3D Mass-Spring System*; Technical Report RR-LIRIS-2007-004, LIRIS UMR 5205 CNRS/INSA de Lyon/Université Claude Bernard Lyon 1/Université Lumiére Lyon 2/École Centrale de Lyon; 2007. Available online: https://link.springer.com/article/10.1007/s11548-008-0271-0 (accessed on 1 December 2020).
22. Lloyd, B.A.; Szekely, G.; Harders, M. Identification of Spring Parameters for Deformable Object Simulation. *IEEE Trans. Vis. Comput. Graph.* **2007**, *13*, 1081–1094. [CrossRef]
23. Ostoja-Starzewski, M. Lattice models in micromechanics. *Appl. Mech. Rev.* **2002**, *55*, 35–60. [CrossRef]
24. Chen, H.; Lin, E.; Liu, Y. A novel Volume-Compensated Particle method for 2D elasticity and plasticity analysis. *Int. J. Solids Struct.* **2014**, *51*, 1819–1833. [CrossRef]
25. Chen, H.; Liu, Y. A Nonlocal Lattice Particle Framework for Modeling of Solids. In *ASME International Mechanical Engineering Congress and Exposition*; American Society of Mechanical Engineers: New York, NY, USA, 2016; p. V001T03A001. [CrossRef]
26. Golec, K. Hybrid 3D Mass Spring System for Soft Tissue Simulation. Ph.D. Theses, Université de Lyon, Lyon, France, 2018.
27. Hardy, R.J. Formulas for determining local properties in molecular-dynamics simulations—Shock waves. *J. Chem. Phys.* **1982**, *76*, 622–628. [CrossRef]
28. Zimmerman, J.A.; WebbIII, E.B.; Hoyt, J.J.; Jones, R.E.; Klein, P.A.; Bammann, D.J. Calculation of stress in atomistic simulation. *Model. Simul. Mater. Sci. Eng.* **2004**, *12*, S319. [CrossRef]
29. Sheinman, M.; Broedersz, C.P.; MacKintosh, F.C. Nonlinear effective-medium theory of disordered spring networks. *Phys. Rev. E* **2012**, *85*, 021801. [CrossRef] [PubMed]
30. Banks, M.K.; Hazel, A.L.; Riley, G.D. Quantitative Validation of Physically Based Deformable Models in Computer Graphics. In *Workshop on Virtual Reality Interaction and Physical Simulation*; Andrews, S., Erleben, K., Jaillet, F., Zachmann, G., Eds.; The Eurographics Association: Genoa, Italy, 2018. [CrossRef]

chemengineering

Fortran Coarray Implementation of Semi-Lagrangian Convected Air Particles within an Atmospheric Model

Soren Rasmussen [1,*], Ethan D. Gutmann [2], Irene Moulitsas [1] and Salvatore Filippone [3]

1   Centre for Computational Engineering Sciences, Cranfield University, Bedford MK43 0AL, UK; i.moulitsas@cranfield.ac.uk
2   National Center for Atmospheric Research, Boulder, CO 80305, USA; gutmann@ucar.edu
3   Department of Civil and Computer Engineering, Università di Roma "Tor Vergata", 32249 Rome, Italy; salvatore.filippone@uniroma2.it
*   Correspondence: s.rasmussen@cranfield.ac.uk

Abstract: This work added semi-Lagrangian convected air particles to the Intermediate Complexity Atmospheric Research (ICAR) model. The ICAR model is a simplified atmospheric model using quasi-dynamical downscaling to gain performance over more traditional atmospheric models. The ICAR model uses Fortran coarrays to split the domain amongst images and handle the halo region communication of the image's boundary regions. The newly implemented convected air particles use trilinear interpolation to compute initial properties from the Eulerian domain and calculate humidity and buoyancy forces as the model runs. This paper investigated the performance cost and scaling attributes of executing unsaturated and saturated air particles versus the original particle-less model. An in-depth analysis was done on the communication patterns and performance of the semi-Lagrangian air particles, as well as the performance cost of a variety of initial conditions such as wind speed and saturation mixing ratios. This study found that given a linear increase in the number of particles communicated, there is an initial decrease in performance, but that it then levels out, indicating that over the runtime of the model, there is an initial cost of particle communication, but that the computational benefits quickly offset it. The study provided insight into the number of processors required to amortize the additional computational cost of the air particles.

Keywords: convection; air parcels; atmospheric model; high-performance computing; Coarray Fortran; PGAS

Citation: Rasmussen, S.; Gutmann, E.D.; Moulitsas, I.; Filippone, S. Fortran Coarray Implementation of Semi-Lagrangian Convected Air Particles within an Atmospheric Model. *ChemEng* **2021**, *5*, 21. https://doi.org/10.3390/chemengineering5020021

Academic Editor: Alessio Alexiadis

Received: 16 January 2021
Accepted: 19 April 2021
Published: 6 May 2021

Publisher's Note: MDPI stays neutral with regard to jurisdictional claims in published maps and institutional affiliations.

1. Introduction

The Intermediate Complexity Atmospheric Research (ICAR) model is an atmospheric model that balances the complexity of more accurate, high-order modeling schemes with the resources required to model larger scale scenarios [1]. It is currently in use at the National Center for Atmospheric Research (NCAR) in the United States and has been used for simulations such as a year-long precipitation simulation of a 169 × 132 km section of the Rocky Mountains and precipitation patterns in the European Alps and the South Island of New Zealand [2–4]. ICAR has a three-dimensional model that handles a three-dimensional wind field, advection of heat and moisture. The ICAR model initializes the domain and tracks the changes of variables such as temperature, humidity, and pressure. The model is able to emulate realistic terrain by handling hills and mountains within the environment.

This paper aimed to examine the implications of adding semi-Lagrangian convected parcels that move through the existing three-dimensional Cartesian grid of the ICAR application. The ICAR model was implemented using Coarray Fortran, a partitioned global address space (PGAS) parallel programming model, to handle the communication required for the domain decomposition of the problem [5–7]. The PGAS model functions as a global memory address space, such that each process or thread has local memory that is accessible by all other processes or threads [8]. Coarrays offer high-level syntax to easily provide the benefits of the PGAS model and will be examined more later (Section 1.1).

This work performed an in-depth analysis of the performance and communication implications of adding semi-Lagrangian convected air parcels on top of a Cartesian atmospheric model. Within the atmospheric community, air parcel is the term used to refer to pockets of air, but due to the nature of the parcels used here and for sake of unity, parcels are referred to as particles.

Other work in the field has been done examining aspects of using Fortran coarrays and the performance cost of various semi-Lagrangian schemes. The Integrated Forecasting System (IFS), which is used by the European Centre for Medium Range Weather Forecasts (ECMWF), has examined the performance benefits of using coarrays and OpenMP within their large-scale model [9,10]. Similar work has been done within the Yi-He Global Spectral Model (YHGSM), which examined the benefits of changing the communication of the semi-Lagrangian scheme from two-sided to one-sided Message Passing Interface (MPI) communication [11]. MPI and two-sided and one-sided communication are discussed in Section 1.1.

A framework called Moist Parcel-in-Cell (MPIC) has been developed for convection of Lagrangian particles [12,13]. These papers focused on the convection physics and accuracy of the MPIC modeling. MPIC uses OpenMP for parallelization, and the research group has worked to integrate it with the Met-Office/NERC cloud model (MONC) infrastructure, which uses the MPI for communication [14]. Work that has closely looked at the results of scaling coarrays and Message Passing Interface (MPI) have done so looking at the Cellular Automata (CA) problem [15,16]. The ICAR model has also previously been used to examine the performance and ease-of-programmability comparisons between the MPI and coarrays [17]. This study showed that re-coding coarrays to an asynchronous two-sided MPI implementation, while having little impact on additional lines for the physics files, required 226 extra lines of code in the module that controlled the communication. This equated to ∼91% additional lines.

The use of particles in atmospheric science is more commonly applied to trajectory analysis using offline models such as the Hybrid Single-Particle Lagrangian Integrated Trajectory (HYSPLIT) model [18]. In the current implementation, the particles are directly coupled within the model analogous to the approach taken by [19] for chemical transport studies. While it is beyond the scope of the current study, future work could use this particle implementation to add a simple atmospheric chemistry module to ICAR to permit the study of atmospheric chemistry in this computationally efficient model.

The novelty of the work presented here is the in-depth analysis of the performance and communication cost of adding semi-Lagrangian air particles to the coarray ICAR model. We analyzed a variety of parameters, such as wind speed and moisture, to understand their effects on performance. We also examined the use of Fortran coarrays with two different communication back-ends, OpenCoarrays with one-sided MPI puts, and Cray's SHMEM-based communication. This helps showcase the performance of HPC clusters from two different manufactures, as well as the performance of a Free and Open-Source Software (FOSS) library with OpenCoarrays and the proprietary Cray SHMEM communication backend.

1.1. Fortran Coarrays

The performance of code has traditionally increased due to advancements in hardware or algorithms, but as time passed and the physical limits of hardware were being approached, this meant the end of Moore's law and Dennard scaling [20]. The move to parallel programming with multicore processors and off-loading to devices such as GPUs has meant that performance gains could still be achieved. Computing with MPI has been a traditional way to use multiple processors locally and communicate data across nodes. The MPI functions on the Single-Program Multiple-Data (SPMD) paradigm, where the user writes a single program that gets run by every processing element, called rank within the MPI, at the same time. Similar to the MPI, coarrays also follow the SPMD paradigm, but refer to the processing elements as images, instead of ranks. In the SPMDP paradigm, data are often split up amongst images, as in Figure 1, so that each processor can work on a chunk of the problem before communicating the results back to a main node.

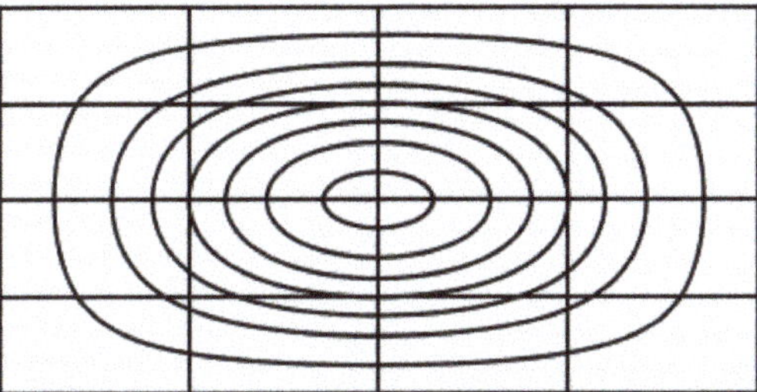

Figure 1. Overhead of the hill contour, distributed across 16 images.

Fortran coarrays started out as a language extension and were formally added to the Fortran Standard in 2008 [5,6]. Coarrays are similar in concept to the MPI, but offer the advantage of a simplified syntax with square brackets [17]. A variable without square brackets will always refer to the current image's copy of a coarray variable. Adding square brackets, such as var[i], will allow any image to access or change the *i*th image's var variable data. For instance, the following code allows the first image to send the local data in the cloud_moisture variable to the second image's rain variable Figure 2.

```
real, allocatable :: rain[:]
...
allocate(rain[*])
if (this_image() .eq. 1) then
  rain[2] = cloud_moisture
end if
sync all
```

Figure 2. Coarray example of variable rain.

In this instance, the coarray variables are scalars, but they can also be arrays and even derived types. Coarrays allow blocking synchronization with the sync all command and offer the typical range of collective communication, sum, reduce, etc.

To perform the same action in the MPI, one could use two-sided or one-sided communication. Two-sided communication is communication where a sender and a receiver both need to make a function call for a message to be communicated. Two-sided communication can be done using blocking MPI_Send and MPI_Recv or non-blocking MPI_Isend and MPI_Irecv. One-sided communication is a Remote Memory Access (RMA) method to allow a single rank to perform communication between multiple ranks. For example, either a sender calls a put to send a message or a receiver calls a get to retrieve the information. This MPI method of calling puts and gets is equivalent to the PGAS of coarrays.

The example above could be done with an MPI_Put. A code snippet showing the equivalent method with one-sided MPI communication is shown below Figure 3.

```
use mpi_f08
type(MPI_Win) :: window
integer(kind=MPI_ADDRESS_KIND) :: wsize
integer :: disp_unit, ierr
real :: rain, cloud_moisture

...
call MPI_Win_create(rain, wsize, disp_unit, MPI_INFO_NULL, &
                    MPI_COMM_WORLD, window, ierr)
call MPI_Win_fence(0, window, ierr)
if (rank .eq. 0) then
  call MPI_Put(cloud_moisture, 1, MPI_REAL, 1, disp_unit, 1, &
               MPI_REAL, window, ierr)
end if
call MPI_Win_fence(0, window, ierr)
...
```

Figure 3. One-sided MPI example.

The ICAR model uses coarrays to handle domain decomposition; see Figure 1. The ICAR mini-app used here has two different topologies, one that contains a single hill, which is split among the available images, and one that is a flat plain. Figure 4 shows a side view of the hill with particles, and Figure 1 shows a top-down view of the hill's contour lines and how the distribution of the domain across 16 images would occur. The physics in the ICAR model has a variety of parameters that need to be updated using numerical analysis and stencils. The stencils have halo regions of depth one that are contained in variables of the *exchangeable_t* type. The type *exchangeable_t* has been created based on object-oriented principles. Each variable of type *exchangeable_t* has a local three-dimensional array that represents parameters such as water vapor, cloud water mass, or potential temperature. It contains halo regions for the north, south, east, and west directions and has procedures, such as `put_north` and `retrieve_south_halo`, to help facilitate the halo communication. The ICAR model uses coarrays to update halo regions; see Figure 5. These halo regions are coarrays that perform "puts" during the update phase. The update phase occurs between the initialization step and the start of the computation and after each computational time step.

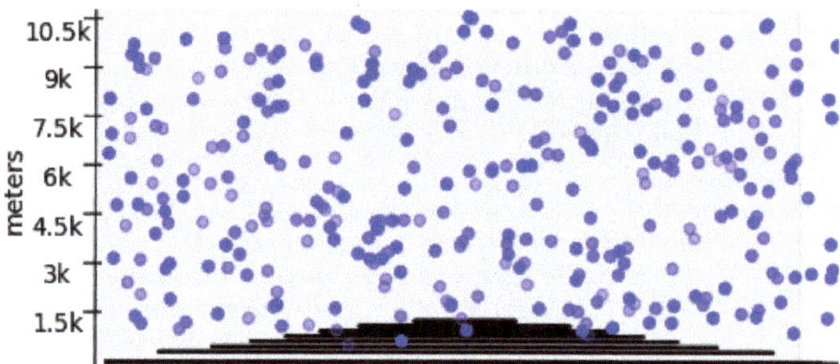

Figure 4. Side view of hill and air particles.

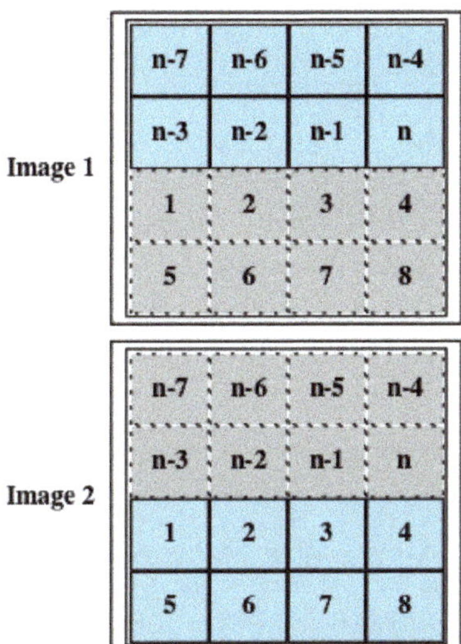

Figure 5. Halo regions to deal with trilinear interpolation.

To handle the movement of the particles, a number of additions needed to be made to the ICAR application. A new type `convection_exchangeable_t` was created that was based on the `exchangeable_t` type. The new type `convection_exchangeable_t` changed the coarray halo arrays to coarray buffers. These are buffers into which the neighboring images will directly put particles that need to be transferred across the boundary. When a particle's horizontal x or y values are greater than the image's local grid, they are loaded into an "input" coarray buffer of the neighboring region. Then, at the end of the particle computation phase, the buffer holding the particles is iterated over and unloaded into the image's array of particles. This is a key difference from the original ICAR model, where there is an explicit compute phase and then a communication phase of the halo regions. With the particles, there is an explicit compute phase and unloading phase.

A complication that occurred from adding semi-Lagrangian particles to the code is that a particle could exist between the boundaries with a halo region of one. Figure 6 indicates how an air particle would look as it exists between the boundaries of Image 1 and 2. A halo region of one would solve this issue because the particle could access the data points 2,3,n-1,n-2.The complication occurs because horizontal forces are applied to the particle, and further calculations for its pressure, saturated mixing ratio, etc., are needed. This sometimes results in a particle in Image 2 needing the points n-1, n-2, n-5, and n-6. This is why the boundary regions within the `convection_exchangeable_t` type were all changed from one to two. While the original code only had halo regions in the north, south, east, and west directions, four additional coarray buffers were needed to handle the diagonal movement of the particles. Since the u and v wind fields work simultaneously, they can move the particle diagonally. If a particle is near the "x" and "y" boundary at the same time, there is a possibility that the particle will jump to a diagonal neighbor.

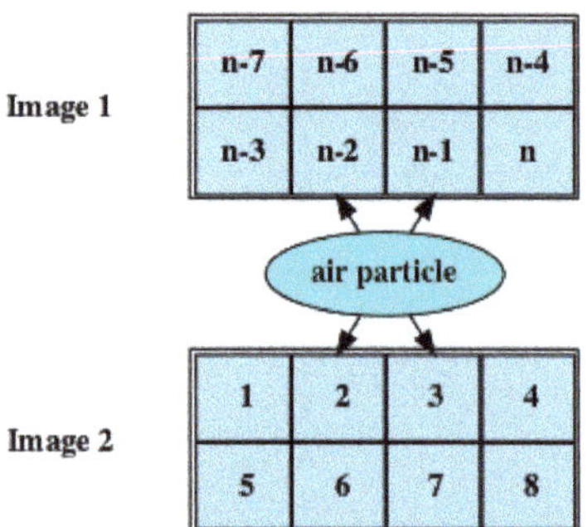

Figure 6. Lagrangian air particle existing on the boundary.

1.2. Convection

As the introduced semi-Lagrangian convected air particles warm, they rise and cool until their momentum brings them to an area where the surrounding region is cooler and they start to fall. As they fall, the particles warm until they reach an area where they are warmer than the surrounding environment, and they begin to rise. This oscillation causes a number of transformations within the air particle. As the air particle is rising, as long as its relative humidity is below 100%, it is considered a dry air particle and will cool at the adiabatic lapse rate with a constant mixing ratio and potential temperature [21]. Potential temperature is defined as the temperature a particle of air would be at if it were brought to a standard reference pressure P_0, defined as 1000 hPa. The equation used for calculating the potential temperature is $\theta = T(\frac{P_0}{P})^{R/c_p}$, where T is the absolute temperature of the air particle with pressure P. $R/c_p = 0.286$, where R is the gas constant of air and c_p is the specific heat of air at a constant pressure [22,23].

When the relative humidity increases over 100%, the air particle is saturated, and the water vapor within the particle will be converted into liquid cloud water. This phase change causes a change in temperature and potential temperature while pressure stays constant. As the particle falls, this process is reversed.

The air particles are implemented in a semi-Lagrangian manner. They are kept within an unordered buffer and operate essentially independently of the environment. This is allowable because as air particles rise, they do so in an adiabatic manner. It is standard practice to assume that the convected air particle's heat exchange with the surrounding environment is so small that it can be neglected [23].

2. Materials and Methods

In accordance with MDPI research data policies, the source code used and data sets created for this paper are publicly available [24,25].

2.1. Air Particle Physics

The framework for handling the physics of the air particle is roughly broken into the steps depicted in Figure 7. An overview of the process will be discussed before going over the equations of state. The first calculations, shown in the Move Particle Section, are to find the buoyancy force in the z-direction and the horizontal forces from the wind. This calculates the trajectory of the air particle and where it should be placed. Once the particle is moved, the changes in the physical properties of the particle are calculated. This process is depicted in the lapse rate section of the figure. First, the dry lapse rate is calculated, and if the relative humidity stays below 100%, the process continues. The particle is checked to see if it has moved beyond the image's boundaries and communicated if so. If the relative humidity goes above 100%, then an iterative process is started to find the correct amount of phase and lapse rate change.

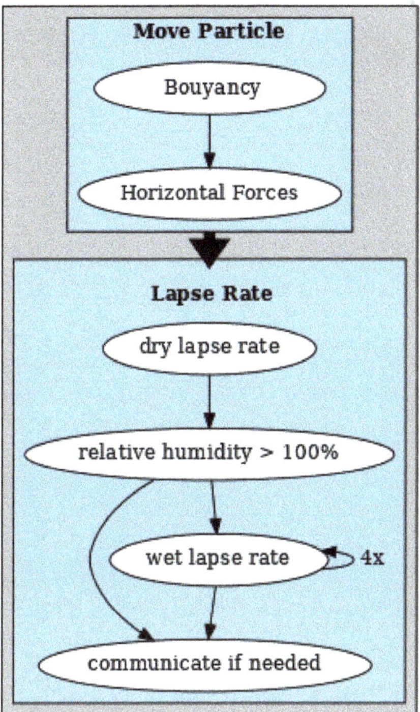

Figure 7. Overview of the computational phase of a particle.

The two equations of state used during this process are presented here. The equation of state used to formulate important properties of the air particle is the ideal gas law, where P is pressure, V is volume, n is the amount of moles, R is the ideal gas constant, and T is temperature.

$$PV = nRT$$

The other equation of state is the first law of thermodynamics and is used such that the change of internal energy within a closed system is equal to the quantity of energy from heat, minus the amount of thermodynamic work done on its surroundings [26], where U is the internal energy of a closed system, Q is the heat, and W is the thermodynamic work.

$$\Delta U = Q - W$$

From these equations of state, the following formulas have been derived and are commonly used within the meteorological community. The subsequent equation is how the buoyancy force for movement of the air particle is calculated. The downward force on the air particle is $\rho g V$, where ρ is the particle's density, g is gravity, and V is the volume [23]. When an air particle rises, it displaces an equal volume of ambient air, so the downward force on the displaced air is $\rho' g V$, where ρ' is the displaced air's density. The upward force is the same for both the air particle and the displaced air and equal to $-V(\delta p / \delta z)$. For the air particles, this is the only force in the z-direction, and so, the particle's equation of motion is:

$$\frac{d^2}{dt^2} = F_b = g\left(\frac{T - T'}{T}\right)$$

where T is the particle's temperature and T' is the environment's.

The force is applied at a constant rate through the time period, where d is displacement, $a = F_b$ is acceleration, and v_0 is the previous time step's velocity.

$$d = v_0 t + \frac{1}{2} a t^2$$

2.1.1. Lapse Rate

After the buoyancy force has been applied and the air particle has been moved to its new elevation, the change in temperature is calculated. The following equation is applied to calculate the new pressure:

$$p = p_0 - z_d \times \frac{g \times p_0}{287.05 \times T_0}$$

where the gravity variable is $g = 9.81$, T_0 and p_0 are the original temperature and pressure, and z_d is the distance the particle is displaced in the z-direction.

For the change in temperature of a dry air particle, the Exner (Π) function is used. The Exner function is defined as the following, where p is pressure, R_d is the gas constant of dry air, c_p is the heat capacity of dry air, T is temperature, and θ is the potential temperature:

$$\Pi = \left(\frac{p}{p_0} \right)^{R_d / c_p} = \frac{T}{\theta}$$

Rearranging, the new temperature is obtained from $T = \frac{\Pi}{\theta}$.

The next step is to check if the relative humidity of the air particle has exceeded 100%. This saturated mixing ratio is obtained from a function within the ICAR code that calculates it from temperature and pressure arguments. The relative humidity is calculated by dividing the particle's current water vapor by the saturated mixing ratio and multiplying by 100 to get the percentage. If the relative humidity is above 100%, then the extra humidity is taken and changed to cloud water. If the relative humidity is below 100% and there is cloud water in the particle, the cloud water will be converted back into water vapor so as to keep the relative humidity as close to 100% as possible. The change in temperature is calculated as follows, where q is heat, C_p is the specific heat of air at constant pressure, T is temperature, and P is pressure. C_p is calculated from $C_p = (1004 * (1 + 1.84 * v))$, where v is the mass of water vapor divided by the mass of dry air [22].

$$\Delta q = C_p \times \Delta T - \frac{\Delta P}{\rho}$$

For this, a specific latent heat of 2.5×10^6 J kg^{-1} is used for the particle. There is no change in pressure during the moist adiabatic process, so the $\frac{\Delta P}{\rho}$ expression is zero. The potential temperature does not change during the dry adiabatic process, but for the moist adiabatic process, the Exner function is used to calculate the new potential temperature.

2.1.2. Orographic Lifting

In addition to forces applied in the z-direction, a wind field is applied for movement in the x- and y-direction. As the air particle moves through the environment, orographic lifting will occur when the particle hits a barrier and has to rise to flow over the barrier. The ICAR application keeps track of the altitude of the ground, and when an air particle is moving in the x- and y-direction due to wind flow, the air particles will rise or fall accordingly. Bilinear interpolation is used in the x- and y-direction to estimate the altitude of the ground. The orographic lifting is simplistic in that the altitude of the air particle is changed the exact amount as the change in altitude of the ground floor.

2.2. Methodology

Within this section, a detailed account of how the air particles are handled is given. It is the purpose of this study to implement and add the fundamental components of functioning semi-Lagrangian air particles to the ICAR mini-app. Every air particle is initialized by randomly choosing a position within a coarray image's subdomain. The program uses Fortran's intrinsic random number generator with a seed of -1. The intrinsic `random_init` is called with both values equal to true. This is to ensure that each image has distinct random values when calling the function and that each time the program is run as a whole, the random numbers will be the same to ensure reproducibility between runs.

At initialization, an air particle of type *convection_particle* is created using the values of the environment at its current position. The *convection_particle* type has variables to keep track of the particle's x, y, z position within the image's grid, the height of the particle in meters, the u, v, w values of the wind field, the velocity, pressure, temperature, potential temperature, water vapor, and relative humidity. For orographic lifting, the height of the ground below the particle is also kept. After the initial x, y, z position is randomly generated, trilinear interpolation is used to gather the values of *convection_particle*'s types. See Figure 8 for an example of trilinear interpolation. For example, the initial x, y, z position would be the red C in Figure 8. Calculating C requires eight regular grid points from the surrounding particle's area, and five of those points are listed in the figure as $c_{000}, c_{001}, c_{100}, c_{110}, c_{111}$. Note, when using trilinear interpolation, one needs to be sure that if $\exists i \in \{x, y, z\}$, such that $floor(i) = ceiling(i)$, then bilinear interpolation is used. For example, if c_{000} and c_{001} were equal, then $floor(x) = ceiling(x)$ would be true. In that case, the trilinear interpolation to calculate the value in a cube would have to be switched to a bilinear interpolation, which finds the value in a square. Likewise, if $\exists i \in \{x, y\}$ such that $floor(i) = ceiling(i)$, then bilinear interpolation needs to be substituted for linear interpolation. If in the rare case that $\forall i \in \{x, y\}$, it is true that $floor(i) = ceiling(i)$, there is no need for interpolation. Finding the right level of interpolation is done to avoid divide by zero errors from $floor(i) - ceiling(i) = 0$.

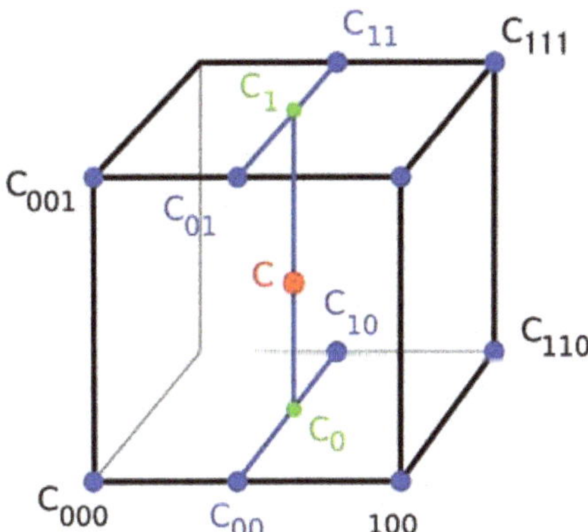

Figure 8. Trilinear interpolation [27].

When running performance tests, to keep the communication costs of each image as similar as possible, any particle that goes beyond the original domain is wrapped around to the other side. In Figure 1, a birds-eye view of the domain is shown and how it would be split up amongst 16 images. If a particle were to leave the top of that image heading north, it would continue to head north from the corresponding bottommost image. This allows for the particle count over the entire domain to remain constant. Likewise, if convection causes a particle to go off the top or bottom of the z-axis, a replacement particle is created.

There were two different types of initialization of atmospheric and environment variables that were tested in our model. One was an ideal case, where the potential temperature θ was set at 300 K for the entire model. The other setup was done by using a real atmospheric sounding of Seattle. The soundings were extracted from the University of Wyoming's weather website; a site with atmospheric soundings from locations around the world [28]. This was done using Python libraries BeautifulSoup, to extract the sounding data, and Pandas to handle the data manipulation and preprocessing. The atmospheric sounding was interpolated to produce data at every meter of elevation and written to a file for later use. The atmospheric sounding of Seattle gave the real-world potential temperature at elevations above 1600 m. This was beyond our model's upper limit, so the extra data were unused in this case, but could be used if the domain of the model were expanded. Potential temperatures below the sounding's lowest elevation were set at 300 K.

2.3. Hardware and Software

It has been noted in the literature that different hardware platforms, coupled with different software environments and compiler vendors, can result in performance variation [29]. For our experiments, we used two different HPC systems and employed two different compilers, thus covering both open-source and proprietary software implementations.

Cheyenne was one of two HPC clusters used. Cheyenne is a 5.34 petaflop SGI machine run by the National Center for Atmospheric Research (NCAR). The SGI ICE XA cluster has 145,152 Intel Xeon cores. Each node is dual socket with two Intel Xeon E5-2697V4 (Broadwell) processors, and a Mellanox EDR InfiniBand interconnect is used. The project uses the `caf` compiler wrapper provided by the Sourcery Institute with the OpenCoarrays library. The OpenCoarrays library is an open-source project that provides the coarray implementation for GNU gfortran [30]. The underlying compiler that `caf` wraps is GNU gfortran 10.0.0 with flags `-fcoarray=lib` and includes paths to the OpenCoarray library. `caf` was used with flags `-cpp -O3` for preprocessing and optimization. The project used OpenCoarrays 2.9.0, which itself was built with MPICH 3.3.2. OpenCoarrays handles the communication requirements of coarrays with a number of possible communication backends. This project uses the MPI backend with ones-sided RMA puts and gets, though it is possible to use OpenSHMEM.

A Cray cluster was used with nodes of two Intel Xeon Broadwell chips for a total of 44 cores. The Cray Fortran compiler Version 11.0.0 was used with flags `-e F -h caf -O3` for preprocessing, coarrays, and optimization. Cray's underlying communication layer for Fortran coarrays is the SHMEM library, standing for Symmetric Hierarchical MEMory [31]. SHMEM was created to implement one-sided PGAS routines. When accessing intranode coarray memory, the program quickly ran out of the default allocation of memory. This led to `lib-4205 : UNRECOVERABLE library error: The program was unable to request more memory space`. This issue was fixed with the help of an environment variable that influences the amount of available memory. By setting `PGAS_MEMINFO_DISPLAY` to one, the system report information on the PGAS library was displayed. It showed that the symmetric heap was equal to 65 MB per process. By increasing the memory by setting `XT_SYMMETRIC_HEAP_SIZE` to 500M, the program was able to procure the memory it needed and run successfully.

When compiling with Cray, the flags `-Rbcps` were initially used. They allowed for run-time checks of array bounds (b), array conformance checking (c), check array allocation status and other items (p), and character substring bounds (s). After initial development, these flags were removed. Flags that were also used included `-e F`, which turns on the preprocessor expansion of macros in Fortran source lines. This was required to handle macros for assertions, changing Cray output to files, and to delay PGAS synchronization.

If the `_CRAYFTN` macro was defined, the following code would be called:

call assign (" **assign** –S on –y on p:\%.txt", ierrr)

This would ensure repeat data values, such as "8 8 8", would be output with spaces between them. This is not the normal Cray behavior, which would output those values as "3*8", presumably for the sake of saving file space. While helpful for memory and time saving, it does make final data analysis more difficult. The additional flag `-h caf` was used, which allows the compiler to recognize coarray syntax. This is a default flag, but when it is defined on the command line, the macro `_CRAY_COARRAY` gets set to one [32]. An important directive to use when using coarrays is `!DIR$ PGAS DEFER_SYNC`. This ensures that the synchronization of data is delayed until a synchronization point [33,34]. This can be verified using Cray's profiling tools to see if the PGAS library calls are the non-blocking routine `__pgas_put_nbi`.

2.3.1. Profiling Tools

Basic performance measurements were done using hand-coded `system_clock` timers. When more sophisticated profiling and analysis was needed, HPE Cray's CrayPat performance analysis tool and the Tuning and Analysis Utility (TAU) were used. TAU was developed by the University of Oregon Performance Research Lab, the LANL Advanced Computing Laboratory, and The Research Center Julich at ZAM, Germany [35]. TAU Commander is TAU's tool that packages all the capabilities of TAU in a simple package for users. Depending on the user's needs, TAU Commander builds the required tools and allows for flexibility in choosing the combination of profiling, tracing, and other counters required. This project, with its use of CMake and the OpenCoarray Library, required a few tweaks to get everything working.

TAU's Fortran compiler wrapper `taucaf` was used as the Fortran compiler when building the ICAR application. The wrapper was missing the OpenCoarray library though, so the system paths needed to be added, and the `libcaf` library had to be loaded. A simple way to achieve this is to call "`caf -show`" from the terminal. The `caf` compiler is itself a wrapper to call `gfortran`, so this option will expand the full command needed to use the OpenCoarray Library. By copying that whole output, without the `${@}`, and then adding it to the `CMAKE_Fortran_FLAGS` Cmake variable, the project will be able to load and link everything correctly. The `caf` compiler is a bash script to implement the wrapper correctly, and the `${@}` is a special parameter that expands the positional arguments, every argument after the script's name [36]. The users can inform TAU of which information that they want in a few ways. One is using environment variables, such as `TAU_COMM_MATRIX`, to profile the communication matrix or running `tau_exec` with command line arguments. This was used in conjunction with the other default values, which turn on sampling and generate a performance profile.

Craypat was designed by Cray to allow the instrumentation of an application without recompilation, only through linking [37]. Using the original binary, `pat_build` was used with the `-g caf` flag to create a new executable that would produce profiling data when run. There are many additional profiling flags that can be passed; this project would sometimes add `heap` for a closer look at the memory behavior. The new binary can then be run using aprun in the same usual fashion "`aprun -n $num_procs ./test-ideal+pat`".

3. Results and Discussions

3.1. Validation

Validation of the model was done by ensuring that the air particles showed the same properties as real-world data and other modeling equations. The domain size used for the validation runs had dimensions $nx = 20, ny = 20, nz = 30$. Each cell corresponds to a length, width, and height of 500 m, and this is consistent throughout the modeling. To test the dry air particles, the hill was removed from the model, and the atmospheric sounding of Seattle was used. During the dry particle runs, the water vapor in the environment and particles was then set to zero to remove all moisture. For the wet particle runs, the water vapor in the particles was set to the saturated mixing ratio of the environment where its coordinates are. This means that if the particle stayed at the current temperature and pressure, the relative humidity would be 100%. Given that particles were initialized with an initial upward velocity of 5 m/s, the wet particles would become saturated immediately. The runs were done for 200 time steps of 20 s each for the ICAR model and time steps of one every second for the particles, totaling 4000 steps for the particles.

The results of the dry and wet runs are shown in Figure 9a,b, respectively. Both figures show a behavior that would be expected for a working model. The potential temperature of a particle experiencing dry adiabatic lifting is constant, which is what occurs in Figure 9a. The potential temperature only changes during the moist adiabatic lapse rate, during the process of the phase shift from water vapor to rain water, or vice versa. The third graph in Figure 9b confirms this behavior. Additionally, the trajectories of the temperature and pressure of the saturated and unsaturated particles follow what would be expected [28].

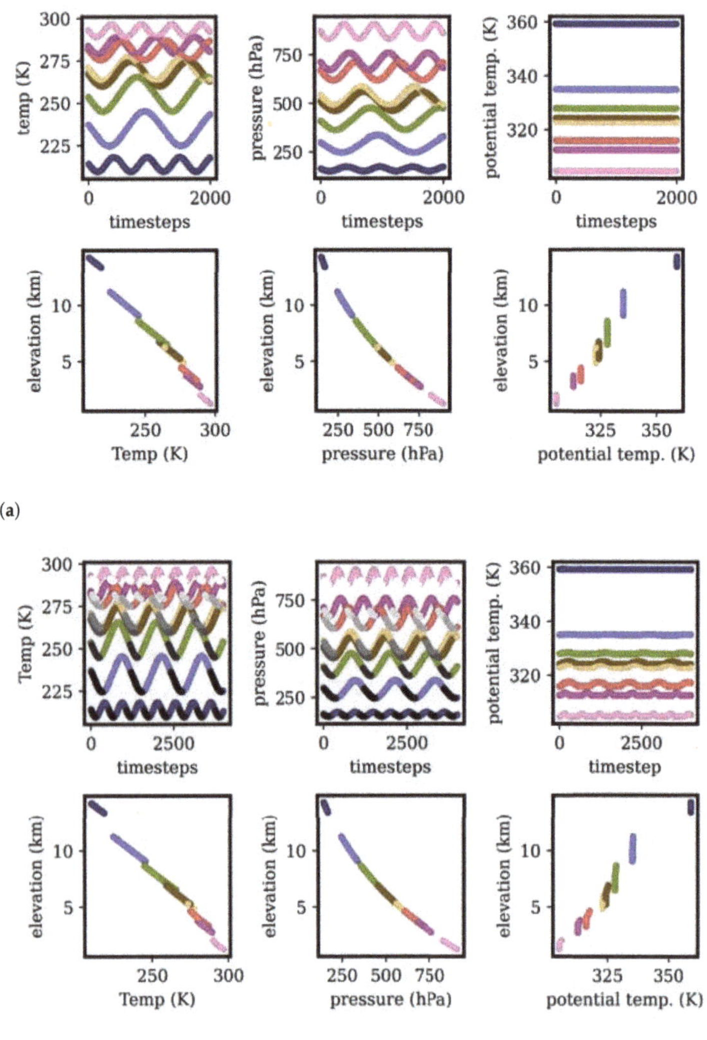

Figure 9. Particle attributes over time and varying elevation for: (**a**) eight dry air particles (**b**) eight saturated air particles; note that the grayscale indicates the particles are saturated.

3.2. Halo Depth

The addition of air particles to the ICAR model required the increase of the halo depth from one to two, and this study wanted to quantify the effects of increasing halo depth size.

Increasing the halo depth size results in creating more data that need to be communicated during the halo exchange. Another benefit of understanding the performance cost is that different numerical methods, possibly used for greater precision, could require larger stencils that would also increase the halo depth size. In Figure 10, the performance is shown as the depth of the halo region is increased using grid dimensions of size $500 \times 500 \times 30$ and $2000 \times 2000 \times 30$. For those two sizes, this graph shows the scaling of one node using 44 cores and two nodes using 22 cores each. The reason for increasing the node count while keeping the work per image constant is to summarize the cost of intranode communication with halo depth variance. Figure 10 shows that while increased halo depth affects run-time, splitting the work across nodes in this case had minimal effect on performance.

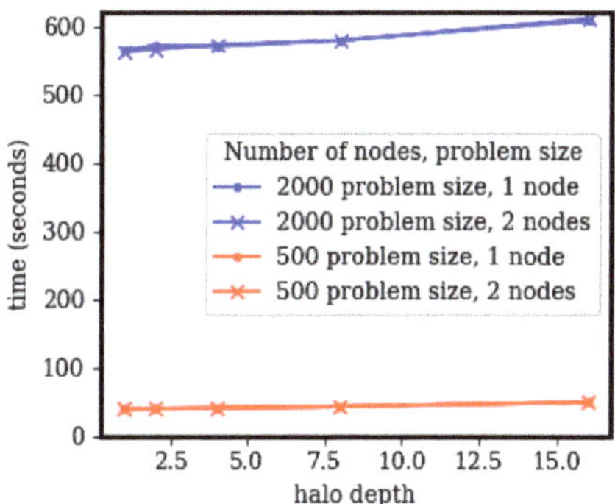

Figure 10. Effects of changing halo depth using a Cray machine.

3.3. Particle Count and Wind Speed

In addition to halo depth, the wind speed and particle count are variables that could possibly affect performance. As wind speed increases the amount of particles to communicate across, image boundaries will naturally increase. The particles are moving faster and will cross an image boundary in a shorter time frame. A larger number of particles would also translate to a larger amount of particles being transferred across image boundaries. Additionally, as the number of particles increases, it is intuitive that the computation cost to extract environmental information for the particle physics would also grow. The graph in Figure 11 attempts to isolate the effects of adding particles to the model by looking at the performance cost of increasing the particle count of two types of particles. Additionally, the graphs in Figure 12 attempt to examine the effects of wind on the number of particles communicated, how that communication looks, and the performance.

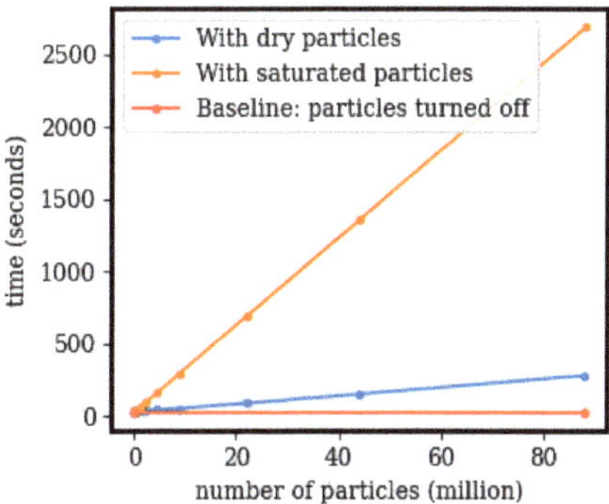

Figure 11. Performance effects of increasing the particle count using a Cray machine.

Figure 11 shows the computation costs of saturated and unsaturated particles as the number of particles increases. For these runs, wind speed was changed to zero, so no communication cost was measured. This figure was run on a single node of the Cray machine with 44 images. The number of particles per image ranged from one to two million, for a total of 88 million particles. The domain was $200 \times 200 \times 30$, and the problem was run for 200 time steps, with 19 additional time steps of 1 s for the particles. There was clearly linear growth for both dry and saturated air particles. The cost of saturated particles grew at a faster rate, with the time cost of 88 million particles being ∼9.5 times the dry particles. One-point-one million saturated particles took ∼1.7 times, and the dry particles and 2.2 million saturated particles took ∼2.5 times. Therefore, around 1.5 million particles, the cost of saturated particles doubled the cost of dry particles. The additional performance cost of the saturated air particles was intuitive since the anytime a particle's relative humidity was over 100%, an iterative process was used to calculate the exact change in temperature and water mass.

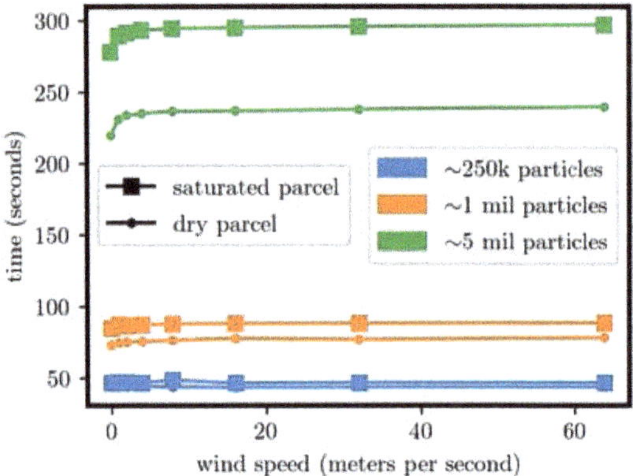

(a)

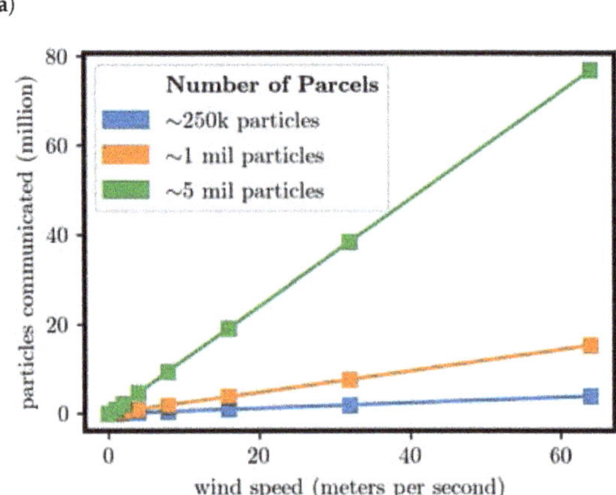

(b)

Figure 12. On a Cray machine. (**a**) The effects of wind speed with varying particle count using dry and saturated particles. (**b**) The number of total particles communicated with the change in wind.

To examine the effects of wind speed on the model, we first turn to Figure 12a. This figure was created using a problem domain of $500 \times 500 \times 30$ over 200 time steps. The runs were done with initially dry or saturated particles and a total particle count of around 250 thousand, 1 million, and 5 million particles. The 250 thousand particles were chosen as the starting point since that would equate to a particle per horizontal grid cell. In all of the varying particle count, the results showed the expected behavior of saturated particles taking more time than unsaturated. As the particle count grew, so did the difference between dry and saturated. This was also an intuitive result since each saturated particle had to perform an iterative process to correctly convert water vapor and cloud water. What was more unexpected was that after an initial performance hit from increased communication, there was negligible cost as the wind speed increased. This was most clearly shown in the five million particle run, where from 0 m/s to 1 m/s, there was a 4.03% performance cost, but from 4 to 8 m/s, there was a 0.45% increase.

Examining Figure 12b shows a linear increase in the number of particles communicated when the wind speed was increased in the model. For five million particles, a change of 0 to 1 m/s in wind speed resulted in zero total particles communicated to 1,048,073. That initial cost of one million particles communicated took an extra 4.03% of the initial time. Compare that with the 0.45% cost when moving from 4 to 8 m/s, even though 4,819,692 additional particles were communicated. This indicated that there was a cost to the communication phase of the particles, but that it did not matter how often they were communicated.

All the previous runs were done on one node, meaning that the communication would all be inter-node. This choice was made due to the exponential growth of runs that would have to be done if this were scaled out to include more nodes. To get a feel for the intra- vs. inter-node communication though, additional runs were done holding the image count constant, but adding two and four nodes. The runs were done over 200 time steps with a wind speed of 8 m/s and with dry air particles. Table 1 shows that running a total of 44 images with one or two nodes leads to approximately the same performance while the jump to four nodes results in a 29.5% performance increase.

Table 1. Inter- vs. intra-node communication.

Nodes	Images	Time (s.)	Average Time (s) PGAS	Percent of Time PGAS	Average Time (s) __pgas_sync_all	Percent of Time __pgas_sync_all	Average Maximum Memory Usage (MBs)
1	44	235.58	12.37	5.0%	6.47	2.6%	100
2	44	234.93	—	—	—	—	—
4	44	165.55	24.49	13.0%	10.88	5.8%	468.8
2	88	117.61	—	—	—	—	—
4	176	59.6	—	—	—	—	—

This jump in performance was a surprising result and required further investigation. Cray's Performance Tools were used to look closely at the data to understand what was occurring. The percent of time spent using PGAS functions was as expected much higher for the four node version, 13.0% vs. 5.0%. While the largest PGAS function in the single-node version was sync all, at 2.6%, for the four-node version it was co_broadcast at 6.0%. The co_broadcast was used in the initial setup phase of the model to communicate physics calculations to all images. The table conveys that sync all within the retrieve function used by the particles took up most of the wait time, but the actual communication of the data did not. The __pgas_put_nbi, used by the particle communication function, took 0.0% and 0.1% of the time for the single- and four-node runs. From testing, it appeared that in __pgas_put_nbi, nbi likely stood for the non-blocking interface, since it was called when a non-blocking put was specifically requested with the !DIR$ PGAS DEFER_SYNC directive. Two things should be noted from this part of the analysis. First, the particle communication in this method took almost no time, and it was the syncs during the unloading of the communication buffers that took the time. Second, despite PGAS taking much longer on the four-node version, it still ran 29.5% faster. It could possibly be from the larger amount of memory each image used, but further investigation would be needed. This would fall in line with what has been noted before, that increasing memory per core will increase performance before reaching saturation [38].

Analysis was done on the distribution of the number of particles communicated. Figure 13a was created using a base wind speed of 8.0 m/s on a $500 \times 500 \times 30$ domain. This figure exhibits how with varying particle counts, the distribution of how many times a particle is communicated will stay relatively constant. The consistency in the distribution suggests we can fix the particle count, without loss of generality, to perform the runs done in Figure 13b. Figure 13b was made with a particle count of 250 thousand. Figure 13b shows the probability distribution of the number of times a particle is communicated as the wind speed changes. Again, this confirmed that particles were being communicated as expected. These figures and Figure 12a,b express that despite linear growth in the number of particles communicated with an increase in wind speed, only the initial increase had a noticeable effect on the runtime.

Particle Communication Distribution

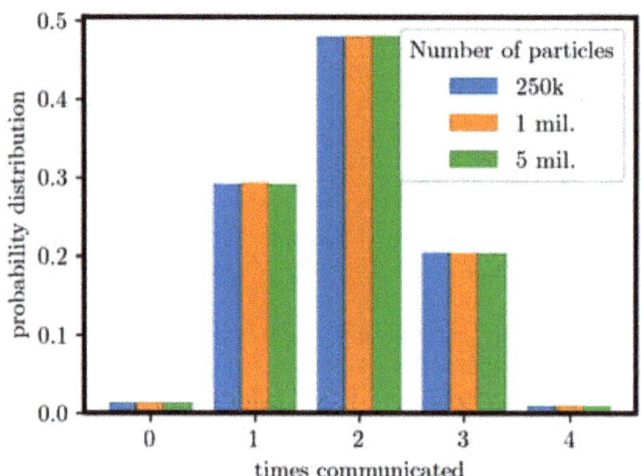

(a)

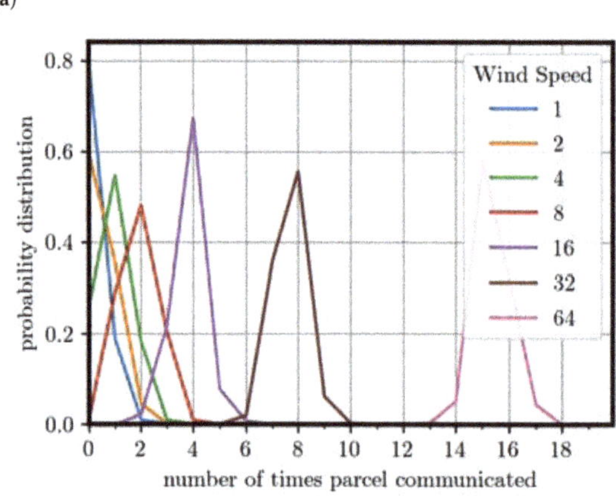

(b)

Figure 13. (a) Distribution of particles communicated with the change in particle count. (b) Distribution of particles communicated with the change in wind speed.

3.4. Scaling and Speedup

As is common practice, we present both strong and weak scaling results to evaluate the performance. Strong scaling suggests that the overall problem size is fixed, and performance is assessed by increasing the number of images. In weak scaling, the size of the problem is fixed per image; therefore, the overall problem size is scaled along with the number of images. The following strong scaling runs were done on a $500 \times 500 \times 30$ and $2000 \times 2000 \times 30$ domain size on both the CRAYand SGI HPC systems. The weak scaling runs were done using 12 thousand and 768 thousand grid points per process. Both strong and weak scaling were run for 200 time steps with each time step being 20 s. The convected air particles' time steps were run every second.

The strong scaling experiments, Figure 14a,b, were created using one image and doubling until 36 images. After that, the full processor count of the nodes was used, which for the Cray cluster was 44 processes times the node count and for the SGI cluster 36 processes times the node count. The number of nodes was doubled every time, up to 16 nodes with a total of 704 processes.

Strong Scaling

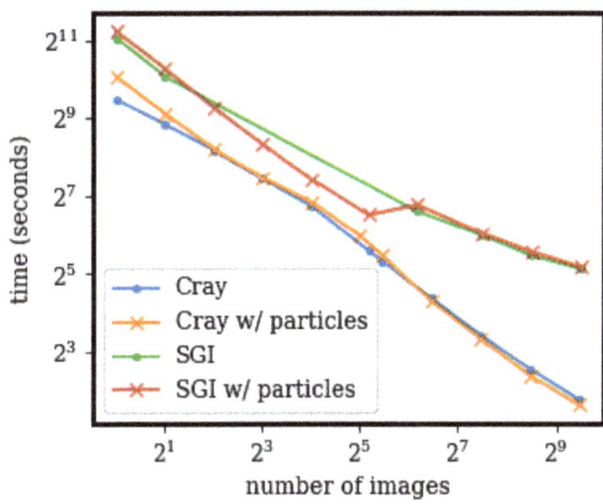

(a)

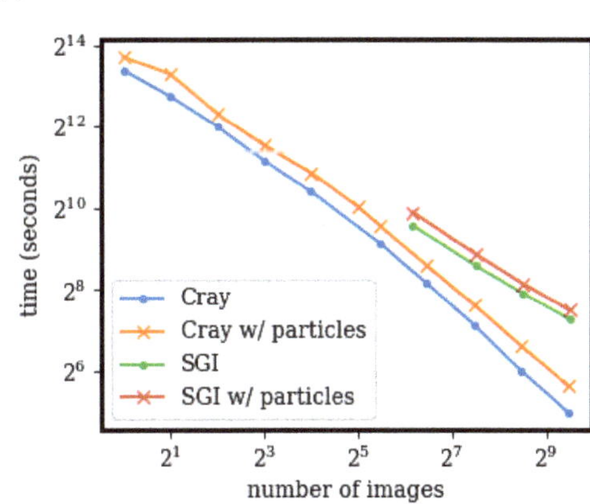

(b)

Figure 14. Log-log graphs of the domains of (a) grid dimensions $500 \times 500 \times 30$ and (b) grid dimensions $2000 \times 2000 \times 30$.

In Figure 14a, the air particles are one per horizontal grid cell. This graph shows that the addition of particles did not really affect the runtime and scaled at the same pace as the particle-less runs. Figure 14b shows that the addition of particles had a consistent cost that did not disappear. However, it also shows that the addition of particles did not affect the scaling on the larger domain size. The strong scaling efficiency, defined as $\frac{T_1}{n \times t_n} \times 100$, is shown in Figure 15a,b. T_1 is the time to complete the work using one processor; n is the number of processors used; and t_n is the amount of time n processors took to complete. Figure 15a shows that the strong scaling efficiency for the SGI machine did not seem to be affected by the addition of particles. On the Cray machine, the efficiency of the run with particles actually increased. This actually switched around with the larger problem domain, where the addition of particles made the efficiency slightly worse. This seemed to indicate that as the problem size grew, the efficiency cost of the addition of particles would be less and less.

Strong Scaling Efficiency

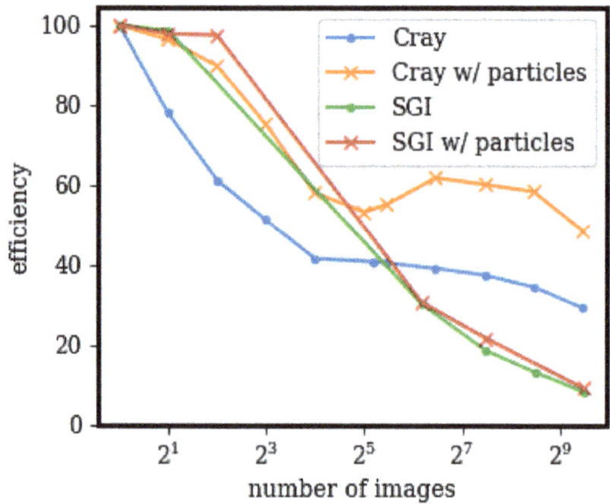

(**a**)

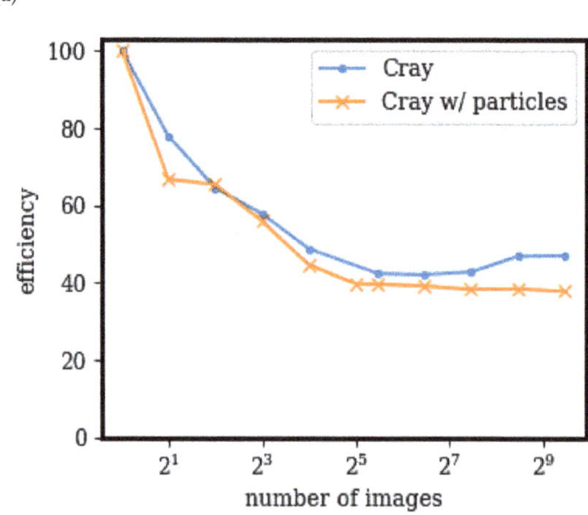

(**b**)

Figure 15. (**a**) Grid dimensions $500 \times 500 \times 30$. (**b**) Grid dimensions $2000 \times 2000 \times 30$.

In Figure 16a,b, the weak scaling results are shown. When the runs were done with particles, a particle per cell was used. Figure 16a starts at the size $20 \times 20 \times 30$, and Figure 16b starts at $160 \times 160 \times 30$. In both figures, the runtime increases until the node processes are all used, at which point the scaling generally levels out. This first increase indicated better cache use and lower communication cost of a single processor. Figure 16a's runtimes are rather low, ranging from 1.15 to 3.2 s. This might be the reason the weak scaling curve never flattened, but oscillated up and down. Figure 16b on the other hand has a weak scaling curve that flattens as all of the first node's 36 or 44 processors are being used and stays flat. Thus, it can be said that linear scaling was achieved and that the ICAR application with and without particles should scale well with larger processor counts. This matched the strong scaling results as well.

Weak Scaling

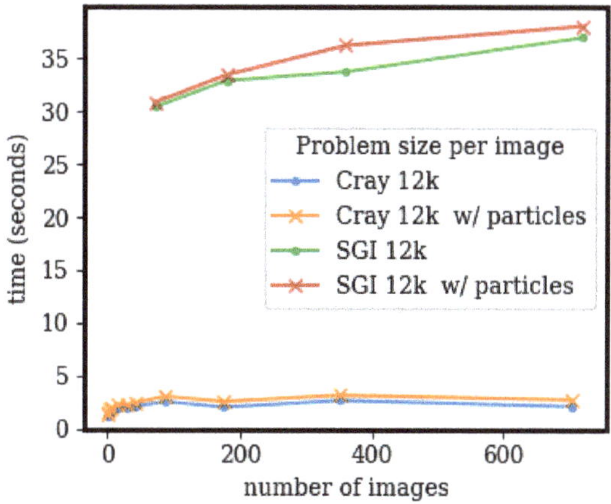

(a)

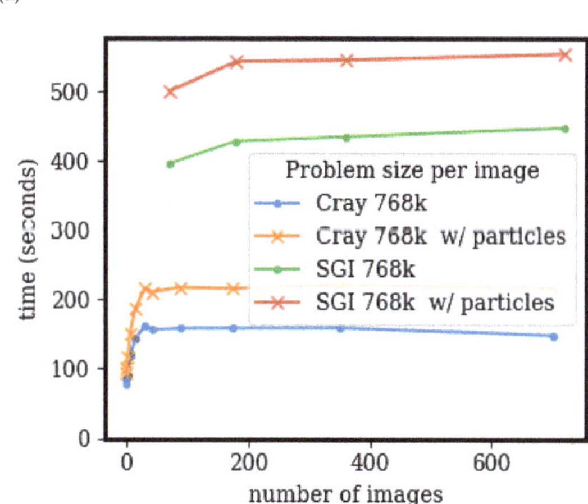

(b)

Figure 16. (a) Twelve-thousand grid cells per process; (b) 768 k grid cells per process.

4. Conclusions

The strong and weak scaling graphs showed that the ICAR model scales well and that the addition of the semi-Lagrangian convected air particles does not affect the scaling. The weak scaling indicated that the framework of communicating the convected air particles immediately upon leaving an image's boundary is an effective way to take advantage of scaling. The strong scaling efficiency suggested that there is better efficiency as the problem size grows. This would need to be tested on larger problem sizes, but the scaling for $2000 \times 2000 \times 30$ was better than $500 \times 500 \times 30$. Furthermore, at $2000 \times 2000 \times 30$, Figure 15b suggested that the scaling efficiency levels off for runs with and without particles, which is desirable.

Investigating the particle communication showed that there was an initial performance hit from communicating the air particles. After that original hit though, increasing the number of times a particle was communicated across boundaries had a minimal effect on the performance. The cost of the communication of the air particles really comes from the cost of the synchronization points. The synchronization time of the air particles took 2.6% and 5.8% of the time for one and four nodes. Communicating the particles across the boundaries took 0.0% and 0.1% percent of the time; note this percentage and the rounding were done by the performance tool.

This work showed that there was a performance cost of adding air particles, but that it did not affect the HPC scalability of the original program. Scaling to a large number of images is a very important feature of any scientific code, even more so in our case where the underlying problem is that of downscaling atmospheric models. Hence, if the introduction of semi-Lagrangian convected air particles did not scale well, our approach would conflict with the primary goal of the underlying model.

Within the ICAR model, it indicates the benefit of spending more time in the future to make some of the processing more efficient. Currently, the air particle physics are calculated every second, and comparing that with the environment time step being 20 s, there is room for improvement. If additional features of air particles are deemed helpful for the ICAR model, it shows that they should not affect scaling; for example, the extra computation from calculating the coriolis force due to the Earth's rotation. Additionally, it would be beneficial to have the particles interact more with the surrounding environment. Currently, the environment is used to calculate buoyancy forces and saturation ratios, but once cloud water is converted, the particle never "rains".

Author Contributions: Conceptualization, S.R., E.D.G., I.M., and S.F.; data curation, S.R.; formal analysis, S.R. and E.D.G.; investigation, S.R. and E.D.G.; methodology, S.R., E.D.G., I.M. and S.F.; software, S.R. and E.D.G.; supervision, E.D.G., I.M. and S.F.; validation, S.R. and E.D.G.; visualization, S.R.; writing—original draft, S.R.; writing—review and editing, S.R., E.D.G., I.M. and S.F. All authors read and agreed to the published version of the manuscript.

Funding: The National Center for Atmospheric Research is sponsored by the National Science Foundation. Funding support provided by the U.S. Army Corps of Engineers. The author Soren Rasmussen is funded by a fellowship provided by Sourcery Institute.

Institutional Review Board Statement: Not applicable.

Data Availability Statement: Data available in a publicly accessible repository. The data presented in this study are openly available in scrasmussen/icar_data at 10.5281/zenodo.4734314.

Acknowledgments: We would like to acknowledge the high-performance computing support from Cheyenne (doi:10.5065/D6RX99HX) provided by NCAR's Computational and Information Systems Laboratory, sponsored by the National Science Foundation. We would also like to acknowledge the high-performance computing support from Swan provided by Cray Inc.

Conflicts of Interest: The author Soren Rasmussen is funded by a fellowship provided by Sourcery Institute. The funders had no role in the design of the study; in the collection, analyses, or interpretation of data; in the writing of the manuscript; nor in the decision to publish the results. Cray Inc. had no role in the design of the study; in the collection, analyses, or interpretation of data; in the writing of the manuscript; nor in the decision to publish the results.

References

1. Gutmann, E.; Barstad, I.; Clark, M.; Arnold, J.; Rasmussen, R. The intermediate complexity atmospheric research model (ICAR). *J. Hydrometeorol.* **2016**, *17*, 957–973. [CrossRef]
2. Bernhardt, M.; Härer, S.; Feigl, M.; Schulz, K. Der Wert Alpiner Forschungseinzugsgebiete im Bereich der Fernerkundung, der Schneedeckenmodellierung und der lokalen Klimamodellierung. *Österreichische-Wasser-Und Abfallwirtsch.* **2018**, *70*, 515–528. [CrossRef]
3. Horak, J.; Hofer, M.; Maussion, F.; Gutmann, E.; Gohm, A.; Rotach, M.W. Assessing the added value of the Intermediate Complexity Atmospheric Research (ICAR) model for precipitation in complex topography. *Hydrol. Earth Syst. Sci.* **2019**, *23*, 2715–2734. [CrossRef]
4. Horak, J.; Hofer, M.; Gutmann, E.; Gohm, A.; Rotach, M.W. A process-based evaluation of the Intermediate Complexity Atmospheric Research Model (ICAR) 1.0. 1. *Geosci. Model Dev.* **2021**, *14*, 1657–1680. [CrossRef]
5. Numrich, R.W.; Reid, J. Co-Array Fortran for parallel programming. In *ACM Sigplan Fortran Forum*; ACM: New York, NY, USA, 1998; Volume 17:2, pp. 1–31.
6. ISO/IEC. *Fortran Standard 2008*; Technical report, J3; ISO/IEC: Geneva, Switzerland, 2010.
7. Coarfa, C.; Dotsenko, Y.; Mellor-Crummey, J.; Cantonnet, F.; El-Ghazawi, T.; Mohanti, A.; Yao, Y.; Chavarría-Miranda, D. An evaluation of global address space languages: Co-array fortran and unified parallel C. In Proceedings of the tenth ACM SIGPLAN Symposium on Principles and Practice of Parallel Programming, Chicago, IL, USA, 15–17 June 2005; pp. 36–47.
8. Stitt, T. *An Introduction to the Partitioned Global Address Space (PGAS) Programming Model*; Connexions, Rice University: Houston, TX, USA, 2009.
9. Mozdzynski, G.; Hamrud, M.; Wedi, N. A partitioned global address space implementation of the European centre for medium range weather forecasts integrated forecasting system. *Int. J. High Perform. Comput. Appl.* **2015**, *29*, 261–273. [CrossRef]
10. Simmons, A.; Burridge, D.; Jarraud, M.; Girard, C.; Wergen, W. The ECMWF medium-range prediction models development of the numerical formulations and the impact of increased resolution. *Meteorol. Atmos. Phys.* **1989**, *40*, 28–60. [CrossRef]
11. Jiang, T.; Guo, P.; Wu, J. One-sided on-demand communication technology for the semi-Lagrange scheme in the YHGSM. *Concurr. Comput. Pract. Exp.* **2020**, *32*, e5586. [CrossRef]
12. Dritschel, D.G.; Böing, S.J.; Parker, D.J.; Blyth, A.M. The moist parcel-in-cell method for modelling moist convection. *Q. J. R. Meteorol. Soc.* **2018**, *144*, 1695–1718. [CrossRef]
13. Böing, S.J.; Dritschel, D.G.; Parker, D.J.; Blyth, A.M. Comparison of the Moist Parcel-in-Cell (MPIC) model with large-eddy simulation for an idealized cloud. *Q. J. R. Meteorol. Soc.* **2019**, *145*, 1865–1881. [CrossRef]
14. Brown, N.; Weiland, M.; Hill, A.; Shipway, B.; Maynard, C.; Allen, T.; Rezny, M. A highly scalable met office nerc cloud model. *arXiv* **2020**, arXiv:2009.12849.
15. Shterenlikht, A.; Cebamanos, L. Cellular automata beyond 100k cores: MPI vs. Fortran coarrays. In Proceedings of the 25th European MPI Users' Group Meeting, Barcelona, Spain, 23 September 2018; pp. 1–10.
16. Shterenlikht, A.; Cebamanos, L. MPI vs Fortran coarrays beyond 100k cores: 3D cellular automata. *Parallel Comput.* **2019**, *84*, 37–49. [CrossRef]
17. Rasmussen, S.; Gutmann, E.D.; Friesen, B.; Rouson, D.; Filippone, S.; Moulitsas, I. Development and Performance Comparison of MPI and Fortran Coarrays within an Atmospheric Research Model. Presented at the Workshop 2018 IEEE/ACM Parallel Applications Workshop, Alternatives To MPI (PAW-ATM), Dallas, TX, USA, 16 November 2018.
18. Stein, A.; Draxler, R.R.; Rolph, G.D.; Stunder, B.J.; Cohen, M.; Ngan, F. NOAA's HYSPLIT atmospheric transport and dispersion modeling system. *Bull. Am. Meteorol. Soc.* **2015**, *96*, 2059–2077. [CrossRef]
19. Ngan, F.; Stein, A.; Finn, D.; Eckman, R. Dispersion simulations using HYSPLIT for the Sagebrush Tracer Experiment. *Atmos. Environ.* **2018**, *186*, 18–31. [CrossRef]
20. Esmaeilzadeh, H.; Blem, E.; Amant, R.S.; Sankaralingam, K.; Burger, D. Dark silicon and the end of multicore scaling. In Proceedings of the 2011 38th Annual International Symposium on Computer Architecture (ISCA), San Jose, CA, USA, 4–8 June 2011; pp. 365–376.
21. Moisseeva, N.; Stull, R. A noniterative approach to modelling moist thermodynamics. *Atmos. Chem. Phys.* **2017**, *17*. [CrossRef]
22. Stull, R.B. *Practical Meteorology: An Algebra-Based Survey of Atmospheric Science*; University of British Columbia: Vancouver, BC, Canada, 2018.
23. Yau, M.K.; Rogers, R.R. *A Short Course in Cloud Physics*; Elsevier: Amsterdam, The Netherlands, 1996.
24. Rasmussen, S.; Gutmann, E. Coarray ICAR Fork. [Code]. Available online: github.com/scrasmussen/coarray_icar/releases/tag/v0.1 (accessed on 14 January 2021).
25. Rasmussen, S.; Gutmann, E. ICAR Data. [Dataset]. Available online: github.com/scrasmussen/icar_data/releases/tag/v0.0.1 (accessed on 14 January 2021).
26. Mandl, F. *Statistical Physics*; Wiley: Hoboken, NJ, USA, 1971.
27. Marmelad. 3D Interpolation. Available online: https://en.wikipedia.org/wiki/Trilinear_interpolation#/media/File:3D_interpolation2.svg (access on 14 January 2021).

28. University of Wyoming. Upper Air Soundings. Available online: weather.uwyo.edu/upperair/sounding.html (accessed on 10 November 2020).
29. Sharma, A.; Moulitsas, I. MPI to Coarray Fortran: Experiences with a CFD Solver for Unstructured Meshes. *Sci. Program.* **2017**, *2017*, 3409647. [CrossRef]
30. Fanfarillo, A.; Burnus, T.; Cardellini, V.; Filippone, S.; Nagle, D.; Rouson, D. OpenCoarrays: Open-source transport layers supporting coarray Fortran compilers. In Proceedings of the 8th International Conference on Partitioned Global Address Space Programming Models, Eugene, OR, USA, 6–10 October 2014; pp. 1–11.
31. Feind, K. Shared memory access (SHMEM) routines. *Cray Res.* **1995**, *53*, 303–308.
32. HPE Cray. *Cray Fortran Reference Manual*; Technical Report; Cray Inc.: Seattle, DC, USA, 2018.
33. Shan, H.; Wright, N.J.; Shalf, J.; Yelick, K.; Wagner, M.; Wichmann, N. A preliminary evaluation of the hardware acceleration of the Cray Gemini interconnect for PGAS languages and comparison with MPI. *ACM Sigmetrics Perform. Eval. Rev.* **2012**, *40*, 92–98. [CrossRef]
34. Shan, H.; Austin, B.; Wright, N.J.; Strohmaier, E.; Shalf, J.; Yelick, K. Accelerating applications at scale using one-sided communication. In Proceedings of the Conference on Partitioned Global Address Space Programming Models (PGAS'12), Santa Barbara, CA, USA, 10–12 October 2012.
35. Shende, S.S.; Malony, A.D. The TAU parallel performance system. *Int. J. High Perform. Comput. Appl.* **2006**, *20*, 287–311. [CrossRef]
36. Ramey, C.; Fox, B. Bash 5.0 Reference Manual. 2019. Available online: gnu.org/software/bash/manual/ (accessed on 12 May 2020).
37. Kaufmann, S.; Homer, B. *Craypat-Cray X1 Performance Analysis Tool*; Cray User Group: Seattle, DC, USA, 2003; pp. 1–32.
38. Zivanovic, D.; Pavlovic, M.; Radulovic, M.; Shin, H.; Son, J.; Mckee, S.A.; Carpenter, P.M.; Radojković, P.; Ayguadé, E. Main memory in HPC: Do we need more or could we live with less? *ACM Trans. Archit. Code Optim. (TACO)* **2017**, *14*, 1–26. [CrossRef]

chemengineering

MDPI

Article

How to Modify LAMMPS: From the Prospective of a Particle Method Researcher

Andrea Albano [1,*]**, Eve le Guillou** [2]**, Antoine Danzé** [2]**, Irene Moulitsas** [2]**, Iwan H. Sahputra** [1,3]**, Amin Rahmat** [1]**, Carlos Alberto Duque-Daza** [1,4]**, Xiaocheng Shang** [5]**, Khai Ching Ng** [6]**, Mostapha Ariane** [7] **and Alessio Alexiadis** [1,*]

[1] School of Chemical Engineering, University of Birmingham, Birmingham B15 2TT, UK;
    halimits@yahoo.com (I.H.S.); A.Rahmat@bham.ac.uk (A.R.); C.A.Duque-Daza@bham.ac.uk (C.A.D.-D.)
[2] Centre for Computational Engineering Sciences, Cranfield University, Bedford MK43 0AL, UK;
    Eve.M.Le-Guillou@cranfield.ac.uk (E.l.G.); A.Danze@cranfield.ac.uk (A.D.); i.moulitsas@cranfield.ac.uk (I.M.)
[3] Industrial Engineering Department, Petra Christian University, Surabaya 60236, Indonesia
[4] Department of Mechanical and Mechatronic Engineering, Universidad Nacional de Colombia,
    Bogotá 111321, Colombia
[5] School of Mathematics, University of Birmingham, Edgbaston, Birmingham B15 2TT, UK;
    X.Shang.1@bham.ac.uk
[6] Department of Mechanical, Materials and Manufacturing Engineering, University of Nottingham Malaysia,
    Jalan Broga, Semenyih 43500, Malaysia; KhaiChing.Ng@nottingham.edu.my
[7] Department of Materials and Engineering, Sayens-University of Burgundy, 21000 Dijon, France;
    Mostapha.Ariane@u-bourgogne.fr
* Correspondence: axa1220@student.bham.ac.uk or aaalbano@gmail.com (A.A.);
    a.alexiadis@bham.ac.uk (A.A.)

check for
updates

Citation: Albano, A.; le Guillou, E.; Danzé, A.; Moulitsas, I.; Sahputra, I.H.; Rahmat, A.; Duque-Daza, C.A.; Shang, X.; Ching Ng, K.; Ariane, M.; et al. How to Modify LAMMPS: From the Prospective of a Particle Method Researcher. *ChemEng* **2021**, *5*, 30. https://doi.org/10.3390/chemengineering5020030

Academic Editors: Mark P. Heitz and Andrew S. Paluch

Received: 11 January 2021
Accepted: 26 May 2021
Published: 13 June 2021

Abstract: LAMMPS is a powerful simulator originally developed for molecular dynamics that, today, also accounts for other particle-based algorithms such as DEM, SPH, or Peridynamics. The versatility of this software is further enhanced by the fact that it is open-source and modifiable by users. This property suits particularly well Discrete Multiphysics and hybrid models that combine multiple particle methods in the same simulation. Modifying LAMMPS can be challenging for researchers with little coding experience. The available material explaining how to modify LAMMPS is either too basic or too advanced for the average researcher. In this work, we provide several examples, with increasing level of complexity, suitable for researchers and practitioners in physics and engineering, who are familiar with coding without been experts. For each feature, step by step instructions for implementing them in LAMMPS are shown to allow researchers to easily follow the procedure and compile a new version of the code. The aim is to fill a gap in the literature with particular reference to the scientific community that uses particle methods for (discrete) multiphysics.

Keywords: LAMMPS; particle method; discrete multiphysics

1. Introduction

LAMMPS, acronym for Large-scale Atomic/Molecular Massively Parallel Simulator, was originally written in F77 by Steve Plimpton [1] in 1993 with the goal of having a large-scale parallel classical Molecular Dynamic (MD) code. The project was a Cooperative Research and Development Agreement (CRADA) between two DOE labs (Sandia and LLNL) and three companies (Cray, Bristol Myers Squibb, and Dupont). Since the initial release LAMMPS has been improved and expanded by many researchers who implemented many mesh-free computational methods such as Perydynamics, Smoothed particle hydrodynamics (SPH), Discrete Elemet Method (DEM) and many more [2–11].

Such a large number of computational methods within the same simulator allows researchers to easily combine them for the simulation of complex phenomena. In particular, our research group has used during the years LAMMPS in a variety of settings that go from classic Molecular Dynamics [12–16], to Discrete Multiphysics simulations of cardiovascular flows [17–20], Modelling drug adsorption in human organs [21–24], Cavitation [25–27], multiphase flow containing cells or capsules [28–31], solidification/dissolution [32–34], material properties [35,36] and even epidemiology [37] and coupling particles methods with Artificial Intelligence [38–40]. An example of a Discrete Multiphysics simulation run with the basic LAMMPS's code is shown in Appendix A.

Thanks to its modular design open source nature and its large community, LAMMPS has been conceived to be modified and expanded by adding new features. In fact, about 95% of its source code is add-on file [41]. However, this can be a tough challenge for researcher with no to little knowledge of coding. The LAMMPS user manual [41] describes the internal structure and algorithms of the code with the intent of helping researcher to expand LAMMPS. However, due to the lack of examples of implementation and validation, the document can be hard to read for user who are not programmers. In fact, the available material is either very basic [41] or requires advanced programming skills [42,43].

The aim of this work is to provide several step-by-step examples with increasing level of complexity that can fill the gap in the middle to help and encourage researchers to use LAMMPS for discrete multiphysics and expand it with new adds on to the code that could fit their needs. In fact, most of the available material focuses on Molecular Dynamics (MD) and implicitly assumes that the reader's background is in MD rather than other particle methods such as SPH or DEM. On the contrary, this paper is dedicated to the particle community and highlights how LAMMPS can be used and modified for methods other than MD. This goal fits particularly well with the scope of this Special Issue on "Discrete Multiphysics: Modelling Complex Systems with Particle Methods" In particular, it relates to some of the topics of the Special Issue such by exploring the potential of LAMMPS for coupling particle methods, and by sharing some "tricks of the trade" on how to modify its code that cannot be found anywhere else in the literature.

In Section 2 LAMMPS structure and hierarchy are explained introducing the concept of style. Following the LAMMPS authors advice, to avoid writing a new style from scratch, Sections 3–6 new styles are developed using existing style are as reference. Finally, in Section 7, all the steps to write a class from scratch are shown.

2. LAMMPS Structure

After initial releases in F77 and F90, LAMMPS is now written in C++, an object oriented language that allows any programmer to exploit the class programming paradigm. The declaration of a class, including the signature of the instance variables and functions (or methods), which can be accessed and used by creating an instance of that class. The data and functions within a class are called members of the class. The definition (or implementation) of a member function can be given inside or outside the class definition.

A class has private, public, and protected sections which contain the corresponding class members.

- The private members, defined before the keyword public, cannot be accessed from outside the class. They can only be accessed by class or "friend" functions, which are declared as having access to class members, without themselves being members. All the class members are private by default.
- The public members can be accessed from outside the class anywhere within the scope of the class object.
- The protected members are similar to private members but they can be accessed by derived classes or child classes while private members cannot.

2.1. Inheritance

An important concepts in object-oriented programming is that of inheritance. Inheritance allows to define a class in terms of another class and the new class inherits the members of the existing class. This existing class is called the base (or parent) class, and the new class is referred to as a subclass, or child class, or derived class.

The idea of inheritance implements the "is a" relationship. For example, Mammal IS-A Animal, Dog IS-A Mammal hence Dog IS-A Animal as well.

The inheritance relationship between the parent and the derived classes is declared in the derived class with the following syntax:

Listing 1: C++ syntax for classes inheritance

```
class name_child_class: access_specifier name_parent_class
{ /*...*/ };
```

The type of inheritance is specified by the access-specifier, one of public, protected, or private. If the access-specifier is not used, then it is private by default, but public inheritance is commonly used: public members of the base class become public members of the derived class and protected members of the base class become protected members of the derived class. A base class's private members are never accessible directly from a derived class, but can be accessed through calls to the public and protected members of the base class.

2.2. Virtual Function

The signature of a function f must be declared with a virtual keyword in a base class C to allow its definition (implementation), or redefinition, in a derived class D. Then, when a derived class D object is used as an element of the base class C, and f is called, the derived class's implementation of the function is executed.

There is nothing wrong with putting the virtual in front of functions inside of the derived classes, but it is not required, unless it is known for sure that the class will not have any children who would need to override the functions of the base class. A class that declares or inherits a virtual function is called a polymorphic class.

2.3. LAMMPS Inheritance and Class Syntax

A schematic representation of the LAMMPS inheritance tree is shown in Figure 1: LAMMPS is the top-level class for the entire code, then all the core classes, highlighted in blue, inherit all the constructors, destructors, assignment operator members, friends and private members declared and defined in LAMMPS. The core classes perform LAMMPS fundamental actions. For instance, the Atom class collects and stores all the per-atom, or per-particle, data while Neighbor class builds the neighbor lists [41].

The style classes, highlighted in reds, inherit all the constructors, destructors, assignment operator members, friends and private members declared and defined in LAMMPS and in the corresponding core class. The style classes are also virtual parents class of many child classes that implement the interface defined by the parent class. For example, the fix style has around 100 child classes.

Each style is composed of a pair of files:

- namestyle.h
  The header of the style, where the class style is defined and all the objects, methods and constructors are declared.
- namestyle.cpp
  Where all the objects, methods and constructors declared in the class of style are defined.

When a new style is written both `namestyle.h` and `namestyle.cpp` files need to be created.

Each "family" style has its own set of methods, declared in the header and defined in the cpp file, in order to define the scope of the style. For example, the pair style are classes that set the formula(s) LAMMPS uses to compute pairwise interactions while bond style set the formula(s) to compute bond interactions between pairs of atoms [41].

Each pair style has some recurrent functions such as `compute`, `allocate` and `coeff`. Although the final scope of those functions can differ for different styles, they all share a similar role within the classes.

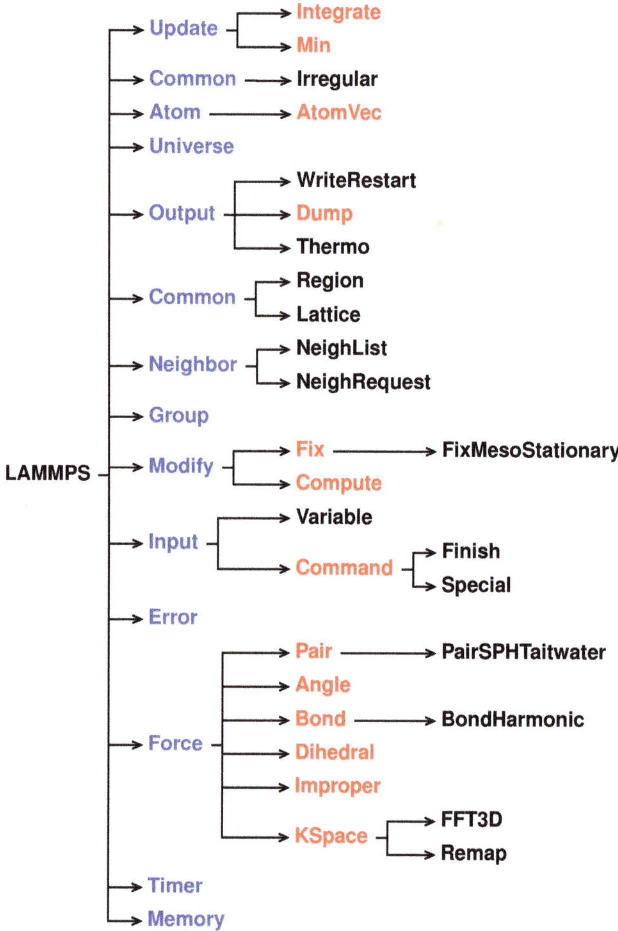

Figure 1. Class hierarchy within LAMMPS source code.

An example of a pair style, sph/taitwater, header in LAMMPS is shown in Listing 2.

Listing 2: Header file of sph/taitwater pair style (pair_sph_taitwater.h)

```cpp
class PairSPHTaitwater: public Pair{// class definition, accessibility and Inheritance
 public:    // access specifier: public
   // public methods
   PairSPHTaitwater(class LAMMPS *); // Constructors
   virtual ~PairSPHTaitwater(); // Destructors
   virtual void compute(int, int);
   void settings(int, char **);
   void coeff(int, char **);
   virtual double init_one(int, int);
   virtual double single(int, int, int, int, double, double, double, double &);

 protected:  // access specifier: protected
   double *rho0, *soundspeed, *B;
   double **cut,**viscosity;
   int first;
// protected methods
   void allocate();
};
```

All the class members are defined in the cpp file. Taking sph/taitwater pair style as reference, each method declared in Listing 2 will be defined and commented in the next sections. Although this can be style-specific, the aim is to give an overview of how the methods are defined in the cpp in LAMMPS. Albeit different style has different methods, the understanding gained can be transferred into others style, as shown in Sections 3 and 6.

2.3.1. Constructor

Any class usually include a member function called constructors. The constructor is mechanically invoked when an object of the class is created. This allows the class to initialise members or allocate storage. Unlike the other member of the class, the constructor name must match the name of the class and it does not have a return type.

Listing 3: Constructor definition in sph/taitwater pair style (pair_sph_taitwater.cpp)

```
1  PairSPHTaitwater::PairSPHTaitwater(LAMMPS *lmp) : Pair(lmp)
2  {
3    restartinfo = 0;
4    first = 1;
5  }
```

2.3.2. Destructor

The role of destructors is to de-allocate the allocated dynamic memory, see Section 2.3.8, being mechanically invoked just before the end of the class lifetime. Similarly to constructors, destructors does not have a return type and have the same name as the class name with a tilde ($\sim$) prefix.

Listing 4: Destructors definition in sph/taitwater pair style (pair_sph_taitwater.cpp)

```
1  PairSPHTaitwater::~PairSPHTaitwater() {
2    if (allocated) { /// check if the pair style uses allocate, see Section 2.8
3    ///   cleanup the memory used by allocate, see Section 2.8
4      memory->destroy(setflag);
5      memory->destroy(cutsq);
6      memory->destroy(cut);
7      memory->destroy(rho0);
8      memory->destroy(soundspeed);
9      memory->destroy(B);
10     memory->destroy(viscosity);
11   }
12 }
```

2.3.3. compute

compute is virtual member of the pair style and is one of the most relevant functions in a number of classes in LAMMPS. For instance, in pair style classes is used to compute pairwise interaction of the specific pair style. This can seen in the commented Listing 5, where the force applied on a pair of neighboring particles is derived using the Tait equation, lines 131–151. In compute all the local parameters needed to compute the pairwise interaction are declared and defined within the method.

Listing 5: compute definition in sph/taitwater pair style (pair_sph_taitwater.cpp)

```
1  void PairSPHTaitwater::compute(int eflag, int vflag) {
2
3  /// start variables and pointer declaration
4    int i, j, ii, jj, inum, jnum, itype, jtype;
5    double xtmp, ytmp, ztmp, delx, dely, delz, fpair;
6
7    int *ilist, *jlist, *numneigh, **firstneigh;
8    double vxtmp, vytmp, vztmp, imass, jmass, fi, fj, fvisc, h, ih, ihsq;
9    double rsq, tmp, wfd, delVdotDelR, mu, deltaE;
10 // end
11
12   if (eflag || vflag)
```

```
13      ev_setup(eflag, vflag);
14    else
15      evflag = vflag_fdotr = 0;
16
17  /// others variables and pointers declaration and initialisation
18    double **v = atom->vest; // pass the value of the pointer that points to a pointers
19    // pointing to the first element of velocity vector of the particles
20    double **x = atom->x; // pass the value of the pointer that points to a pointers
21    // pointing to the first element of position vector of the particles
22    double **f = atom->f; // pass the value of the pointer that points to a pointers
23    // pointing to the first element of force vector of the particles
24    double *rho = atom->rho; // pass the value of the pointer that points
25    // to the density vector of the particles
26    double *mass = atom->mass; // pass the value of the pointer that points
27     // to the mass vector of the particles
28    double *de = atom->de; // pass the value of the pointer that points
29    // to the change of internal energy of the particles
30    double *drho = atom->drho; // pass the value of the pointer that points
31    // to the change of density of the particles
32    int *type = atom->type; // pass the value of the pointer that points to the type of the
          particles
33    int nlocal = atom->nlocal; // pass the value of the numbers of owned and ghost atoms on
          this proc
34    int newton_pair = force->newton_pair;  // pass the value of the Newton's 3rd law
          settings
35  /// end
36
37
38    // check consistency of pair coefficients
39
40    if (first) {
41      for (i = 1; i <= atom->ntypes; i++) {
42        for (j = 1; i <= atom->ntypes; j++) {
43          if (cutsq[i][j] > 1.e-32) {
44            if (!setflag[i][i] || !setflag[j][j]) {
45              if (comm->me == 0) {
46                printf(
47                    "SPH particle types %d and %d interact with cutoff=%g,
48                    but not all of their single particle properties are set.\n",
49                    i, j, sqrt(cutsq[i][j]));
50            } } } } }
51      first = 0;
52    }
53
54
55    inum = list->inum; // pass the value of number of I atoms neighbors are stored for
56    ilist = list->ilist; // pass the value of the pointer pointing to the local indices of I
          atoms
57    numneigh = list->numneigh; // pass the address of a pointer pointing to the number of J
          neighbors
58    //  for each I atom
59    firstneigh = list->firstneigh; // pass the value of a pointer that points to pointer
60    // pointing to 1st J int value of each I atom
61
62
63
64
65    for (ii = 0; ii < inum; ii++) { // loop for each i particles stored in inum
66      i = ilist[ii]; // pass the index of the i particle
67      xtmp = x[i][0]; // pass the x position of the i particle
68      ytmp = x[i][1]; // pass the y position of the i particle
69      ztmp = x[i][2]; // pass the z position of the i particle
70      vxtmp = v[i][0]; // pass the x velocity of the i particle
71      vytmp = v[i][1]; // pass the y velocity of the i particle
72      vztmp = v[i][2]; // pass the z velocity of the i particle
73      itype = type[i]; // pass the type of the i particle
74      jlist = firstneigh[i]; // pass the 1st J int value of each I atom
75      jnum = numneigh[i]; //pass number of J neighbors for each I atom
76
77      imass = mass[itype]; // pass the mass of the i particle
78
```

```
// compute force of atom i with Tait EOS
tmp = rho[i] / rho0[itype];
fi = tmp * tmp * tmp;
fi = B[itype] * (fi * fi * tmp - 1.0) / (rho[i] * rho[i]);
// end

for (jj = 0; jj < jnum; jj++) {   // loop over neighbours list of particle i
  j = jlist[jj]; // pass the index of the j particle
  j &= NEIGHMASK;

  delx = xtmp - x[j][0]; // x distance between particles i and j
  dely = ytmp - x[j][1]; // y distance between particles i and j
  delz = ztmp - x[j][2]; // z distance between particles i and j
  rsq = delx * delx + dely * dely + delz * delz; // squared distance between particles
  i and j
  jtype = type[j]; // pass the type of the i particle
  jmass = mass[jtype]; // pass the mass of the j particle

  if (rsq < cutsq[itype][jtype]) { // check if i and j are neighbor

    h = cut[itype][jtype]; // pass the smoothing length
    ih = 1.0 / h; // calculate the inverse, divisions are computationally expensive
    ihsq = ih * ih; // squared inverse

    wfd = h - sqrt(rsq);

    if (domain->dimension == 3) {
      // Lucy Kernel, 3d
      wfd = -25.0669035369735515383e0 * wfd * wfd * ihsq * ihsq * ihsq * ih;
    } else {
      // Lucy Kernel, 2d
      wfd = -19.0985931710274440292e0 * wfd * wfd * ihsq * ihsq * ihsq;
    }

    // compute force of atom j with Tait EOS
    tmp = rho[j] / rho0[jtype];
    fj = tmp * tmp * tmp;
    fj = B[jtype] * (fj * fj * tmp - 1.0) / (rho[j] * rho[j]);
    // end

    // dot product of velocity delta and distance vector
    delVdotDelR = delx * (vxtmp - v[j][0]) + dely * (vytmp - v[j][1])
        + delz * (vztmp - v[j][2]);

    // artificial viscosity (Monaghan 1992)
    if (delVdotDelR < 0.) {
      mu = h * delVdotDelR / (rsq + 0.01 * h * h);
      fvisc = -viscosity[itype][jtype] * (soundspeed[itype]
          + soundspeed[jtype]) * mu / (rho[i] + rho[j]);
    } else {
      fvisc = 0.;
    }

    fpair = -imass * jmass * (fi + fj + fvisc) * wfd; // total pair force
    deltaE = -0.5 * fpair * delVdotDelR; // internal energy increment

// change in force in each direction for particle i
    f[i][0] += delx * fpair;
    f[i][1] += dely * fpair;
    f[i][2] += delz * fpair;

    //change in density for particle i
    drho[i] += jmass * delVdotDelR * wfd;

    // change in internal energy for particle i
    de[i] += deltaE;

    if (newton_pair || j < nlocal) {
    // change in force in each direction for particle j
```

```
147              f[j][0] -= delx * fpair;
148              f[j][1] -= dely * fpair;
149              f[j][2] -= delz * fpair;
150
151              de[j] += deltaE; // change in internal energy for particle j
152
153              drho[j] += imass * delVdotDelR * wfd; // change in density for particle j
154            }
155
156          if (evflag)
157            ev_tally(i, j, nlocal, newton_pair, 0.0, 0.0, fpair, delx, dely, delz);
158        }
159      }
160    }
161
162    if (vflag_fdotr) virial_fdotr_compute();
163 }
```

2.3.4. `settings`

`settings` is a public void function that reads the input script checking that all the arguments of the pair style are declared. If arguments are present, `settings` stores them so they can be used by `compute`. Examples for no arguments pair style and arguments pair style input script with the corresponding `settings` are listed below:

- No arguments pair style: sph/taitwater
  As described in the SPH for LAMMPS manual [6], the command line to invoke the sph/taitwater pair style is shown in Listing 6.

  Listing 6: Command line to invoke sph/taitwater pair style

```
1 pair_style sph/taitwater
```

  In this pair style there is just a string defining the pair style, sph/taitwater, with no arguments. For this reason in `settings`, Listing 7, when the `if` statement is true (number of arguments other than zero) an error is produced.

  Listing 7: `setting` definition in sph/taitwater pair style (pair_sph_taitwater.cpp)

```
1 void PairSPHTaitwater::settings(int narg, char **arg) {
2   if (narg != 0)  ///  check the number of arguments
3     error->all(FLERR,"Illegal number of setting arguments for pair_style sph/
       taitwater");
4 }
```

- Arguments pair syle: sph/rhosum
  As described in the SPH for LAMMPS manual [6], the command line to invoke the sph/rhosum pair style is shown in Listing 8.

  Listing 8: Command lines to invoke sph/rhosum pair style

```
1 pair_style sph/rhosum Nstep
```

  In this pair style there is a string defining the pair style, sph/rhosum, plus one argument, Nstep. For this reason in `settings`, Listing 9, when the `if` statement is true (number of arguments other than one) an error is produced. When the `if` statement is false `settings` assigns the value of Nstep in the variable nstep, line 5, by using the `inumeric` function defined in the force class.

  Listing 9: `setting` definition in sph/rhosum pair style (pair_sph_rhosum.cpp)

```
1 void PairSPHRhoSum::settings(int narg, char **arg) {
2   if (narg != 1) ///  check the number of arguments
3     error->all(FLERR,
4         "Illegal number of setting arguments for pair_style sph/rhosum");
5   nstep = force->inumeric(FLERR,arg[0]); // store the variable in the position 0 (Nstep)
       into nstep
6 }
```

2.3.5. `coeff`

Similar to `setting`, `coeff` is a public void function that reads and set the coefficients used in by `compute` of the pair style. For each *i j* pair is possible to set different coefficients. The coefficients are passed in the input file with the command line pair coeff, see Listing 10. As before, examples for different pair coeff input script and the corresponding `coeff` are listed below:

- **sph/taitwater**
  As described in the SPH for LAMMPS manual [6], the command line to invoke sph/taitwater pair coeff is shown in Listing 10.

  Listing 10: Command line to invoke sph/taitwater pair coeff

  ```
  pair_coeff I J rho_0 c_0 alpha h
  ```

  In total there are six arguments. Thus, in `coeff`, Listing 11, when `if` statement is true (number of arguments other than six) an error is produced. When the `if` statement is false `coeff` assigns the type of particles I and J plus the value of `rho_0`, `c_0`, `alpha` and `h` in from the string to the variables by using the `numeric` function defined in force class. At last, within the double `for` loop from line 19 to 32, the variables are assigned for each particles.

  Listing 11: `coeff` definition in sph/taitwater pair style (pair_sph_taitwater.cpp)

  ```cpp
  void PairSPHTaitwater::coeff(int narg, char **arg) {
    if (narg != 6) ///  check the number of arguments
      error->all(FLERR,
          "Incorrect args for pair_style sph/taitwater coefficients");
    if (!allocated) ///  check if allocate has been called
      allocate(); ///  call allocate, see section 2.8

    int ilo, ihi, jlo, jhi;
    force->bounds(arg[0], atom->ntypes, ilo, ihi);
    force->bounds(arg[1], atom->ntypes, jlo, jhi);

    ///  store the variables in the position 2--5
    double rho0_one = force->numeric(FLERR,arg[2]);
    double soundspeed_one = force->numeric(FLERR,arg[3]);
    double viscosity_one = force->numeric(FLERR,arg[4]);
    double cut_one = force->numeric(FLERR,arg[5]);
    /// B_one is a constant used in tait EOS inside compute, see section 2.3
    double B_one = soundspeed_one * soundspeed_one * rho0_one / 7.0;

    ///  assign the coefficient to the corresponding particle (i)
    ///  and to the pair of particles (i,j)
    int count = 0;
    for (int i = ilo; i <= ihi; i++) {
      rho0[i] = rho0_one;
      soundspeed[i] = soundspeed_one;
      B[i] = B_one;
      for (int j = MAX(jlo,i); j <= jhi; j++) {
        viscosity[i][j] = viscosity_one;
        cut[i][j] = cut_one;

        setflag[i][j] = 1;

        count++;
      }
    }
    if (count == 0) ///  check if the arguments have been assigned
      error->all(FLERR,"Incorrect args for pair coefficients");
  }
  ```

- **sph/rhosum**
  As described in the SPH for LAMMPS manual [6], the syntax to invoke the command is shown in Listing 12.

Listing 12: Command lines to invoke sph/rhosum pair style

```
1  pair_coeff I J h
```

In this case there are three arguments. Thus, in the `coeff`, Listing 13, when the `if` statement is true (number of arguments other than six) an error is produced. When the error is not produced function assigns the type of particles I and J plus the value of *h* in the string to the variable `cut_one`, line 11, by using `bounds` and `numeric` function defined in force class. At last, within the double `for` loop from line 14 to 20, the variables are assigned for each particles.

Listing 13: `coeff` definition in sph/rhosum pair style (pair_sph_rhosum.cpp)

```
1   void PairSPHRhoSum::coeff(int narg, char **arg) {
2     if (narg != 3) /// check the number of arguments
3       error->all(FLERR,"Incorrect number of args for sph/rhosum coefficients");
4     if (!allocated) /// check if allocate has been called
5       allocate(); /// call allocate, see section 2.8
6
7     int ilo, ihi, jlo, jhi;
8     force->bounds(arg[0], atom->ntypes, ilo, ihi);
9     force->bounds(arg[1], atom->ntypes, jlo, jhi);
10
11    double cut_one = force->numeric(FLERR,arg[2]);
12
13  /// assign the coefficient to the pair of particles (i,j)
14    int count = 0;
15    for (int i = ilo; i <= ihi; i++) {
16      for (int j = MAX(jlo,i); j <= jhi; j++) {
17        cut[i][j] = cut_one;
18        setflag[i][j] = 1;
19        count++;
20      }
21    }
22
23    if (count == 0) /// check if the arguments have been assigned
24      error->all(FLERR,"Incorrect args for pair coefficients");
25  }
```

2.3.6. `init_one`

`init_one` check if all the pair coefficients for a given *i j* pair have been assigned. If they were assigned the methods ensure the symmetry of the matrix.

Listing 14: `init_one` definition in sph/taitwater pair style (pair_sph_taitwater.cpp)

```
1   double PairSPHTaitwater::init_one(int i, int j) {
2   /// check if the coefficient of the pair of particles (i,j) were assigned
3     if (setflag[i][j] == 0) {
4       error->all(FLERR,"Not all pair sph/taitwater coeffs are set");
5     }
6   /// ensure the matrix symmetry
7     cut[j][i] = cut[i][j];
8     viscosity[j][i] = viscosity[i][j];
9
10    return cut[i][j];
11  }
```

2.3.7. `single`

In `single` the force and energy of a single pairwise interaction, or single bond or angle (in case of bond or angle style), between two atoms is evalutated. The method is specifically invoked by the command line compute pair/local (or compute bond/local) to calculate properties of individual pair, or bond, interactions [41].

Listing 15: `single` definition in sph/taitwater pair style (pair_sph_taitwater.cpp)

```cpp
double PairSPHTaitwater::single(int i, int j, int itype, int jtype,
    double rsq, double factor_coul, double factor_lj, double &fforce) {
  fforce = 0.0;

  return 0.0;
}
```

2.3.8. allocate

`allocate` is a protected void function that allocates dynamic memory. The dynamic memory allocation is used when the amount of memory needed depends on user input. As explained before, at the end of the lifetime of the class, the destructors will de-allocate the memory the memory used by `allocate`.

Listing 16: `allocate` definition in sph/taitwater pair style (pair_sph_taitwater.cpp)

```cpp
void PairSPHTaitwater::allocate() {
  allocated = 1;   ///  confirm that allocated has been called
  int n = atom->ntypes;   /// assign the value of the number of types

  memory->create(setflag, n + 1, n + 1, "pair:setflag");
  for (int i = 1; i <= n; i++)
    for (int j = i; j <= n; j++)
      setflag[i][j] = 0;

/// allocate the memory for the arguments of the pair style
  memory->create(cutsq, n + 1, n + 1, "pair:cutsq");
  memory->create(rho0, n + 1, "pair:rho0");
  memory->create(soundspeed, n + 1, "pair:soundspeed");
  memory->create(B, n + 1, "pair:B");
  memory->create(cut, n + 1, n + 1, "pair:cut");
  memory->create(viscosity, n + 1, n + 1, "pair:viscosity");
}
```

3. Kelvin–Voigt Bond Style

We can use what we learned in the previous section to generate a new dissipative bond potential that can be used to model viscoelastic materials. The Kelvin–Voigt model [44] is used to model viscoelastic material as a purely viscous damper and purely elastic spring connected in parallel as shown in Figure 2.

Since the two components of the model are arranged in parallel, the strain in each component is identical:

$$\varepsilon_{tot} = \varepsilon_{spring} = \varepsilon_{damper}. \tag{1}$$

On the other hand, the total stress σ_{tot} will be split into σ_{spring} and σ_{damper} to have $\varepsilon_{spring} = \varepsilon_{damper}$. Thus we have

$$\sigma_{tot} = \sigma_{spring} + \sigma_{damper}. \tag{2}$$

Combining Equations (1) and (2) with the constitutive relation for both the spring and the dumper, $\sigma_{spring} = k\varepsilon$ and $\sigma_{damper} = b\dot{\varepsilon}$, is possible to write that

$$\sigma = k\varepsilon(t) + b\frac{d\varepsilon(t)}{dt} = k\varepsilon(t) + b\dot{\varepsilon}, \tag{3}$$

where k is the elastic modulus and b is the coefficient of viscosity. Equation (3) relates stress to strain and strain rate for a Kelvin–Voigt material [44].

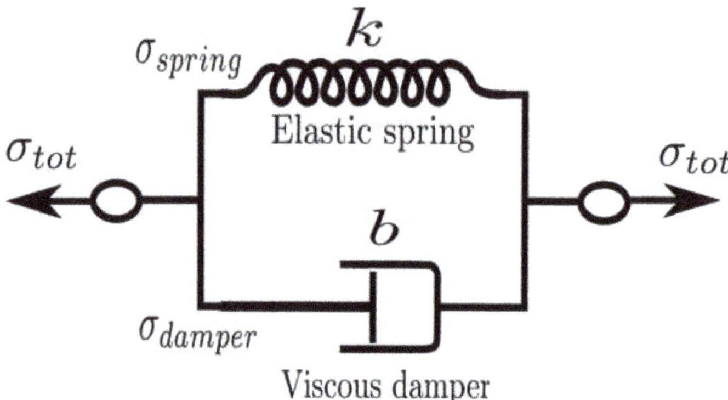

Figure 2. Schematic representation of Kelvin–Voigt model [44].

Similarly to bond test to write a new pair style called bond kv we take the bond harmonic pair style as reference. The new pair style is declared and initialised in bond_kv.h and bond_kv.cpp saved in the /src/MOLECULE directory and its hierarchy is shown in Figure 3.

Figure 3. Class hierarchy of the new bond style.

3.1. Validation

The bond kv pair style has been validated by Sahputra et al. [45] in their Discrete Multiphysics model for encapsulate particles with a soft outer shell.

3.2. bond_kv.cpp

All the functions will be the same as in the reference bond harmonic. However, in our new bond kv, we need to substitute the "BondHarmonic" text by a new "BondKv" text, as can be seen in Listings 17 and 18. Form now on, when we show a side-by-side comparison between the reference and the modified file, we highlight in yellow the modified lines and in red the deleted lines.

Listing 17: Original script (bond_harmonic.cpp)

```
1  #include "math.h"
2  #include "stdlib.h"
3  #include "bond_harmonic.h"
4  #include "atom.h"
5  #include "neighbor.h"
6  #include "domain.h"
7  #include "comm.h"
8  #include "force.h"
9  #include "memory.h"
10 #include "error.h"
11
12 using namespace LAMMPS_NS;
13
14 BondHarmonic::BondHarmonic(LAMMPS *lmp) : Bond(lmp)
15 {}
16 BondHarmonic::~BondHarmonic()
17 { ... }
18 void BondHarmonic::compute(int eflag, int vflag)
19 { ... }
20 void BondHarmonic::allocate()
21 { ... }
22 void BondHarmonic::coeff(int narg, char **arg)
```

```
23 { ... }
24 double BondHarmonic::equilibrium_distance(int i)
25 { ... }
26 void BondHarmonic::write_restart(FILE *fp)
27 { ... }
28 void BondHarmonic::read_restart(FILE *fp)
29 { ... }
30 void BondHarmonic::write_data(FILE *fp)
31 { ... }
32 double BondHarmonic::single(int type, double rsq,
33   int i, int j, double &fforce)
34 { ... }
```

Listing 18: Modified script (bond_kv.cpp)

```
 1 #include "math.h"
 2 #include "stdlib.h"
 3 #include "bond_kv.h"
 4 #include "atom.h"
 5 #include "neighbor.h"
 6 #include "domain.h"
 7 #include "comm.h"
 8 #include "force.h"
 9 #include "memory.h"
10 #include "error.h"
11
12 using namespace LAMMPS_NS;
13
14 BondKv::BondKv(LAMMPS *lmp) : Bond(lmp)
15   {}
16 BondKv::~BondKv()
17 { ... }
18 void BondKv::compute(int eflag, int vflag)
19 { ... }
20 void BondKv::allocate()
21 { ... }
22 void BondKv::coeff(int narg, char **arg)
23 { ... }
24 double BondKv::equilibrium_distance(int i)
25 { ... }
26 void BondKv::write_restart(FILE *fp)
27 { ... }
28 void BondKv::read_restart(FILE *fp)
29 { ... }
30 void BondKv::write_data(FILE *fp)
31 { ... }
32 double BondKv::single(int type, double rsq,
33   int i, int j, double &fforce)
34 { ... }
```

Compared to the bond harmonic we are introducing a new parameter, *b*, from the input file. For this reason we need to modify destructor, compute, allocate, coeff, write_restart and read_restart. Following the order of function initialisation, see Listing 18, the destructor is modified as shown in Listing 20.

Listing 19: Original destructor (bond_harmonic.cpp)

```
1 BondHarmonic::~BondHarmonic()
2 {
3   if (allocated) {
4     memory->destroy(setflag);
5     memory->destroy(k);
6     memory->destroy(r0);
7   }
8 }
```

Listing 20: Modified destructor (bond_kv.cpp)

```cpp
BondKv::~ BondKv()
{
  if (allocated) {
    memory->destroy(setflag);
    memory->destroy(k);
    memory->destroy(r0);
    memory->destroy(b);  /* dashpot/damper costant */
  }
}
```

The next function to modify is `compute`. The strain rate, $\dot{\varepsilon}$, can also be seen as the speed of deformation. To use it within the new pair style we need to declared and initialised the velocities of each particles, see Listing 22.

Listing 21: Original compute (bond_harmonic.cpp)

```cpp
void BondTest::compute(int eflag, int vflag)
{
  int i1,i2,n,type;
  double delx,dely,delz,ebond,fbond;
  double rsq,r,dr,rk;

  ebond = 0.0;
  if (eflag || vflag) ev_setup(eflag,vflag);
  else evflag = 0;

  double **x = atom->x;
  double **f = atom->f;
  }
}
```

Listing 22: Modified compute (bond_kv.cpp)

```cpp
void BondTest::compute(int eflag, int vflag)
{
    int i1,i2,n,type;
  double delx,dely,delz,ebond,fbond;
  double rsq,r,dr,rk

/* declaration of new variables */
  double delv_x, delv_y, delv_z;
  double dir_vx1, dir_vy1, dir_vz1, dir_vx2, dir_vy2,
    dir_vz2, dir_vx1, dir_vx1;
  double rsq_x, rsq_y, rsq_z;
/* end declaration of new variables */



  ebond = 0.0;
  if (eflag || vflag) ev_setup(eflag,vflag);
  else evflag = 0;

  double **x = atom->x;
  double **f = atom->f;
  double **v = atom->v; /* delcaration and inizalitaizon
   of a new pointer*/
  }
}
```

Moreover, inside the loop `for (n = 0; n < nbondlist; n++)` of the original `compute`, we need to add a new set of lines between the lines to calculate the spring force and the lines to calculate force and energy increment. Those lines calculate velocities and directions to compute the dashpot forces, see Listing 23.

Now is possible to write the new expression of the force applied to pair of atoms.

Listing 23: Modified compute (bond_kv.cpp)

```cpp
/* dashpot velocities and directions */
dev_x = v[ i1 ][ 0 ] - v[ i2 ][ 0 ];
dev_y = v[ i1 ][ 1 ] - v[ i2 ][ 1 ];
dev_z = v[ i1 ][ 2 ] - v[ i2 ][ 2 ];
rsq_vx = dev_x * dev_x;
rsq_vy = dev_y * dev_y;
rsq_vz = dev_z * dev_z;
velx = sqrt( rsq_vx );
vely = sqrt( rsq_vy );
velz = sqrt( rsq_vz );

if ( v[ i1 ][ 0 ] >= 0.0 ) dir_vx1 = 1;
else dir_vx1 = -1;

if ( v[ i1 ][ 1 ] >= 0.0 ) dir_vy1 = 1;
else dir_vy1 = -1;

if ( v[ i1 ][ 2 ] >= 0.0 ) dir_vz1 = 1;
else dir_vz1 = -1;

if ( v[ i2 ][ 0 ] >= 0.0 ) dir_vx2 = 1;
else dir_vx2 = -1;

if ( v[ i2 ][ 1 ] >= 0.0 ) dir_vy2 = 1;
else dir_vy2 = -1;

if ( v[ i2 ][ 2 ] >= 0.0 ) dir_vz2 = 1;
else dir_vz2 = -1;
```

Listing 24: Original compute (bond_harmonic.cpp)

```cpp
    if (newton_bond || i1 < nlocal) {
      f[i1][0] += delx*fbond;
      f[i1][1] += dely*fbond;
      f[i1][2] += delz*fbond;
    }

    if (newton_bond || i2 < nlocal) {
      f[i2][0] -= delx*fbond;
      f[i2][1] -= dely*fbond;
      f[i2][2] -= delz*fbond;
    }

    if (evflag) ev_tally(i1,i2,nlocal,
    newton_bond,ebond,fbond,delx,dely,delz);
  }
}
```

Listing 25: Modified compute (bond_kv.cpp)

```cpp
/// eq 3 implementation for each force component
    if (newton_bond || i1 < nlocal) {
      f[i1][0] += (delx*fbond) - (dir_vx1*b[type]*velx);
      f[i1][1] += (dely*fbond) - (dir_vy1*b[type]*vely);
      f[i1][2] += (delz*fbond) - (dir_vy1*b[type]*velz);
    }

    if (newton_bond || i2 < nlocal) {
      f[i2][0] -= (delx*fbond) - (dir_vx2*b[type]*velx);
      f[i2][1] -= (dely*fbond) - (dir_vy2*b[type]*vely);
      f[i2][2] -= (delz*fbond) - (dir_vz2*b[type]*velz);
    }

    if (evflag) ev_tally(i1,i2,nlocal,
    newton_bond,ebond,fbond,delx,dely,delz);
  }
}
```

With the introduction of a new parameter in the pair style we need to make a new dynamic memory allocation by modifying `allocate`.

Listing 26: Original `allocate` (bond_harmonic.cpp)

```cpp
void BondHarmonic::allocate()
{
  allocated = 1;
  int n = atom->nbondtypes;

  memory->create(k,n+1,"bond:k");
  memory->create(r0,n+1,"bond:r0");

  memory->create(setflag,n+1,"bond:setflag");
  for (int i = 1; i <= n; i++) setflag[i] = 0;
}
```

Listing 27: Modified `allocate` (bond_kv.cpp)

```cpp
void BondKv::allocate()
{
  allocated = 1;
  int n = atom->nbondtypes;

  memory->create(k,n+1,"bond:k");
  memory->create(r0,n+1,"bond:r0");
  memory->create(b,n+1,"bond:b"); // new line to
  // dynamically allocate b

  memory->create(setflag,n+1,"bond:setflag");
  for (int i = 1; i <= n; i++) setflag[i] = 0;
}
```

The viscosity of the damper, *b*, is given by the user in the input file. For this reason, we also need to modify `coeff`.

Listing 28: Original `coeff` (bond_harmonic.cpp)

```cpp
void BondHarmonic::coeff(int narg, char **arg)
{
  if (narg != 3) error->all(FLERR,"Incorrect args for
  bond coefficients");
  if (!allocated) allocate();

  int ilo,ihi;
  force->bounds(arg[0],atom->nbondtypes,ilo,ihi);

  double k_one = force->numeric(FLERR,arg[1]);
  double r0_one = force->numeric(FLERR,arg[2]);

  int count = 0;
  for (int i = ilo; i <= ihi; i++) {
    k[i] = k_one;
    r0[i] = r0_one;
    setflag[i] = 1;
    count++;
  }

  if (count == 0) error->all(FLERR,"Incorrect args for
  bond coefficients");
}
```

Listing 29: Modified coeff (bond_kv.cpp)

```cpp
void BondKv::coeff(int narg, char **arg)
{
  if (narg != 4) error->all(FLERR,"Incorrect args for
  bond coefficients");
  if (!allocated) allocate();

  int ilo,ihi;
  force->bounds(arg[0],atom->nbondtypes,ilo,ihi);

  double k_one = force->numeric(FLERR,arg[1]);
  double r0_one = force->numeric(FLERR,arg[2]);
  double b_one = force->numeric(FLERR,arg[3]);
// to allocate in b_one the 3rd argument of bond_coeff
  int count = 0;
  for (int i = ilo; i <= ihi; i++) {
    k[i] = k_one;
    r0[i] = r0_one;
    b[i] = b_one;
// to allocate the value stored in b_one used in compute
    setflag[i] = 1;
    count++;
  }

  if (count == 0) error->all(FLERR,"Incorrect args for
  bond coefficients");
}
```

This pair style also has the `write_restart` and `read_restart` functions that have to be modified. They basically, write and read geometry file that can be used as a support file in the input file.

Listing 30: Original `write_restart` and `read_restart` (bond_harmonic.cpp)

```cpp
void BondHarmonic::write_restart(FILE *fp)
{
  fwrite(&k[1],sizeof(double),atom->nbondtypes,fp);
  fwrite(&r0[1],sizeof(double),atom->nbondtypes,fp);
}
/*------------*/
void BondHarmonic::read_restart(FILE *fp)
{
  allocate();

  if (comm->me == 0) {
    fread(&k[1],sizeof(double),atom->nbondtypes,fp);
    fread(&r0[1],sizeof(double),atom->nbondtypes,fp);
  }
  MPI_Bcast(&k[1],atom->nbondtypes,MPI_DOUBLE,0,world);
  MPI_Bcast(&r0[1],atom->nbondtypes,MPI_DOUBLE,0,world);

  for (int i = 1; i <= atom->nbondtypes; i++)
  setflag[i] = 1;
}
```

Listing 31: Modified `write_restart` and `read_restart` (bond_kv.cpp)

```cpp
void BondKv::write_restart(FILE *fp)
{
  fwrite(&k[1],sizeof(double),atom->nbondtypes,fp);
  fwrite(&r0[1],sizeof(double),atom->nbondtypes,fp);
  fwrite(&b[1],sizeof(double),atom->nbondtypes,fp);
}
/*------------*/
void BondKv::read_restart(FILE *fp)
{
  allocate();
```

```
12    if (comm->me == 0) {
13      fread(&k[1],sizeof(double),atom->nbondtypes,fp);
14      fread(&r0[1],sizeof(double),atom->nbondtypes,fp);
15      fread(&b[1],sizeof(double),atom->nbondtypes,fp);
16    }
17    MPI_Bcast(&k[1],atom->nbondtypes,MPI_DOUBLE,0,world);
18    MPI_Bcast(&r0[1],atom->nbondtypes,MPI_DOUBLE,0,world);
19    MPI_Bcast(&n[1],atom->nbondtypes,MPI_DOUBLE,0,world);
20
21
22    for (int i = 1; i <= atom->nbondtypes; i++)
23    setflag[i] = 1;
24 }
```

3.3. bond_kv.h

In the header of the new pair style we need to substitute the "BondHarmonic" text by a new "BondKv" text as well as declare a new protected member in the class, the pointer to b.

Listing 32: Original header (bond_harmonic.h)

```
1  #ifdef BOND_CLASS
2
3  BondStyle(harmonic,BondHarmonic)
4
5  #else
6
7  #ifndef LMP_BOND_HARMONIC_H
8  #define LMP_BOND_HARMONIC_H
9
10 #include "stdio.h"
11 #include "bond.h"
12
13 namespace LAMMPS_NS {
14
15 class BondHarmonic : public Bond {
16  public:
17    BondHarmonic(class LAMMPS *);
18    virtual ~BondHarmonic();
19    virtual void compute(int, int);
20    void coeff(int, char **);
21    double equilibrium_distance(int);
22    void write_restart(FILE *);
23    void read_restart(FILE *);
24    void write_data(FILE *);
25    double single(int, double, int, int, double &);
26
27  protected:
28    double *k,*r0;
29
30    void allocate();
31 };
32 }
33 #endif
34 #endif
```

Listing 33: Modified header (bond_kv.h)

```
1  #ifdef BOND_CLASS
2
3  BondStyle(kv,BondKv)
4
5  #else
6
7  #ifndef LMP_BOND_KV_H
8  #define LMP_BOND_KV_H
9
10 #include "stdio.h"
```

```cpp
11  #include "bond.h"
12
13  namespace LAMMPS_NS {
14
15  class BondKv : public Bond {
16   public:
17    BondKv(class LAMMPS *);
18    virtual ~BondKv();
19    virtual void compute(int, int);
20    void coeff(int, char **);
21    double equilibrium_distance(int);
22    void write_restart(FILE *);
23    void read_restart(FILE *);
24    void write_data(FILE *);
25    double single(int, double, int, int, double &);
26
27   protected:
28    double *k,*r0, *b; // new pointer
29
30    void allocate();
31  };
32  }
33  #endif
34  #endif
```

3.4. Invoking kv Pair Style

Now the new pair style is completed. To run LAMMPS with the new style we need to compile it and then invoke it by writing the command lines in shown in Listing 34 in the input file.

Listing 34: Command lines to invoke the kv pair style

```
1  bond_style kv
2  bond_coeff K r0 b
```

4. Noble–Abel Stiffened-Gas Pair Style

In the SPH framework is possible to determine all the particles properties by solving the particle form of the continuity equation [6,26]

$$\frac{d\rho_i}{dt} = \sum_j m_j \mathbf{v}_{ij} \nabla_j W_{ij};$$

(4)

the momentum equation [6,26]

$$m_i \frac{d\mathbf{v}_i}{dt} = \sum_j m_i m_j \left(\frac{P_i}{\rho_i} + \frac{P_i}{\rho_i} + \Pi_{ij} \right) \nabla_j W_{ij};$$

(5)

and the energy conservation equation [6,26]

$$m_i \frac{de_i}{dt} = \frac{1}{2} \sum_j m_i m_j \left(\frac{P_i}{\rho_i} + \frac{P_i}{\rho_i} + \Pi_{ij} \right) : \mathbf{v}_{ij} \nabla_j W_{ij} - \sum_j \frac{m_i m_j}{\rho_i \rho_j} \frac{(\kappa_i + \kappa_j)(T_i - T_j)}{r_{ij}^2} \mathbf{r}_{ij} \cdot \nabla_j W_{ij}.$$

(6)

However, to be able to solve this set of equations an Equation of State (EOS) linking the pressure P and the density ρ is needed [46]. In the user-SPH package of LAMMPS one EOS is used for the liquid (Tait's EOS) and one for gas phase (ideal gas EOS). In this section we will implement a new EOS for the liquid phase. Note that with similar steps is also possible to implement a new gas EOS.

Le Métayer and Saurel [47] combined the "Noble–Abel" and the "Stiffened-Gas" EOS proposing a new EOS called Noble–Abel Stiffened-Gas (NASG), suitable for multiphase flow. The expression of the EOS does not change with the phase considered. For each phases, the pressure and temperature are calculated as function of density and specific internal energy, e.g.,

$$P(\rho, e) = (\gamma - 1)\frac{(e - q)}{\left(\frac{1}{\rho} - b\right)} - \gamma P_\infty,\tag{7}$$

and temperature-wise

$$T(\rho, e) = \frac{e - q}{C_v} - \left(\frac{1}{\rho} - b\right)\frac{P_\infty}{C_v},\tag{8}$$

where P, ρ, e, and q are, respectively, the pressure, the density, the specific internal energy, and the heat bond of the corresponding phase. γ, P_∞, q, and b are constant coefficients that defines the thermodynamic properties of the fluid.

For this new pair style, called sph/nasgliquid, we take as a reference the sph/taitwater pair style declared and initialised in pair_sph_taitwater.h and pair_sph_taitwater.cpp files in the directory /src/USER-SPH. All the files regarding sph/nasgliquid must be saved in the /src/USER-SPH directory and its hierarchy is shown in Figure 4..

Figure 4. Class hierarchy of the new bond style.

4.1. Validation

The sph/nasgliquid pair style has validated by Albano and Alexiadis [26] to study the Rayleigh collapse of an empty cavity.

4.2. pair_sph_nasgliquid.cpp

All the functions will be the same as in the reference sph/taitwater. However, in our new sph/nasgliquid, we need to substitute the "PairSPHTaitwater" text in "PairSPHNasgliquid", as can be seen in Listings 35 and 36.

Listing 35: Original script (pair_sph_taitwater.cpp)

```cpp
#include <cmath>
#include <cstdlib>
#include "pair_sph_taitwater.h"
#include "atom.h"
#include "force.h"
#include "comm.h"
#include "neigh_list.h"
#include "memory.h"
#include "error.h"
#include "domain.h"

using namespace LAMMPS_NS;

PairSPHTaitwater::PairSPHTaitwater(LAMMPS *lmp) :
Pair(lmp)
{...}
PairSPHTaitwater::~PairSPHTaitwater()
{...}
void PairSPHTaitwater::compute(int eflag, int vflag)
{...}
void PairSPHTaitwater::allocate()
```

```
22 {...}
23 void PairSPHTaitwater::settings(int narg, char **/*arg*/)
24 {...}
25 void PairSPHTaitwater::coeff(int narg, char **arg)
26 {...}
27 double PairSPHTaitwater::init_one(int i, int j)
28 {...}
```

Listing 36: Modified script (pair_sph_nasgliquid.cpp)

```
1  #include <cmath>
2  #include <cstdlib>
3  #include "pair_sph_nasgliquid.h"
4  #include "atom.h"
5  #include "force.h"
6  #include "comm.h"
7  #include "neigh_list.h"
8  #include "memory.h"
9  #include "error.h"
10 #include "domain.h"
11
12 using namespace LAMMPS_NS;
13
14 PairSPHNasgliquid:: PairSPHNasgliquid(LAMMPS *lmp) :
15 Pair(lmp)
16 {...}
17 PairSPHNasgliquid::~ PairSPHNasgliquid()
18 {...}
19 void PairSPHNasgliquid::compute(int eflag, int vflag)
20 {...}
21 void PairSPHNasgliquid::allocate()
22 {...}
23 void PairSPHNasgliquid::settings(int narg, char **/*arg*/)
24 {...}
25 void PairSPHNasgliquid::coeff(int narg, char **arg)
26 {...}
27 double PairSPHNasgliquid::init_one(int i, int j)
28 {...}
```

For the sph/nasgliquid we need to pass a total of 12 arguments from the input file,
while they were only six for sph/taitwater. For this reason we need to modify destructor,
compute, allocate, settings and coeff. Following the order of function initialisation, see
Listing 36, the destructor is modified as shown in Listing 38.

Listing 37: Original destructor (pair_sph_taitwater.cpp)

```
1  PairSPHTaitwater::~PairSPHTaitwater() {
2    if (allocated) {
3      memory->destroy(setflag);
4      memory->destroy(cutsq);
5      memory->destroy(cut);
6      memory->destroy(rho0);
7      memory->destroy(soundspeed);
8      memory->destroy(B);
9      memory->destroy(viscosity);
10   }
11 }
```

Listing 38: Modified destructor (pair_sph_nasgliquid.cpp)

```
1  PairSPHNasgliquid::~PairSPHNasgliquid() {
2    if (allocated) {
3      memory->destroy(setflag);
4      memory->destroy(cutsq);
5      memory->destroy(cut);
6      memory->destroy(soundspeed);
7      memory->destroy(B);
8      memory->destroy(CP);
```

```
 9    memory->destroy(CV);
10    memory->destroy(gamma);
11    memory->destroy(P00);
12    memory->destroy(b);
13    memory->destroy(q);
14    memory->destroy(q1);
15    memory->destroy(viscosity);
16  }
17 }
```

In the NASG EOS the pressure is function of both density, ρ, and internal energy, e. For this reason, we need to declare more pointers and variables in compute compared to the reference pair style, see line 6 and 20 in Listing 40.

Listing 39: Original compute (pair_sph_taitwater.cpp)

```
 1 void PairSPHTaitwater::compute(int eflag, int vflag) {
 2   int i, j, ii, jj, inum, jnum, itype, jtype;
 3   double xtmp, ytmp, ztmp, delx, dely, delz, fpair;
 4
 5   int *ilist, *jlist, *numneigh, **firstneigh;
 6   double vxtmp, vytmp, vztmp, imass, jmass,
 7   fi, fj, fvisc, h, ih, ihsq;
 8   double rsq, tmp, wfd, delVdotDelR, mu, deltaE;
 9
10   if (eflag || vflag)
11     ev_setup(eflag, vflag);
12   else
13     evflag = vflag_fdotr = 0;
14
15   double **v = atom->vest;
16   double **x = atom->x;
17   double **f = atom->f;
18   double *rho = atom->rho;
19   double *mass = atom->mass;
20   double *de = atom->de;
21   double *drho = atom->drho;
22   int *type = atom->type;
23   int nlocal = atom->nlocal;
24   int newton_pair = force->newton_pair;
```

Listing 40: Modified compute (pair_sph_nasgliquid.cpp)

```
 1 void PairSPHNasgliquid::compute(int eflag, int vflag) {
 2   int i, j, ii, jj, inum, jnum, itype, jtype;
 3   double xtmp, ytmp, ztmp, delx, dely, delz, fpair;
 4
 5   int *ilist, *jlist, *numneigh, **firstneigh;
 6   double vxtmp, vytmp, vztmp, imass, jmass,
 7   fi, fj, fvisc, h, ih, ihsq, iirho, ijrho;
 8   double rsq, tmp, wfd, delVdotDelR, mu, deltaE;
 9
10   if (eflag || vflag)
11     ev_setup(eflag, vflag);
12   else
13     evflag = vflag_fdotr = 0;
14
15   double **v = atom->vest;
16   double **x = atom->x;
17   double **f = atom->f;
18   double *rho = atom->rho;
19   double *mass = atom->mass;
20   double *de = atom->de;
21   double *e = atom->e;
22   double *drho = atom->drho;
23   int *type = atom->type;
24   int nlocal = atom->nlocal;
25   int newton_pair = force->newton_pair;
```

Another modification for `compute` regards the expression of the force applied to the *i*-th, see Listing 42, and *j*-th, see Listing 44, particle.

Listing 41: Original `compute` (pair_sph_taitwater.cpp)

```
// compute pressure of atom i with Tait EOS
tmp = rho[i]/rho0[itype];
fi = tmp * tmp * tmp;
fi = B[itype] * (fi * fi * tmp - 1.0)/ (rho[i] * rho[i]);
```

Listing 42: Modified `compute` (pair_sph_nasgliquid.cpp)

```
// compute pressure of atom i with NASG EOS
tmp = e[i] / imass;
iirho= 1.0/rho[i];
iirho= iirho - b[itype];
fi = (( tmp - q[itype]) * B[itype] / iirho);
fi = fi - gamma[itype] * P00[itype];
fi = fi / (rho[i] * rho[i]);
```

Listing 43: Original `compute` (pair_sph_taitwater.cpp)

```
// compute pressure  of atom j with Tait EOS
tmp = rho[j] / rho0[jtype];
fj = tmp * tmp * tmp;
fj = B[jtype] * (fj * fj * tmp - 1.0) / (rho[j] * rho[j]);
```

Listing 44: Modified `compute` (pair_sph_nasgliquid.cpp)

```
// compute pressure  of atom j with NASG EOS
tmp = e[j] / jmass;
ijrho= 1/rho[j];
ijrho= ijrho - b[jtype];
fj = ((tmp - q[jtype])* B[jtype]/ijrho);
fj = fj - gamma[jtype]*P00[jtype];
fj = fj / (rho[j] * rho[j]);
```

With the introduction of a new parameter in the pair style we need to make a new dynamic memory allocation by modifying `allocate`.

Listing 45: Original `allocate` (pair_sph_taitwater.pp)

```
void PairSPHTaitwater::allocate() {
  allocated = 1;
  int n = atom->ntypes;

  memory->create(setflag, n + 1, n + 1, "pair:setflag");
  for (int i = 1; i <= n; i++)
    for (int j = i; j <= n; j++)
      setflag[i][j] = 0;

  memory->create(cutsq, n + 1, n + 1, "pair:cutsq");
  memory->create(rho0, n + 1, "pair:rho0");
  memory->create(soundspeed, n + 1, "pair:soundspeed");
  memory->create(B, n + 1, "pair:B");
  memory->create(cut, n + 1, n + 1, "pair:cut");
  memory->create(viscosity,n + 1,n + 1,"pair:viscosity");
}
```

Listing 46: Modified `allocate` (pair_sph_nasgliquid.cpp)

```
void PairSPHNasgliquid::allocate() {
  allocated = 1;
  int n = atom->ntypes;

  memory->create(setflag, n + 1, n + 1, "pair:setflag");
  for (int i = 1; i <= n; i++)
    for (int j = i; j <= n; j++)
      setflag[i][j] = 0;
```

```
 9
10    memory->create(cutsq, n + 1, n + 1, "pair:cutsq");
11    memory->create(soundspeed, n + 1, "pair:soundspeed");
12    memory->create(B, n + 1, "pair:B");
13    memory->create(CP, n + 1, "pair:CP");
14    memory->create(CV, n + 1, "pair:CV");
15    memory->create(gamma, n + 1, "pair:gamma");
16    memory->create(P00, n + 1, "pair:P00");
17    memory->create(b, n + 1, "pair:b");
18    memory->create(q, n + 1, "pair:q");
19    memory->create(q1, n + 1, "pair:q1");
20    memory->create(cut, n + 1, n + 1, "pair:cut");
21    memory->create(viscosity,n + 1,n + 1,"pair:viscosity");
22 }
```

The 12 arguments used in the pair style are passed by the used in the input file. For this reason, we also have to modify `coeff`.

Listing 47: Original `coeff` (pair_sph_taitwater.cpp)

```
 1 void PairSPHTaitwater::coeff(int narg, char **arg) {
 2   if (narg != 6)
 3     error->all(FLERR,
 4         "Incorrect args for pair_style sph/taitwater
 5         coefficients");
 6   if (!allocated)
 7     allocate();
 8   int ilo, ihi, jlo, jhi;
 9   force->bounds(FLERR,arg[0], atom->ntypes, ilo, ihi);
10   force->bounds(FLERR,arg[1], atom->ntypes, jlo, jhi);
11   double rho0_one = force->numeric(FLERR,arg[2]);
12   double soundspeed_one = force->numeric(FLERR,arg[3]);
13   double viscosity_one = force->numeric(FLERR,arg[4]);
14   double cut_one = force->numeric(FLERR,arg[5]);
15   double B_one=soundspeed_one*soundspeed_one*rho0_one/7.0;
16   int count = 0;
17   for (int i = ilo; i <= ihi; i++) {
18     rho0[i] = rho0_one;
19     soundspeed[i] = soundspeed_one;
20     B[i] = B_one;
21     for (int j = MAX(jlo,i); j <= jhi; j++) {
22       viscosity[i][j] = viscosity_one;
23       cut[i][j] = cut_one;
24       setflag[i][j] = 1;
25       count++; } }
```

Listing 48: Modified `coeff` (pair_sph_nasgliquid.cpp)

```
 1 void PairSPHNasgliquid::coeff(int narg, char **arg) {
 2   if (narg != 12)
 3     error->all(FLERR,
 4         "Incorrect args for pair_style sph/nasgliquid
 5         coefficients");
 6   if (!allocated)
 7     allocate();
 8   int ilo, ihi, jlo, jhi;
 9   force->bounds(FLERR,arg[0], atom->ntypes, ilo, ihi);
10   force->bounds(FLERR,arg[1], atom->ntypes, jlo, jhi);
11   double soundspeed_one = force->numeric(FLERR,arg[2]);
12   double viscosity_one = force->numeric(FLERR,arg[3]);
13   double cut_one = force->numeric(FLERR,arg[4]);
14   double CP_one = force->numeric(FLERR,arg[5]);
15   double CV_one = force->numeric(FLERR,arg[6]);
16   double gamma_one = force->numeric(FLERR,arg[7]);
17   double P00_one = force->numeric(FLERR,arg[8]);
18   double b_one = force->numeric(FLERR,arg[9]);
19   double q_one = force->numeric(FLERR,arg[10]);
20   double q1_one = force->numeric(FLERR,arg[11]);
21   double B_one = (gamma_one - 1);
```

```
22    int count = 0;
23    for (int i = ilo; i <= ihi; i++) {
24      soundspeed[i] = soundspeed_one;
25      B[i] = B_one;
26      CP[i] = CP_one;
27      CV[i] = CV_one;
28      gamma[i] = gamma_one;
29      P00[i] = P00_one;
30      b[i] = b_one;
31      q[i] = q_one;
32      q1[i] = q1_one;
33      for (int j = MAX(jlo,i); j <= jhi; j++) {
34        viscosity[i][j] = viscosity_one;
35        cut[i][j] = cut_one;
36        setflag[i][j] = 1;
37        count++;  }  }
```

4.3. pair_sph_nasgliquid.h

In the header of the new pair style we need to substitute the "PairSPHTaitwater" text in "PairSPHNasgliquid" as well as declare new protected members in the class, the pointers to the new arguments.

Listing 49: Original header (pair_sph_taitwater.h)

```
1  #ifdef PAIR_CLASS
2
3  PairStyle(sph/taitwater,PairSPHTaitwater)
4
5  #else
6
7  #ifndef LMP_PAIR_TAITWATER_H
8  #define LMP_PAIR_TAITWATER_H
9
10 #include "pair.h"
11
12 namespace LAMMPS_NS {
13
14 class PairSPHTaitwater : public Pair {
15  public:
16   PairSPHTaitwater(class LAMMPS *);
17   virtual ~PairSPHTaitwater();
18   virtual void compute(int, int);
19   void settings(int, char **);
20   void coeff(int, char **);
21   virtual double init_one(int, int);
22
23  protected:
24   double *rho0, *soundspeed, *B;
25   double **cut,**viscosity;
26   int first;
27   void allocate();
28 };
29 }
30 #endif
31 #endif
```

Listing 50: Modified header (pair_sph_nasgliquid.h)

```
1  #ifdef PAIR_CLASS
2
3  PairStyle(sph/nasgliquid,PairSPHNasgliquid)
4
5  #else
6
7  #ifndef LMP_PAIR_NASGLIQUID_H
8  #define LMP_PAIR_NASGLIQUID_H
9
```

```
10  #include "pair.h"
11
12  namespace LAMMPS_NS {
13
14  class PairSPHNasgliquid : public Pair {
15   public:
16    PairSPHNasgliquid(class LAMMPS *);
17    virtual ~PairSPHNasgliquid();
18    virtual void compute(int, int);
19    void settings(int, char **);
20    void coeff(int, char **);
21    virtual double init_one(int, int);
22
23   protected:
24    double *soundspeed, *B, *CP, *CV, *gamma, *P00,
25    *b, *q, *q1;
26    double **cut,**viscosity;
27    int first;
28    void allocate();
29  };
30  }
31  #endif
32  #endif
```

4.4. Invoking Sph/Nasgliquid Pair Style

Now the new pair style is completed. To run LAMMPS with the new style we need to compile it and then invoke it by writing the command lines shown in Listing 51 in the input file.

Listing 51: Command lines to invoke the NASG pair style for liquid

```
1  pair_style sph/nasgliquid
2  pair_coeff  I J c_0 alpha h Cv Cp gamma P00 b q q'
```

5. Multiphase (Liquid–Gas) Heat Exchange Pair Style

In LAMMPS thermal conductivity between SPH particles is enabled using the sph/heat-conduction pair style inside the user-SPH package. However, the pair style is designed only for mono phase fluid where the thermal conductivities is constant ($\kappa_i = \kappa$). When more than one phase is present, the heat conduction at the interface can be implemented by using [6,26]

$$m_i \frac{de_i}{dt} = \sum_j \frac{m_i m_j}{\rho_i \rho_j} \frac{(\kappa_i + \kappa_j)(T_i - T_j)}{r_{ij}^2} \mathbf{r_{ij}} \cdot \nabla_j W_{ij}. \tag{9}$$

In the new pair style, called sph/heatgasliquid, one phase is assumed to be liquid with an initial temperature of $T_{l,0}$ and the other is assumed to be and ideal gas. Each time-step the temperature of the fluid is updated as [26].

$$T_l = T_{l,0} + \frac{E_l - E_{l,0}}{C_{p,l}}, \tag{10}$$

where $T_{l,0}$ is the reference temperature, E_0 the internal energy in [J], E_l internal energy [J] at the current time step and $C_{p,l}$ is heat capacity of the fluid in [J K^{-1}]. The temperature of the gas is updated following the ideal EOS [26].

$$T_g = MM \frac{(\gamma - 1)e_g}{R}, \tag{11}$$

where MM is the molar mass [kg kmol^{-1}], e_g is the specific internal energy in [J kg^{-1}], γ is the heat capacity ratio and R is the ideal gas constant in [J K^{-1} kmol^{-1}]. Generally the choice of the reference states $E_{l,0}$ is arbitrary, but if the Equation of State (EOS) used for the phase is function of both density and internal energy of the reference state will be determined by the EOS.

In the sph/heatgasliquid pair style is important to check if the i-th and j-th particles are liquid or gas phase to apply either Equation (10) or Equation (11). This "phase check" is explained in Section 5.2 `compute` function is modified.

For the energy balance the new pair style needs $T_{l,0}$, $E_{l,0}$, $C_{p,l}$ and κ_l for the liquid phase and κ_g for the gas phase. Moreover, for the phase check, the particle types of each phases must be specified. All this informations is passed by the user in the in the input file.

The reference pair style is sph/heatconduction. It is declared and initialised in the pair_sph_heatconduction.cpp pair_sph_heatconduction.cpp files in the directory /src/USER-SPH. All the files regarding sph/heatgasliquid must be saved in the /src/USER-SPH directory and its hierarchy is shown in Figure 5.

Figure 5. Class hierarchy of the new pair style.

5.1. Validation

The sph/heatgasliquid pair style has validated by Albano and Alexiadis [26] to study the role of the heat diffusion in for a gas filled Rayleigh collapse in water.

5.2. pair_sph_heatgasliquid.cpp

All the functions will be the same as in the reference sph/heatconduction. However, in our new sph/heatgasliquid, we need to substitute the "PairSPHHeatConduction" text in "PairSPHHeatgasliquid", as can be seen in Listings 52 and 53.

Listing 52: Original script (pair_sph_heatconduction.cpp)

```
1  #include "math.h"
2  #include "stdlib.h"
3  #include "pair_sph_heatconduction.h"
4  #include "atom.h"
5  #include "force.h"
6  #include "comm.h"
7  #include "memory.h"
8  #include "error.h"
9  #include "neigh_list.h"
10 #include "domain.h"
11
12 using namespace LAMMPS_NS;
13
14 PairSPHHeatConduction::PairSPHHeatConduction(LAMMPS *lmp)
15 : Pair(lmp)
16 { ... }
17 PairSPHHeatConduction::~PairSPHHeatConduction()
18 { ... }
19 void PairSPHHeatConduction::compute(int eflag, int vflag)
20 { ... }
21 void PairSPHHeatConduction::allocate()
22 { ... }
23 void PairSPHHeatConduction::settings(int narg, char **arg)
24 { ... }
25 void PairSPHHeatConduction::coeff(int narg, char **arg)
26 { ... }
27 double PairSPHHeatConduction::init_one(int i, int j)
28 { ... }
29 double PairSPHHeatConduction::single(int i, int j,
30   int itype, int jtype, double rsq, double factor_coul,
31   double factor_lj, double &fforce)
32 { ... }
```

Listing 53: Modified script (pair_sph_heatgasliquid.cpp)

```cpp
#include <cmath>
#include <cstdlib>
#include "pair_sph_heatgasliquid.h"
#include "atom.h"
#include "force.h"
#include "comm.h"
#include "memory.h"
#include "error.h"
#include "neigh_list.h"
#include "domain.h"

using namespace LAMMPS_NS;

PairSPHHeatgasliquid::PairSPHHeatgasliquid(LAMMPS *lmp)
: Pair(lmp)
{ ... }
PairSPHHeatgasliquid::~PairSPHHeatgasliquid()
{ ... }
void PairSPHHeatgasliquid::compute(int eflag, int vflag)
{ ... }
void PairSPHHeatgasliquid::allocate()
{ ... }
void PairSPHHeatgasliquid::settings(int narg, char **arg)
{ ... }
void PairSPHHeatgasliquid::coeff(int narg, char **arg)
{ ...
double PairSPHHeatgasliquid::init_one(int i, int j)
{ ... }
double PairSPHHeatgasliquid::single(int i, int j,
int itype, int jtype, double rsq, double factor_coul,
double factor_lj, double &fforce)
{ ... }
```

For the sph/heatgasliquid we need to pass a total of nine arguments from the input file, while they were only seven for sph/heatconduction. For this reason we need to modify destructor, `compute`, `allocate`, `settings` and `coeff`. Following the order of function initialisation, see Listing 53, the destructor is modified by removing the heat diffusion coefficient, line 6 in Listing 54.

Listing 54: Original destructor (pair_sph_heatconduction.cpp)

```cpp
PairSPHHeatConduction::~PairSPHHeatConduction() {
  if (allocated) {
    memory->destroy(setflag);
    memory->destroy(cutsq);
    memory->destroy(cut);
    memory->destroy(alpha);
  }
}
```

Listing 55: Modified destructor (pair_sph_heatgasliquid.cpp)

```cpp
PairSPHHeatgasliquid::~PairSPHHeatgasliquid() {
  if (allocated) {
    memory->destroy(setflag);
    memory->destroy(cutsq);
    memory->destroy(cut);
  }
}
```

To compute Equation (6) we need to declare more variables in `compute` compared to the reference pair style, see line 4 in Listing 57.

Listing 56: Original compute (pair_sph_heatconduction.cpp)

```cpp
void PairSPHHeatConduction::compute(int eflag, int vflag){
  int i, j, ii, jj, inum, jnum, itype, jtype;
  double xtmp, ytmp, ztmp, delx, dely, delz;
```

Listing 57: Modified compute (pair_sph_heatgasliquid.cpp)

```cpp
void PairSPHHeatgasliquid::compute(int eflag, int vflag){
  int i, j, ii, jj, inum, jnum, itype, jtype;
  double xtmp, ytmp, ztmp, delx, dely, delz;
  double Ti, Tj, ki, kj;  /// new parameters
```

Another important modification is to add the phase check inside `compute`. The phase check has to be implemented for both the *i*-th particle and the *j*-th particle inside the loop over neighbours, `for (ii = 0; ii < inum; ii++)` in the reference pair style. The phase check for the *i*-th particle starts after the assignment of imass, line 3 of Listing 58.

Listing 58: Modified compute (pair_sph_heatgasliquid.cpp)

```cpp
imass = mass[itype];

if (itype == liquidtype)
    {
        Ti= e[i] - el0;
        Ti= Ti/CP1;
        Ti= T01 + Ti;
        ki=kl;
    }
else {
        Ti=0.40*e[i]*18;
        Ti= Ti/imass;
        Ti= Ti/8314.33;
        ki=kg;
    }
```

Similarly, for the *j*-th the phase check start at line 3 of Listing 59.

Listing 59: Modified compute (pair_sph_heatgasliquid.cpp)

```cpp
    jmass = mass[jtype];

    if (jtype == liquidtype)
        {
            Tj= e[j] - el0;
            Tj= Tj/CP1;
            Tj= T01 + Tj;
            kj=kl;
        }
    else {
            Tj=0.40*e[j]*18;
            Tj= Tj/jmass;
            Tj= Tj/8314.33;
            kj=kg;
        }
```

The last change in `compute` is to implement the change in internal energy as shown in Equation (9).

Listing 60: Original compute (pair_sph_heatconduction.cpp)

```cpp
        D = alpha[itype][jtype]; // diffusion coefficient

        deltaE = 2.0 * imass * jmass / (imass+jmass);
        deltaE *= (rho[i] + rho[j]) / (rho[i] * rho[j]);
        deltaE *= D * (e[i] - e[j]) * wfd;

        de[i] += deltaE;
        if (newton_pair || j < nlocal) {
          de[j] -= deltaE;
        }
```

Listing 61: Modified compute (pair_sph_heatgasliquid.cpp)

```cpp
        deltaE = imass * jmass / (rho[i] * rho[j]);   ///
        deltaE *= (ki + kj) * (Ti - Tj) * wfd;    ///
        ///  implementation of eq 3.4
        de[i] += deltaE;
        if (newton_pair || j < nlocal) {
          de[j] -= deltaE;
        }
```

With the introduction of new arguments in the pair style we need to make a new
dynamic memory allocation by modifying `allocate`.

Listing 62: Original allocate (pair_sph_heatconduction.cpp)

```cpp
void PairSPHHeatConduction::allocate() {
  allocated = 1;
  int n = atom->ntypes;

  memory->create(setflag, n + 1, n + 1, "pair:setflag");
  for (int i = 1; i <= n; i++)
    for (int j = i; j <= n; j++)
      setflag[i][j] = 0;

  memory->create(cutsq, n + 1, n + 1, "pair:cutsq");
  memory->create(cut, n + 1, n + 1, "pair:cut");
  memory->create(alpha, n + 1, n + 1, "pair:alpha");
}
```

Listing 63: Modified allocate (pair_sph_heatgasliquid.cpp)

```cpp
void PairSPHHeatgasliquid::allocate() {
  allocated = 1;
  int n = atom->ntypes;

  memory->create(setflag, n + 1, n + 1, "pair:setflag");
  for (int i = 1; i <= n; i++)
    for (int j = i; j <= n; j++)
      setflag[i][j] = 0;

  memory->create(cutsq, n + 1, n + 1, "pair:cutsq");
  memory->create(cut, n + 1, n + 1, "pair:cut");
}
```

The nine arguments used in the pair style are passed by the user in the input file. For
this reason, we also have to modify `coeff`.

Listing 64: Original coeff (pair_sph_heatconduction.cpp)

```cpp
void PairSPHHeatConduction::coeff(int narg, char **arg) {
  if (narg != 4)
    error->all(FLERR,"Incorrect number of args for
    pair_style sph/heatconduction coefficients");
  if (!allocated)
    allocate();
```

```
 8  int ilo, ihi, jlo, jhi;
 9  force->bounds(arg[0], atom->ntypes, ilo, ihi);
10  force->bounds(arg[1], atom->ntypes, jlo, jhi);
11
12  double alpha_one = force->numeric(FLERR,arg[2]);
13  double cut_one   = force->numeric(FLERR,arg[3]);
14
15  int count = 0;
16  for (int i = ilo; i <= ihi; i++) {
17    for (int j = MAX(jlo,i); j <= jhi; j++) {
18  //printf("setting cut[%d][%d] = %f\n", i, j, cut_one);
19      cut[i][j] = cut_one;
20      alpha[i][j] = alpha_one;
21      setflag[i][j] = 1;
22      count++;
23    }
24  }
25
26  if (count == 0)
27    error->all(FLERR,"Incorrect args for pair
28    coefficients");
29 }
```

Listing 65: Modified <code>coeff</code> (pair_sph_heatgasliquid.cpp)

```
 1 void PairSPHHeatgasliquid::coeff(int narg, char **arg) {
 2   if (narg != 9)
 3     error->all(FLERR,"Incorrect number of args for
 4      pair_style sph/heatgasliquid coefficients");
 5   if (!allocated)
 6     allocate();
 7
 8   int ilo, ihi, jlo, jhi;
 9   force->bounds(FLERR,arg[0], atom->ntypes, ilo, ihi);
10   force->bounds(FLERR,arg[1], atom->ntypes, jlo, jhi);
11
12   e10 = force->numeric(FLERR,arg[2]);
13   k1 = force->numeric(FLERR,arg[3]);
14   kg = force->numeric(FLERR,arg[4]);
15   T01 = force->numeric(FLERR,arg[5]);
16   double cut_one = force->numeric(FLERR,arg[6]);
17   CP1 = force->numeric(FLERR,arg[7]);
18   liquidtype = force->numeric(FLERR,arg[8]);
19
20   int count = 0;
21   for (int i = ilo; i <= ihi; i++) {
22     for (int j = MAX(jlo,i); j <= jhi; j++) {
23  //printf("setting cut[%d][%d] = %f\n", i, j, cut_one);
24       cut[i][j] = cut_one;
25       setflag[i][j] = 1;
26       count++;
27     }
28   }
29   if (count == 0)
30     error->all(FLERR,"Incorrect args for pair
31      coefficients");
32 }
```

5.3. *pair_sph_heatgasliquid.h*

In the header of the new pair style we need to substitute the "PairSPHHeatConduction" text in "PairSPHHeatgasliquid" and declare new protected members in the class.

Listing 66: Original header (pair_sph_heatconduction.h)

```
1  #ifdef PAIR_CLASS
2
3  PairStyle(sph/heatconduction,PairSPHHeatConduction)
4
5  #else
6
7  #ifndef LMP_PAIR_SPH_HEATCONDUCTION_H
8  #define LMP_PAIR_SPH_HEATCONDUCTION_H
9
10 #include "pair.h"
11
12 namespace LAMMPS_NS {
13
14 class PairSPHHeatConduction : public Pair {
15  public:
16   PairSPHHeatConduction(class LAMMPS *);
17   virtual ~PairSPHHeatConduction();
18   virtual void compute(int, int);
19   void settings(int, char **);
20   void coeff(int, char **);
21   virtual double init_one(int, int);
22   virtual double single(int, int, int, int, double,
23   double, double, double &);
24
25  protected:
26   double **cut, **alpha;
27   void allocate();
28 };
29 }
30 #endif
31 #endif
```

Listing 67: Modified header (pair_sph_heatgasliquid.h)

```
1  #ifdef PAIR_CLASS
2
3  PairStyle(sph/heatgasliquid,PairSPHHeatgasliquid)
4
5  #else
6
7  #ifndef LMP_PAIR_SPH_HEATGASLIQUID_H
8  #define LMP_PAIR_SPH_HEATGASLIQUID_H
9
10 #include "pair.h"
11
12 namespace LAMMPS_NS {
13
14 class PairSPHHeatgasliquid : public Pair {
15  public:
16   PairSPHHeatgasliquid(class LAMMPS *);
17   virtual ~PairSPHHeatgasliquid();
18   virtual void compute(int, int);
19   void settings(int, char **);
20   void coeff(int, char **);
21   virtual double init_one(int, int);
22   virtual double single(int, int, int, int, double,
23   double, double, double &);
24
25  protected:
26   int liquidtype;
27   double e10, kg, kl, T01, CP1;
28   double **cut;
29   void allocate();
30 };
31 }
32 #endif
33 #endif
```

5.4. Invoking Sph/Heatgasliquid Pair Style

Now the new pair style is completed. To run LAMMPS with the new style we need to compile it and then invoke it by writing the command lines shown in Listing 68 in the input file.

Listing 68: Command lines to invoke the sph/heatgasliquid pair style

```
1 pair_style    sph/heatgasliquid
2 pair_coeff    i j e10 kl kg T10 h Cpl liquidtype
```

6. Full Stationary Fix Style

In LAMMPS a `fix` style is any operation that is applied to the system, usually to a group of particles, during time stepping or minimisation used to alter some property of the sytem [41]. There are hundreds of fixes defined in LAMMPS and new ones can be added. Usually fixes are used for time integration, force constraints, boundary conditions and diagnostics.

In the user-sph package in LAMMPS there is the so called meso/stationary fix used to set boundary condition. With meso/stationary is possible to fix position and velocity for a group of particles, walls as example, but internal energy and density will be updated. In some cases, it is useful to have a fully stationary conditions that maintains constant also the energy and the density. For this new fix, called meso/fullstationary, we take as a reference the meso/stationary fix declared and initialised in fix_meso_stationary.h and fix_meso_stationary.cpp files in the directory /src/USER-SPH. All the files regarding meso/fullstationary must be saved in the /src/USER-SPH directory and its hierarchy is shown in Figure 6.

Figure 6. Class hierarchy of the new fix style.

6.1. Validation

The meso/fullstationary has been used in the validation of the new viscosity class to set the boundary condition of a constant asymmetric heated walls, see Section 7.2.

6.2. fix_meso_fullstationary.cpp

All the functions will be the same as in the reference meso/stationary. However, in our new fullstationary, we need to substitute the "FixMesoStationary" text in "FixMeso-FullStationary", as can be seen in Listings 69 and 70.

Listing 69: Original script (fix_meso_stationary.cpp)

```
1  #include <cstdio>
2  #include <cstring>
3  #include <cmath>
4  #include <cstdlib>
5  #include "fix_meso_stationary.h"
6  #include "atom.h"
7  #include "comm.h"
8  #include "force.h"
9  #include "neighbor.h"
10 #include "neigh_list.h"
11 #include "neigh_request.h"
12 #include "update.h"
13 #include "integrate.h"
14 #include "respa.h"
15 #include "memory.h"
16 #include "error.h"
17 #include "pair.h"
18
```

```
19  using namespace LAMMPS_NS;
20  using namespace FixConst;
21
22  FixMesoStationary:: FixMesoStationary(LAMMPS *lmp,
23  int narg, char **arg) : Fix(lmp, narg, arg)
24  {...}
25  int FixMesoStationary::setmask()
26  {...}
27  void FixMesoStationary::init()
28  {...}
29  void FixMesoStationary::initial_integrate(int /*vflag*/)
30  {...}
31  void FixMesoStationary::final_integrate()
32  {...}
33  void FixMesoStationary::reset_dt()
34  {...}
```

Listing 70: Modified script (fix_meso_fullstationary.cpp)

```
1   #include <cstdio>
2   #include <cstring>
3   #include <cmath>
4   #include <cstdlib>
5   #include "fix_meso_fullstationary.h"
6   #include "atom.h"
7   #include "comm.h"
8   #include "force.h"
9   #include "neighbor.h"
10  #include "neigh_list.h"
11  #include "neigh_request.h"
12  #include "update.h"
13  #include "integrate.h"
14  #include "respa.h"
15  #include "memory.h"
16  #include "error.h"
17  #include "pair.h"
18
19  using namespace LAMMPS_NS;
20  using namespace FixConst;
21
22  FixMesoFullStationary::FixMesoFullStationary(LAMMPS *lmp,
23   int narg, char **arg) : Fix(lmp, narg, arg)
24  {...}
25  int FixMesoFullStationary::setmask()
26  {...}
27  void FixMesoFullStationary::init()
28  {...}
29  void FixMesoFullStationary::initial_integrate
30  (int /*vflag*/)
31  {...}
32  void FixMesoFullStationary::final_integrate()
33  {...}
34  void FixMesoFullStationary::reset_dt()
35  {...}
```

For the meso/fullstationary we need to modify two function: `initial_integrate`, see Listing 72 line 16 and 17, and `final_integrate`, see Listing 74 line 14 and 15.

Listing 71: Original initial_integrate (fix_meso_stationary.cpp)

```cpp
void FixMesoStationary::initial_integrate(int vflag) {

  double *rho = atom->rho;
  double *drho = atom->drho;
  double *e = atom->e;
  double *de = atom->de;
  int *type = atom->type;
  int *mask = atom->mask;
  int nlocal = atom->nlocal;
  int i;

  if (igroup == atom->firstgroup)
    nlocal = atom->nfirst;

  for (i = 0; i < nlocal; i++) {
    if (mask[i] & groupbit) {
      e[i] += dtf * de[i];
// with this line is possible to update internal energy
      rho[i] += dtf * drho[i];
// ... and density every half-step
    }}}
```

Listing 72: Modified initial_integrate (fix_meso_fullstationary.cpp)

```cpp
void FixMesoFullStationary::initial_integrate(int vflag) {

  double *rho = atom->rho;
  double *drho = atom->drho;
  double *e = atom->e;
  double *de = atom->de;
  int *mask = atom->mask;
  int nlocal = atom->nlocal;
  int i;

  if (igroup == atom->firstgroup)
    nlocal = atom->nfirst;

  for (i = 0; i < nlocal; i++) {
    if (mask[i] & groupbit) {
      e[i] +=0; // with this line internal energy
      rho[i] += 0; // ... and density are constant
    }}}
```

Listing 73: Original final_integrate (fix_meso_stationary.cpp)

```cpp
void FixMesoStationary::final_integrate() {

  double *e = atom->e;
  double *de = atom->de;
  double *rho = atom->rho;
  double *drho = atom->drho;
  int *type = atom->type;
  int *mask = atom->mask;
  double *mass = atom->mass;
  int nlocal = atom->nlocal;
  if (igroup == atom->firstgroup)
    nlocal = atom->nfirst;

  for (int i = 0; i < nlocal; i++) {
    if (mask[i] & groupbit) {
      e[i] += dtf * de[i];
      rho[i] += dtf * drho[i];
    }}}
```

Listing 74: Modified final_integrate (fix_meso_fullstationary.cpp)

```
1  vvoid FixMesoFullStationary::final_integrate() {
2
3    double *e = atom->e;
4    double *de = atom->de;
5    double *rho = atom->rho;
6    double *drho = atom->drho;
7    int *mask = atom->mask;
8    int nlocal = atom->nlocal;
9    if (igroup == atom->firstgroup)
10     nlocal = atom->nfirst;
11
12   for (int i = 0; i < nlocal; i++) {
13     if (mask[i] & groupbit) {
14       e[i] += 0; //  with this line internal energy
15       rho[i] += 0; //... and density are constant
16     }}}
```

6.3. *fix_mes_fullstationary.h*

In the header of the new fix we need to substitute the "FixMesoStationary" text in "FixMesoFullStationary".

Listing 75: Original header (pair_sph_heatconduction.h)

```
1  #ifdef FIX_CLASS
2
3  FixStyle(meso/stationary,FixMesoStationary)
4
5  #else
6
7  #ifndef LMP_FIX_MESO_STATIONARY_H
8  #define LMP_FIX_MESO_STATIONARY_H
9
10 #include "fix.h"
11
12 namespace LAMMPS_NS {
13
14 class FixMesoStationary : public Fix {
15  public:
16   FixMesoStationary(class LAMMPS *, int, char **);
17   int setmask();
18   virtual void init();
19   virtual void initial_integrate(int);
20   virtual void final_integrate();
21   void reset_dt();
22
23  private:
24   class NeighList *list;
25  protected:
26   double dtv,dtf;
27   double *step_respa;
28   int mass_require;
29
30   class Pair *pair;
31 };
32 }
33 #endif
34 #endif
```

Listing 76: Modified header (pair_sph_heatgasliquid.h)

```
1  #ifdef FIX_CLASS
2
3  FixStyle(meso/fullstationary,FixMesoFullStationary)
4
5  #else
6
```

```cpp
7  #ifndef LMP_FIX_MESO_FULLSTATIONARY_H
8  #define LMP_FIX_MESO_FULLSTATIONARY_H
9
10 #include "fix.h"
11
12 namespace LAMMPS_NS {
13
14 class FixMesoFullStationary: public Fix {
15  public:
16   FixMesoFullStationary(class LAMMPS *, int, char **);
17   int setmask();
18   virtual void init();
19   virtual void initial_integrate(int);
20   virtual void final_integrate();
21   void reset_dt();
22
23  private:
24   class NeighList *list;
25  protected:
26   double dtv,dtf;
27   double *step_respa;
28   int mass_require;
29
30   class Pair *pair;
31 };
32 }
33 #endif
34 #endif
```

6.4. Invoking Meso/Fullstationary Fix

Now the new fix is completed. To run LAMMPS with the new style we need to compile it and then invoke it by writing the command lines shown in Listing 77 in the input file.

Listing 77: Command lines to invoke the new pair style

```
1  fix ID group-ID meso/fullstationary
```

7. Viscosity Class

Viscosity in the SPH method has been addressed with different solutions [46]. Shock waves, for example, have been a challenge to model due to the arise of numerical oscillations around the shocked region. Monaghan solved this problem with the introduction of the so-called Monaghan artificial viscosity [48]. Artificial viscosity is still used nowadays for energy dissipation and to prevent unphysical penetration for particles approaching each other [25,49]. The SPH package of LAMMPS uses the following artificial viscosity expression [6], within the sph/idealgas and sph/taitwater pair style.

$$\Pi_{ij} = -\alpha h \frac{c_i + c_j}{\rho_i + \rho_j} \frac{\mathbf{v}_{ij} \cdot \mathbf{r}_{ij}}{r_{ij}^2 + \epsilon h^2}, \tag{12}$$

where α is the dimensionless dissipation factor, c_i and c_j are the speed of sound of particle i and j. The dissipation factor, α, can be linked with the real viscosity in term of [6]

$$\alpha = 8 \frac{\mu}{ch\rho}, \tag{13}$$

where c is the speed of sound, ρ the density, μ the dynamic viscosity and h the smoothing length.

The artificial viscosity approach performs well at a high Reynolds number but better solutions are available for laminar flow: Morris et al. [50] approximated and implemented the viscosity momentum term for SPH. The same solution can be found in the sph/taitwater/morris pair style with the expression [6].

$$\sum_j \frac{m_i m_j (\mu_i + \mu_j) \mathbf{v}_{ij}}{\rho_i \rho_j} \left(\frac{1}{r_{ij}} \frac{\partial W_{ij}}{\partial r_i} \right), \tag{14}$$

where μ is the real dynamic viscosity.

In LAMMPS both the dissipation factor and the dynamic viscosity are treated as a constant between a pair of particles when they interact within the smoothing length. In this section we want to make the viscosity a per atom property instead of a pair property only existing within a pair style. Moreover, five temperature dependent viscosity models are added. For this example, no reference file is used; a new class, Viscosity, is implemented in LAMMPS from scratch and its hierarchy is shown in Figure 7.

Figure 7. Class hierarchy of the new class.

7.1. Temperature Dependant Viscosity

In literature multiples empirical models that correlate viscosity with temperature are available [51–53]. In the new viscosity class five different viscosity models have been implemented:

1. Andrade's equation [54]

$$\mu = A exp \left(\frac{B}{T} + CT + DT^2 \right), \tag{15}$$

where μ is the viscosity in [Kg m^{-1} s^{-1}], T is the static temperature in Kelvin, A, B, C and D are fluid-dependent dimensional coefficients available in literature.

2. Arrhenius viscosity by Raman [55,56]

$$\mu = C_1 exp(C_2 / T), \tag{16}$$

where μ is the dynamic viscosity in [Kg m^{-1} s^{-1}], T is the temperature in Kelvin, C_1 and C_2 are fluid-dependent dimensional coefficients available in literature.

3. Sutherland's viscosity [57,58] for gas phase
    Sutherland's law can be expressed as:

$$\mu = \frac{C_1 T^{3/2}}{T + C_2}, \tag{17}$$

where μ is the viscosity in [Kg m^{-1} s^{-1}], T is the static temperature in Kelvin, C_1 and C_2 are dimensional coefficients.

4. Power-Law viscosity law [57] for gas phase
    A power-law viscosity law with two coefficients has the form :

$$\mu = BT^n, \tag{18}$$

where μ is the viscosity in [Kg m^{-1} s^{-1}], T is the static temperature in Kelvin, and B is a dimensional coefficient.

5. Constant viscosity

With constant viscosity both dissipation factor and dynamic viscosity will be constant during the simulation.

When the artificial viscosity is used the dissipation factor of Equation (12) is defined as the arithmetic mean of the dissipation factors of i-th particle and j-th particle.

$$\alpha_{ij} = -\frac{4}{h}\left(\frac{\mu_i}{c_i\rho_i} + \frac{\mu_j}{c_j\rho_j}\right),\tag{19}$$

where α_{ij} is the dissipation factor of the particles pair i and j.

7.2. Validation

In order to validate the new Viscosity class, we will study the effect of asymmetrically heating walls in a channel flow, and more specifically the effect on the velocity field of the fluid. The data obtained with our model will be compared with the analytical solution obtained by Sameen and Govindarajan [59].

The water flows between two walls in the x-direction with periodic conditions. The walls are set at different temperatures T_{cold} and T_{hot}, see Figure 8. Both water and walls are modelled as fluid following the tait EOS. The physical properties of the walls are set constant throughout the simulation using the full stationary conditions described in Section 6.

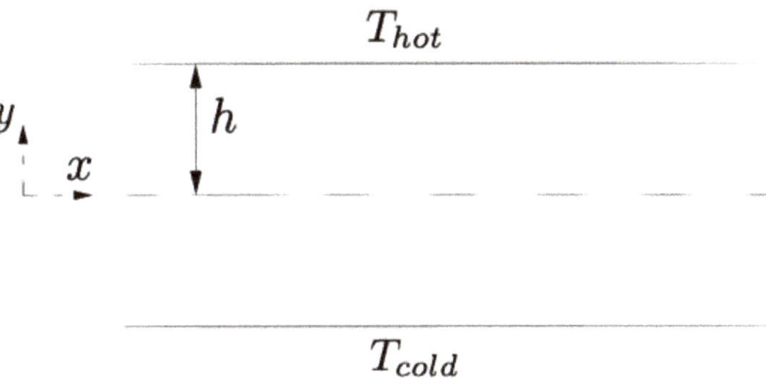

Figure 8. Geometry of the simulation.

To match the condition used by Sameen and Govindarajan we set the cold wall temperature to $T_{cold} = 295$ K and the temperature dependence of the dynamic viscosity described by the Arrhenius model, Equation (16), with $C_1 = 0.000183$ [Ns m^{-2}] and $C_2 = 1879.9$ K [59]. To describe the asymmetric heating Sameen and Govindarajan introduced the parameter m, defined as:

$$m = \frac{\mu_{cold}}{\mu_{ref}}\tag{20}$$

where $\mu_{ref} = \mu_{hot}$ is the viscosity at the hot wall in the case of asymmetric heating and μ_{cold} is the viscosity at the cold wall. By combining (16) and (20), with the given T_{cold}, is possible to express the temperature difference of the walls ΔT as function of m.

Figure 9 shows the viscosity trend for different values of m and the corresponding ΔT. Sometimes, in particle methods, instantaneous data can be noisy (scattered) as can be seen from the blue circles of both Figures 9 and 10.

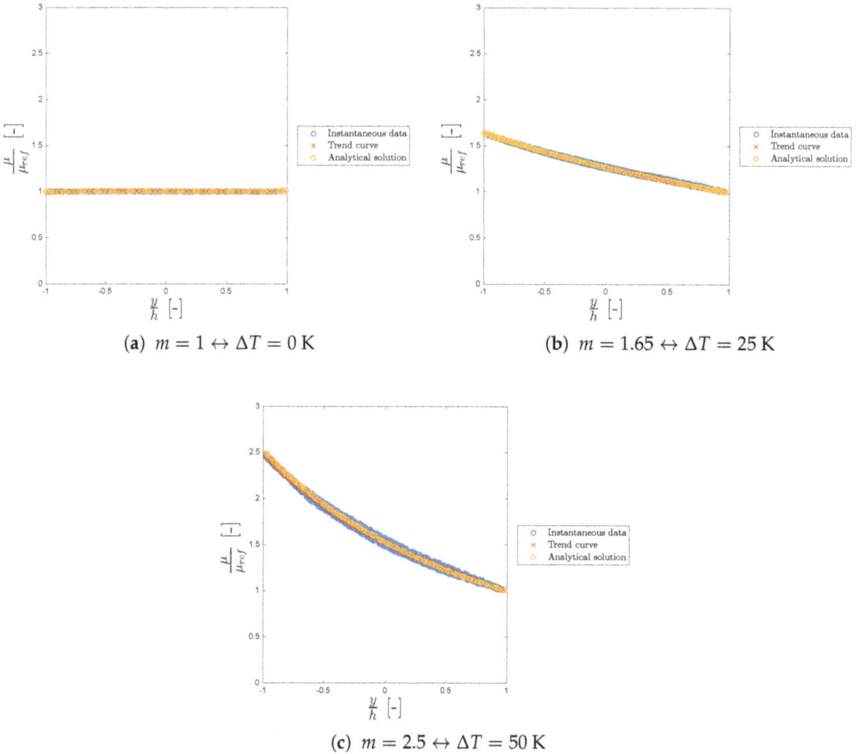

(a) $m = 1 \leftrightarrow \Delta T = 0\,\mathrm{K}$

(b) $m = 1.65 \leftrightarrow \Delta T = 25\,\mathrm{K}$

(c) $m = 2.5 \leftrightarrow \Delta T = 50\,\mathrm{K}$

Figure 9. Dimensionless viscosity profile for different $m = \mu_{cold}/\mu_{ref}$. Blue circles are the instantaneous data in the x direction, the orange curve is the trend curve extrapolated from the instantaneous data, yellow circles are obtained form the analytical solution from Sameen and Govindarajan [59].

In all the cases considered, the model is in good agreement with the work of Sameen and Govindarajan.

Figure 10 shows the dimensionless velocity trend for different values of m.

Again, the model is in good agreement with the analytical solution of Sameen and Govindarajan always laying within the velocity scattered points. In both our model and in the analytical solution the maximum of the velocity shifts to the right as m increases. We can conclude that our model is in good agreement with the literature, showing the typical viscosity and velocity profiles for asymmetric heating confirming the correct functionality of the new viscosity class.

7.3. New Abstract Class: Viscosity

To implement the new viscosity model a new abstract class has been created, called Viscosity. The class has no attribute, and one virtual method: `compute_visc`, that is used to compute the viscosity using one of the Equations (15)–(18). As usual, the Viscosity class is divided in two files, see Listings 78 and 79. As it is an abstract class, it cannot be instantiated. It is used as a base, a mold, to implement the viscosity models. All implemented viscosity classes, such as the ones implementing the Arrhenius viscosity or the Sutherland viscosity, will inherit from this class.

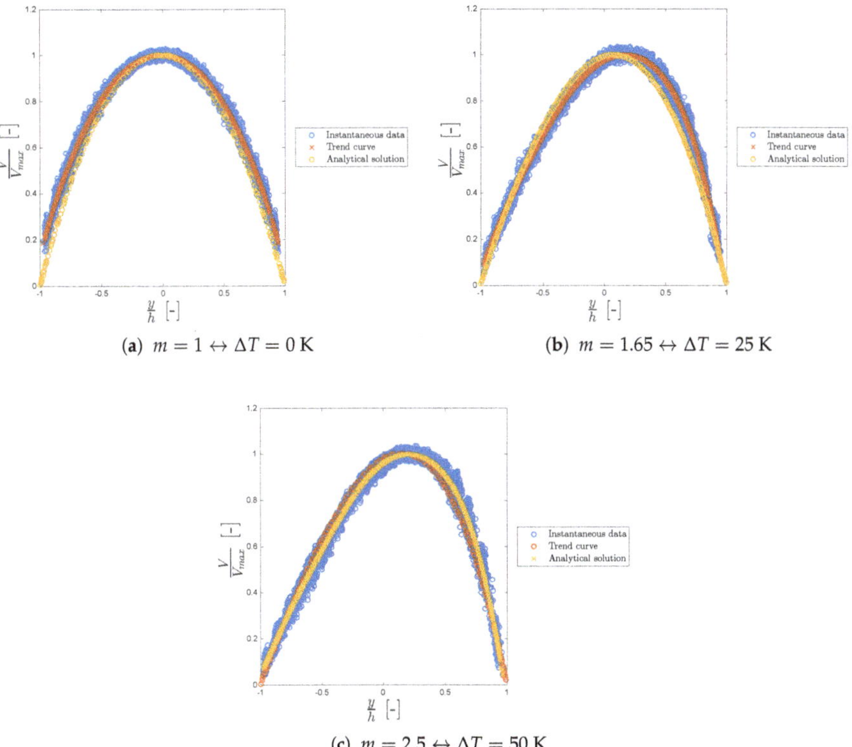

(a) $m = 1 \leftrightarrow \Delta T = 0 \, \text{K}$

(b) $m = 1.65 \leftrightarrow \Delta T = 25 \, \text{K}$

(c) $m = 2.5 \leftrightarrow \Delta T = 50 \, \text{K}$

Figure 10. Dimensionless velocity profile for different $m = \mu_{cold}/\mu_{ref}$. Blue circles are the instantaneous data in the x direction, the orange curve is the trend curve extrapolated from the instantaneous data, yellow circles are obtained form the analytical solution from Sameen and Govindarajan [59].

Listing 78: viscosity.cpp

```cpp
#include "viscosity.h"

using namespace LAMMPS_NS;

Viscosity::Viscosity() {};
```

Listing 79: viscosity.h

```cpp
#ifndef LAMMPS_VISCOSITY_H
#define LAMMPS_VISCOSITY_H

namespace LAMMPS_NS {
    class Viscosity {
        /**
         * Abstract base class for the viscosity attribute.
         * All viscosity types should inherit from this class.
         */
    public:
        Viscosity();
        /**
         * Virtual function.
         * Returns the viscosity, given the temperature.
         */
        virtual double compute_visc(double temperature) = 0;
```

```
17        };
18 }
19 #endif //LAMMPS_VISCOSITY_H
```

This type of base class is called an interface, though as the code is written in C++, there is no actual difference in the implementation. The difference is only in concepts.

This structure allows for a very simple procedure to add a new viscosity type to LAMMPS, as one doesn't have to go through all of the code everytime a new viscosity type is implemented. All that is required is to implement a new viscosity class inheriting from the Viscosity abstract class and modify the `add_viscosity` function. The details of the changes required for those two actions are detailed later in this section.

Another structure one might think of to implement the viscosity abstract base class would be a template. Indeed, templates are more efficient than inherited classes as inherited classes create additional virtual calls when calling the class's methods. However, the choice of which viscosity should be called is made at runtime, and not at compile time, which means the abstract base class would be a better fit. When runtime polymorphism is needed, the structure preferred is an abstract base class.

The abstract class is not the most efficient implementation, but it allows for simplicity of use, which is important considering most of LAMMPS users are not programmers. In this work, we have chosen to sacrifice a bit of efficiency to gain ease of use.

7.4. Implementing a New Viscosity Class

In this section the steps to implement one of Equations (15)–(18) are shown, using the four parameter exponential viscosity as an example.

A new class is created that inherits from the Viscosity abstract class. The new class have as much attributes as the viscosity type has parameters. In this example that means four, as shown in the header in Listing 80.

Listing 80: viscosity_four_parameter_exp.h

```
1  #ifndef LAMMPS_VISCOSITY_FOURPARAMETEREXP_H
2  #define LAMMPS_VISCOSITY_FOURPARAMETEREXP_H
3
4  #include "math.h"
5  #include "viscosity.h"
6  namespace LAMMPS_NS{
7
8  class ViscosityFourParameterExp  : public Viscosity{
9      /**
10      * Implementation of the four parameter exponential viscosity.
11      * This viscosity has four attributes.
12      */
13 private:
14     double A;
15     double B;
16     double C;
17     double D;
18 public:
19     ViscosityFourParameterExp(double A, double B, double C, double D);
20
21     double compute_visc(double temperature) override final;
22
23 };
24 };
25 #endif //LAMMPS_VISCOSITY_FOURPARAMETEREXP_H
```

The constructor therefore should take as arguments the four parameters of the Andrade's equation and initialise the class's attributes with those values. The last step is to implement the `compute_visc` method so it returns the value of the viscosity at the given temperature. The implementation of both those functions is shown in Listing 81.

Listing 81: viscosity_four_parameter_exp.cpp

```cpp
#include "viscosity_four_parameter_exp.h"

using namespace LAMMPS_NS;

ViscosityFourParameterExp::ViscosityFourParameterExp(double A, double B, double C,
    double D) {
    this->A = A;
    this->B = B;
    this->C = C;
    this->D = D;
}

double ViscosityFourParameterExp::compute_visc(double temperature) {
    return  A*exp(B/temperature +C*temperature + D *temperature*temperature);
}
```

Similar steps have to be taken to implement the classes corresponding to the other viscosity models, see the Supplementary material.

7.5. Processing the Viscosity in the Atom Class

In the header of the Atom class we need to include the new viscosity class and declare a new set of public members.

Listing 82: Original header (atom.h)

```cpp
#include "pointers.h"
#include <map>
#include <string>
```

Listing 83: Modified header (atom.h)

```cpp
#include "pointers.h"
#include "viscosity.h"
#include <map>
#include <string>
```

We add two new attributes in the USER-SPH section of the Atom attribute lists: `viscosity`, a pointer to a Viscosity object and `viscosities`, a pointer to an array containing the values of dynamic viscosities for all atoms at the current time step.

Listing 84: Original header (atom.h)

```cpp
// USER-SPH package
  double *rho,*drho,*e,*de,*cv;
  double **vest;
```

Listing 85: Modified header (atom.h)

```cpp
// USER-SPH package
  double *rho,*drho,*e,*de,*cv;
  double **vest;
  Viscosity *viscosity;
  double *viscosities;

```

We want to be able to choose which type of viscosity is being used in the simulation from the input file, using a new command called `viscosity`. Let's discuss the implementation of this feature. First we need to define the `viscosity` command. This is done by modifying the `execute_command` method of the Input class. We then define a new function called `add_viscosity`, whose declaration is shown in Listing 86 and definition in Listing 87. This function will have to be modified each time one wants to create a new viscosity class. In `add_viscosity`, the element arg[0] is the string representing the type of viscosity. For each viscosity class, the method performs the following procedure:

- It checks which type of viscosity is asked to be created using the function strcmp on arg[0] (for Andrade's viscosity it corresponds to line 3 of Listing 87)
- It checks if the number of arguments is coherent with the number of parameter of the viscosity type (line 4–5)
- It scans the coefficients of that viscosity type (line 6–10)
- It creates the appropriate viscosity and initializes the Viscosity attribute (line 11).

This process should be followed for any new implementation.

Listing 86: Modified header (atom.h)

```cpp
void add_viscosity(int narg, char **arg);
```

Listing 87: New add_viscosity function (atom.cpp)

```cpp
void Atom::add_viscosity(int narg, char **arg) {
  if (narg < 1) error->all(FLERR, "Too few arguments for creation of viscosity");
  if (!strcmp(arg[0], "FourParameterExp")) {
    if (narg != 5)
      error->all(FLERR, "Wrong number of arguments for creation of four
      parameter exponential viscosity");
    double A, B, C, D;
    sscanf(arg[1], "%lg", &A);
    sscanf(arg[2], "%lg", &B);
    sscanf(arg[3], "%lg", &C);
    sscanf(arg[4], "%lg", &D);
    this->viscosity = new ViscosityFourParameterExp(A, B, C, D);
    std::cout <<"Viscosity created" <<std::endl;
  } else {
    if (!strcmp(arg[0], "SutherlandViscosityLaw")) {
      if (narg != 3)
        error->all(FLERR, "Wrong number of arguments for creation of Sutherland
    viscosity");
      double A, B;
      sscanf(arg[1], "%lg", &A);
      sscanf(arg[2], "%lg", &B);
      this->viscosity = new SutherlandViscosityLaw(A, B);
      std::cout <<"Viscosity created" <<std::endl;
    } else {
      if (!strcmp(arg[0], "PowerLawGas")) {
        if (narg != 2)
          error->all(FLERR, "Wrong number of arguments for creation of power law
    gas viscosity");
        double B;
        sscanf(arg[1], "%lg", &B);
        this->viscosity = new PowerLawGas(B);
        std::cout <<"Viscosity created" <<std::endl;
      } else {
        if (!strcmp(arg[0], "Arrhenius")) {
          if (narg != 3)
            error->all(FLERR, "Wrong number of arguments for creation of
    Arrhenius viscosity");
          double A;
          double B;
          sscanf(arg[1], "%lg", &A);
          sscanf(arg[2], "%lg", &B);
          this->viscosity = new ViscosityArrhenius(A, B);
          std::cout <<"Viscosity created" <<std::endl;
        } else {
          if (!strcmp(arg[0], "Constant")) {
            if (narg != 2)
              error->all(FLERR, "Wrong number of arguments for creation
              of Constant viscosity");
            double A;
            sscanf(arg[1], "%lg", &A);
            this->viscosity = new ViscosityConstant(A);
            std::cout <<"Viscosity created" <<std::endl;
          } else {
            std::cout <<"Nothing implemented for " << arg[0]<< std::endl;
```

```
52                          }
53                        }
54                      }
55                    }
56                  }
57 }
```

All headers of the new viscosity types implemented in the `add_viscosity` function need to be included in the Atom class, see Listing 88.

Listing 88: New include (atom.cpp)

```cpp
#include <string.h>
#include <iostream>
#include "viscosity_four_parameter_exp.h"
#include "viscosity_sutherland_law.h"
#include "viscosity_power_law_gas.h"
#include "viscosity_arrhenius.h"
#include "viscosity_constant.h"
```

The viscosity attribute is initialised to NULL in the constructor, see Listing 89.

Listing 89: Inside Atom::Atom(LAMMPS *lmp) : Pointers(lmp) (atom.cpp)

```cpp
viscosity = NULL;
```

In the destructor of the Atom class, we add a line to delete the viscosity attribute, see Listing 90.

Listing 90: Inside Atom:: Atom() (atom.cpp)

```cpp
memory->destroy(viscosity);
```

The `extract` function is modified to process the viscosity attribute, see Listing 91.

Listing 91: Modified extract function (atom.cpp)

```cpp
if (strcmp(name, "viscosity") == 0) return (void *) viscosity;
```

7.6. Using compute_Visc in SPH Pair Styles: Tait Water Implementation

The dynamic viscosity is used to compute the artificial viscosity force, that is used in the compute function of the following SPH pair style: sph/idealgas, sph/lj, sph/taitwater and sph/taitwater/morris. In this section the steps to implement compute_visc in sph/taitwater are shown, the others required a similar procedure.

The first function to modify is the destructor, as we don't have to allocate the viscosity parameter anymore.

Listing 92: Original file (pair_sph_taitwater.cpp)

```cpp
PairSPHTaitwater::~PairSPHTaitwater() {
  if (allocated) {
    memory->destroy(setflag);
    memory->destroy(cutsq);
    memory->destroy(cut);
    memory->destroy(rho0);
    memory->destroy(soundspeed);
    memory->destroy(B);
    memory->destroy(viscosity);
  }
}
```

Listing 93: Modified file (pair_sph_taitwater.cpp)

```cpp
PairSPHTaitwater::~PairSPHTaitwater() {
  if (allocated) {
    memory->destroy(setflag);
    memory->destroy(cutsq);
```

```
5    memory->destroy(cut);
6    memory->destroy(rho0);
7    memory->destroy(soundspeed);
8    memory->destroy(B);
9    }
10 }
```

For the same reason as the destructor we need to modify allocate.

Listing 94: Original file (pair_sph_taitwater.cpp)

```
1  void PairSPHTaitwater::allocate() {
2    allocated = 1;
3    int n = atom->ntypes;
4    memory->create(setflag, n + 1, n + 1, "pair:setflag");
5    for (int i = 1; i <= n; i++)
6      for (int j = i; j <= n; j++)
7        setflag[i][j] = 0;
8    memory->create(cutsq, n + 1, n + 1, "pair:cutsq");
9    memory->create(rho0, n + 1, "pair:rho0");
10   memory->create(soundspeed, n + 1, "pair:soundspeed");
11   memory->create(B, n + 1, "pair:B");
12   memory->create(cut, n + 1, n + 1, "pair:cut");
13   memory->create(viscosity, n + 1,n + 1,"pair:viscosity");
14 }
```

Listing 95: Modified file (pair_sph_taitwater.cpp)

```
1  void PairSPHTaitwater::allocate() {
2    allocated = 1;
3    int n = atom->ntypes;
4    memory->create(setflag, n + 1, n + 1, "pair:setflag");
5    for (int i = 1; i <= n; i++)
6      for (int j = i; j <= n; j++)
7        setflag[i][j] = 0;
8    memory->create(cutsq, n + 1, n + 1, "pair:cutsq");
9    memory->create(rho0, n + 1, "pair:rho0");
10   memory->create(soundspeed, n + 1, "pair:soundspeed");
11   memory->create(B, n + 1, "pair:B");
12   memory->create(cut, n + 1, n + 1, "pair:cut");
13 }
```

Inside the compute function of the sph/taitwater pair style we need to declare a new set of variables. Where e is the energy and cv the heat capacity, now needed to calculate the temperature and thus the viscosity.

Listing 96: Original file (pair_sph_taitwater.cpp)

```
1  int *type = atom->type;
2  int nlocal = atom->nlocal;
3  int newton_pair = force->newton_pair;
```

```
1          [linebackgroundcolor={\listyellow{4,5,6,7}},
2          label=820, caption={\small Modified file (pair\textunderscore sph\textunderscore
           taitwater.cpp)}\label{32}]   % Start your code-block

3
4  int *type = atom->type;
5  int nlocal = atom->nlocal;
6  int newton_pair = force->newton_pair;
7  double *e = atom->e;
8  double *cv = atom->cv;
9  Viscosity* viscosity = atom->viscosity;
10 double* viscosities = atom->viscosities;
```

The next modification is inside the loop over the j-th atom when the force induced by the artificial viscosity is calculated inside the pair's `compute` function.

The dynamic viscosities μ_i and μ_j are calculated for each atoms, using the formula implemented in the `compute_visc` method. The temperature for the i-th atom is obtained using $T_i = e_i / cv_i$. It is important to note that using such expression for the energy balance prevents the reference state of the internal energy to be set at 0.

The constant viscosity matrix element is replaced by the formula defined in Equation (19), see Listings 97 and 98.

Listing 97: Original file (pair_sph_taitwater.cpp)

```
// artificial viscosity (Monaghan 1992)
if (delVdotDelR < 0.) {
    mu = h * delVdotDelR / (rsq + 0.01 * h * h);
    fvisc = -viscosity[itype][jtype] * (soundspeed[itype]
    + soundspeed[jtype]) * mu / (rho[i] + rho[j]);
} else {
    fvisc = 0.;
}
```

Listing 98: Modified file (pair_sph_taitwater.cpp)

```
viscosities[i] = viscosity->compute_visc(e[i]/cv[i]);
viscosities[j] = viscosity->compute_visc(e[j]/cv[j]);
// artificial viscosity (Monaghan 1992)
if (delVdotDelR < 0.) {
    mu = h * delVdotDelR / (rsq + 0.01 * h * h);
    fvisc =-4/h*(viscosities[i]/(soundspeed[itype]*rho[i])
    +viscosities[j]/(soundspeed[jtype]*rho[j]))
    *(soundspeed[itype]+ soundspeed[jtype])
    * mu / (rho[i] + rho[j]);
} else {
    fvisc = 0.;
}
```

Viscosity is now a per atom property, this means that we don't have to pass its value then the pair style is invoked. For this reason we need to delete the viscosity related lines inside coeff.

Listing 99: Original coeff (pair_sph_taitwater.cpp)

```
void PairSPHTaitwater::coeff(int narg, char **arg) {
  if (narg != 6)
    error->all(FLERR,
        "Incorrect args for pair_style sph/taitwater
        coefficients");
  if (!allocated)
    allocate();

  int ilo, ihi, jlo, jhi;
  force->bounds(FLERR,arg[0], atom->ntypes, ilo, ihi);
  force->bounds(FLERR,arg[1], atom->ntypes, jlo, jhi);

  double rho0_one = force->numeric(FLERR,arg[2]);
  double soundspeed_one = force->numeric(FLERR,arg[3]);
  double viscosity_one = force->numeric(FLERR,arg[4]);
  double cut_one = force->numeric(FLERR,arg[5]);
  double B_one = soundspeed_one*soundspeed_one*rho0_one/7;

  int count = 0;
  for (int i = ilo; i <= ihi; i++) {
    rho0[i] = rho0_one;
    soundspeed[i] = soundspeed_one;
    B[i] = B_one;
    for (int j = MAX(jlo,i); j <= jhi; j++) {
      viscosity[i][j] = viscosity_one;
      cut[i][j] = cut_one;
      setflag[i][j] = 1;
      count++;
    }
  }
  if (count == 0)
    error->all(FLERR,"Incorrect args for pair
```

```
33    coefficients");
34  }
```

Listing 100: Modified coeff (pair_sph_taitwater.cpp)

```
 1  void PairSPHTaitwater::coeff(int narg, char **arg) {
 2    if (narg != 5)
 3      error->all(FLERR,
 4        "Incorrect args for pair_style sph/taitwater
 5          coefficients");
 6    if (!allocated)
 7      allocate();
 8
 9    int ilo, ihi, jlo, jhi;
10    force->bounds(FLERR,arg[0], atom->ntypes, ilo, ihi);
11    force->bounds(FLERR,arg[1], atom->ntypes, jlo, jhi);
12
13    double rho0_one = force->numeric(FLERR,arg[2]);
14    double soundspeed_one = force->numeric(FLERR,arg[3]);
15    double cut_one = force->numeric(FLERR,arg[4]);
16    double B_one = soundspeed_one*soundspeed_one*rho0_one/7;
17
18    int count = 0;
19    for (int i = ilo; i <= ihi; i++) {
20      rho0[i] = rho0_one;
21      soundspeed[i] = soundspeed_one;
22      B[i] = B_one;
23      for (int j = MAX(jlo,i); j <= jhi; j++) {
24        cut[i][j] = cut_one;
25        setflag[i][j] = 1;
26        count++;
27      }
28    }
29    if (count == 0)
30      error->all(FLERR,"Incorrect args for pair
31        coefficients");
32  }
```

The last modification is in init_one. Again, we delete lines related to the former viscosity attribute.

Listing 101: Original file (pair_sph_taitwater.cpp)

```
1  double PairSPHTaitwater::init_one(int i, int j) {
2    if (setflag[i][j] == 0) {
3      error->all(FLERR,"All pair sph/taitwater coeffs
4        are set");
5    }
6    cut[j][i] = cut[i][j];
7    viscosity[j][i] = viscosity[i][j];
8    return cut[i][j];
9  }
```

Listing 102: Modified file (pair_sph_taitwater.cpp)

```
1  double PairSPHTaitwater::init_one(int i, int j) {
2    if (setflag[i][j] == 0) {
3      error->all(FLERR,"All pair sph/taitwater coeffs
4        are set");
5    }
6    cut[j][i] = cut[i][j];
7    return cut[i][j];
8  }
```

7.7. Running the New Software with Mpirun

At this stage, the software is designed to only run in serial. Changes need to be made to make it run with Message Passing Interface (MPI). This will allow the software to run in parallel: some computations being independent from each other, they can be performed at the same time. Instead of using one processor for a long time, we will use multiple processors for a shorter period. The simulation will therefore take more computing resources but will take a lot shorter to compute. The original SPH module can already be run with MPI however as we have modified the code that is no longer true. We need to make additional changes to the software. All those changes are located in the Atom Vec Meso class of the SPH module.

In LAMMPS, the different MPI processes have to communicate with each other as the computations they perform are not completely independent from each other. They need data from other processes in order to perform their own calculations. They communicate with each other using a buffer that will contain all the necessary data. The buffer is simply an array that we will fill with the data. The different methods for packing and unpacking this buffer are defined in the Atom Vec Meso class. We need to add a new data to transmit: the calculated viscosity.

The first thing to do is to increase the size of the buffers in their initialisation so they can accept the viscosity value, an example is shown in Listings 103 and 104.

Listing 103: Original constructor (atom_vec_meso.cpp)

```
AtomVecMeso::AtomVecMeso(LAMMPS *lmp) : AtomVec(lmp)
{
  molecular = 0;
  mass_type = 1;
  forceclearflag = 1;

// we communicate not only x forward but also vest ..
  comm_x_only = 0; .
// we also communicate de and drho in reverse direction
  comm_f_only = 0;
// 3 + rho + e + vest[3], that means we may
// only communicate 5 in hybrid
  size_forward = 8;
  size_reverse = 5; // 3 + drho + de
  size_border = 12; // 6 + rho + e + vest[3] + cv
  size_velocity = 3;
  size_data_atom = 8;
  size_data_vel = 4;
  xcol_data = 6;

  atom->e_flag = 1;
  atom->rho_flag = 1;
  atom->cv_flag = 1;
  atom->vest_flag = 1;
}
```

Listing 104: Modified constructor (atom_vec_meso.cpp)

```
AtomVecMeso::AtomVecMeso(LAMMPS *lmp) : AtomVec(lmp)
{
  molecular = 0;
  mass_type = 1;
  forceclearflag = 1;

// we communicate not only x forward but also vest ...
  comm_x_only = 0;
// we also communicate de and drho in reverse direction
  comm_f_only = 0;
// 3 + rho + e + vest[3] + viscosities, that means we may
// only communicate 6 in hybrid
  size_forward = 9;
  size_reverse = 5; // 3 + drho + de
// 6 + rho + e + vest[3] + cv + viscosities
  size_border = 13;
  size_velocity = 3;
  size_data_atom = 8;
  size_data_vel = 4;
```

```
20   xcol_data = 6;
21
22   atom->e_flag = 1;
23   atom->rho_flag = 1;
24   atom->cv_flag = 1;
25   atom->vest_flag = 1;
26 }
```

Then, we added the relevant elements of the attribute `viscosities` to the buffer in all the methods handling buffers, an example is shown in Listings 105 and 106.

Listing 105: Original `pack_vec_hybrid` (atom_vec_meso.cpp)

```
1  int AtomVecMeso::pack_comm_hybrid(int n, int *list,
2   double *buf) {
3    //printf("in AtomVecMeso::pack_comm_hybrid\n");
4    int i, j, m;
5
6    m = 0;
7    for (i = 0; i < n; i++) {
8      j = list[i];
9      buf[m++] = rho[j];
10     buf[m++] = e[j];
11     buf[m++] = vest[j][0];
12     buf[m++] = vest[j][1];
13     buf[m++] = vest[j][2];
14   }
15   return m;
16 }
```

Listing 106: Modified `pack_vec_hybrid` (atom_vec_meso.cpp)

```
1  int AtomVecMeso::pack_comm_hybrid(int n, int *list,
2   double *buf) {
3    //printf("in AtomVecMeso::pack_comm_hybrid\n");
4    int i, j, m;
5
6    m = 0;
7    for (i = 0; i < n; i++) {
8      j = list[i];
9      buf[m++] = rho[j];
10     buf[m++] = e[j];
11     buf[m++] = vest[j][0];
12     buf[m++] = vest[j][1];
13     buf[m++] = vest[j][2];
14     buf[m++] = viscosities[j];
15   }
16   return m;
17 }
```

After making those changes for all the methods in the class, the software can be run using mpirun.

7.8. Invoking, Selecting and Computing a Viscosity Object

To compute the new viscosity a new argument was added to the `compute` command: `viscosities`. This allows the user to use the `compute` command to output the dynamic viscosity to the dump file. This can be done by the following command:

```
1 compute          viscosities_peratom all meso/viscosities/atom
```

The implementation of this feature is simple, as it is very similar to other `compute` argument implementation. All that needs to be done is to modify another `compute`'s implementation, such as `compute_meso_rho_atom` so it processes the variable `viscosities` instead of `rho`.

The viscosity used in the simulation can be invoked in the input file, using the following command:

```
1 viscosity      [type of viscosity]     [parameters of the viscosity]
```

The type of viscosity can be chosen from the following list:

- FourParameterExp: the four parameter exponential viscosity law.
- SutherlandViscosityLaw: the Sutherland viscosity law.
- PowerLawGas: the power viscosity law for gases.
- Arrhenius: the Arrhenius viscosity law.
- Constant: a constant viscosity.

For example, to invoke the four parameter exponential viscosity, we can write in the input file:

```
1 viscosity      FourParameterExp C1 C2 C3 C4
```

As stated earlier, this list can easily be extended by the user by modifying the `add_viscosity` function defined earlier.

8. Conclusions

Particle methods are very versatile and can be applied in a variety of applications, ranging from modelling of molecules to the simulation of galaxies. Their power is even amplified when they are coupled together within a discrete multiphysics framework. This versatility matches well with LAMMPS, which is a particle simulator, whose open-source code can be extended with new functionalities. However, modifying LAMMPS can be challenging for researchers with little coding experience and the available support material on how to modify LAMMPS is either too basic or too advanced for the average researcher. Moreover, most of the available material focuses on MD; while the aim of this paper is to support researchers that use other particle methods such as SPH or DEM.

In this work, we present several examples, explained step-by-step and with increasing level of complexity. We begin with simple cases and concluding with more complex ones: Section 3 shows the implementation of the Kelvin–Voigt bond style used to model encapsulate particles with a soft outer shell and validated validated by simulating spherical homogeneous linear elastic and viscoelastic particles [45]; Section 7 show how to implement a new per-atom temperature dependant viscosity property and is validated finding the same viscosity and velocity trend shown by Sameen and Govindarajan [59] in their analytical solution for a channel flow in a asymmetrical heating walls.

The work perfectly fits in the "Discrete Multiphysics: Modelling Complex Systems with Particle Methods" special issue by sharing some in dept know-how and "trick and trades" developed by our group in years of use of LAMMPS. In fact, the aim is to support, in several ways, researchers that use computational particle methods. Often researchers tend to write their own code. The advantage of this approach is that the code is well understood by the researcher and, therefore, easily extendible. However, this sometimes implies reinventing the wheel and countless hours of debugging. Familiarity with a code like LAMMPS, which has an active community of practice and is periodically enriched with new features would be beneficial to this type of researchers allowing them to save considerable time. In the long term, there is another advantage. Modules written for in-house code are hardly sharable. At the moment, the largest portion of the LAMMPS community is dedicated to MD. While this article was under review, for instance, a new book dedicated to modifying LAMMPS came out [60]. However, it focuses only on MD and it does not mention other discrete methods like SPH or DEM. Instead, the aim of this paper is to make LAMMPS more accessible for the Discrete Multiphysics community facilitating sharing reusable code among practitioners in this field.

Supplementary Materials: The codes used in this work are freely available under the GNU General Public License v3 and can be downloaded from the University of Birmingham repository (http://edata.bham.ac.uk/560/).

Author Contributions: Conceptualization, A.A. (Andrea Albano) and A.A. (Alessio Alexiadis); methodology, A.A. (Andrea Albano); validation, A.A. (Andrea Albano), E.l.G., A.D., I.H.S. and A.R.; writing—original draft preparation, A.A. (Andrea Albano), E.l.G., A.D., I.H.S.; writing—review and editing, A.A. (Andrea Albano), A.A. (Alessio Alexiadis), C.A.D.-D., X.S., K.C.N. and M.A.; supervision, A.A. (Alessio Alexiadis), A.R. and I.M.; funding acquisition, A.A. (Alessio Alexiadis). All authors have read and agreed to the published version of the manuscript.

Funding: This work was supported by the US office of Naval Research Global (ONRG) under NICOP Grant N62909-17-1-2051.

Acknowledgments: The authors would like to thank Prof Albano (University of Pisa) for his advice and comments. The computations described in this paper were performed using the University of Birmingham's BlueBEAR HPC service, which provides a High Performance Computing service to the University's research community. See http://www.birmingham.ac.uk/bear for more details.

Conflicts of Interest: The authors declare no conflict of interest.

Abbreviations

The following abbreviations are used in this manuscript:

MS	Molecular Dynamics
DMP	Discrete MultiPhysics
SPH	Smoothed Particle Hydrodynamics
LAMMPS	Large-scale Atomic/Molecular Massively Parallel Simulator
EOS	Equation Of State
LSM	Lattice Spring Model

Appendix A. An Example of Discrete Multiphysics Simulation in LAMMPS

In this section we present a simple case of DMP simulation with LAMMPS. It is an explanatory example deliberately simple for illustrative purposes. It involves only a small number of particles. Sensitivity analysis of the results with the model resolution or other numerical parameters are beyond the scope of this example and not carried out.

The geometry is a 2D tube with an elastic membrane at one end (Figure A1). The tube contains a liquid simulated with the SPH model, Tait EOS and Morris viscosity. The wall is simulated with stationary particles and the membrane with the LSM using Hookean springs. In Figure A1, the liquid particles are red, the wall particles blue and the membrane particles yellow. During the simulation, the fluid is subjected to a force in the x-direction that pushed the particles against the membrane. Because the membrane is elastic, it stretches inflating the right end of the tube like a balloon. The resolution of the membrane is ten times higher than the fluid. This ensures that, as the membrane stretches, fluid particles do not 'leak' in the gaps formed between two consecutive membrane particles. The Lennard Jones potential, truncated to consider only the repulsive part, is used to avoid compenetration between solid and liquid particles. A weaker Lennard Jones potential is used as 'artificial pressure' to avoid excessive compression of the fluid particles.

The initial data file (data.initial) for the geometry was create according to LAMMPS' rules for formatting the Data File [41] and is shared as additional material. In Data File, the fluid particles are called type 1, the wall particles type 2 and the membrane particles type 3. Here we focus on the input file (membrane.lmp), which is also shared in its entirety as additional material. We do not discuss LAMMPS syntax (the reader can refer to LAMMPS User's Guide for this [41]), but only on specific parts of the input file that concern the DMP implementation.

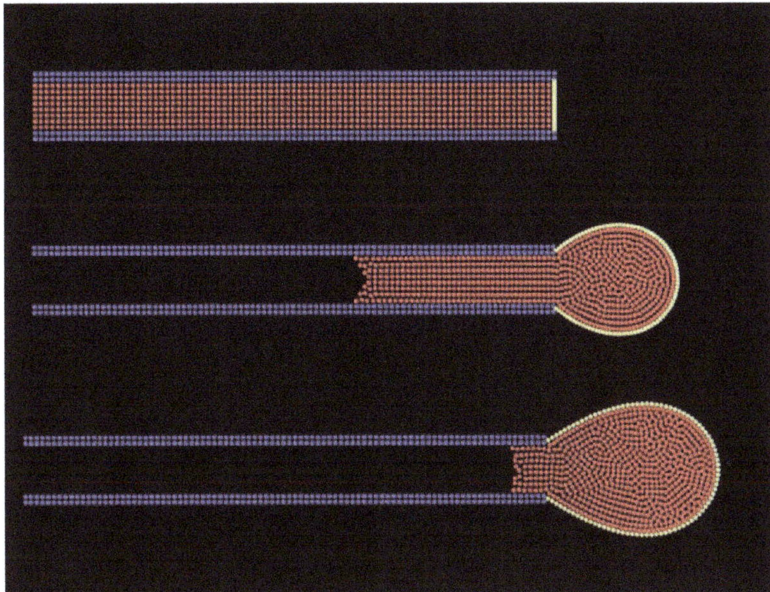

Figure A1. The inflating balloon simulation.

The first section of the input file determines the dimensionality of the problem (2D), the boundary conditions (periodic), the units used (SI), the type of potential used in the simulation (`atom_style `) and the input file that contains the initial position of all the particles

```
1 dimension          2
2 boundary        p p p
3 units             si
4 atom_style  hybrid meso bond angle
5 read_data     data.initial
```

The crucial line for DMP simulations is the `hybrid` keyword of the `atom_style`, which allows for combining different particle models. The keyword `meso` refers to the SPH model and bond, in the case under consideration, to the LSM. The `angle` keyword corresponds to angular springs, but, as it will be clear later, it is not used in this simulation.

The following section contains several variables that are going to be used later on. In particular, the initial particle distance is `dL` and their mass `m`. The resolution of the membrane is `Nt` times higher than the fluid. The initial distance between membrane particle is therefore `db=dL/Nt` and their mass `mM=m/Nt`.

```
 1 variable     dL        equal   0.000111111
 2 variable     m       equal   1.23457e-05
 3 variable     Nt        equal 10
 4 variable     dB        equal ${dL}/${Nt}
 5 variable     mM        equal   ${m}/${Nt}
 6 variable     h       equal 1.5*${dL}
 7 variable     h2        equal ${dL}/${Nt}
 8 variable     c       equal 0.1
 9 variable     mu        equal 1.0e-3
10 variable     rho       equal 1000
11 variable     kA        equal 1.e-8
12 variable     kB        equal 100
13 variable     skin      equal 0.3*${h}
14 variable     epsL      equal 1.e-12
15 variable     epsS      equal 1.e-10
16 variable     sgmL    equal ${dL}
```

```
17 variable     sgmS    equal 0.5*${sgmL}/${Nt}
18 variable     fmax      equal 0.00005
19 variable     ft      equal ramp(0.,${fmax})
```

The section below identifies particles type 1 as a group called fluid, particles type 2 as a group called wall and particles type 3 as a group called membrane. The mass of type 3 particles is assigned (the mass of type 1, 2 was assigned in the data.initial file). The density of all particle is also assigned based on the value rho defined previously.

```
1 group              fluid       type 1
2 group              wall      type 2
3 group              membrane    type 3
4 mass 3 ${mM}
5 set group all meso/rho ${rho}
```

The next section defines the pair potentials for non-bonded particles. In this simulation, we use different styles together (keyword hybrid/overlay). The sph/taitwater/morris pair style, which is used for all pair interactions except 2-2 (i.e., wall particles with themselves); and the Lennard Jones potential lj/cut, which, as explained above, is used both as 'artificial pressure' and to avoid compenetration of solid and fluid particles.

```
1 pair_style hybrid/overlay sph/taitwater/morris lj/cut ${sgmL}
2 pair_coeff 1 * sph/taitwater/morris ${rho} ${c} ${mu} ${h}
3 pair_coeff 2 3 sph/taitwater/morris ${rho} ${c} ${mu} ${h2}
4 pair_coeff 3 3 sph/taitwater/morris ${rho} ${c} ${mu} ${h2}
5
6 pair_coeff 1 * lj/cut ${epsL} ${sgmL}
7 pair_coeff 2 * lj/cut ${epsL} ${sgmL}
8 pair_coeff 1 3 lj/cut ${epsS} ${sgmL}
9 pair_coeff 3 3 lj/cut ${epsS} ${sgmS}
```

After the non-bonded potentials, the script assigns the harmonic potential, with Hook constant kB and equilibrium distance dB, to the bonded particles (i.e., the membrane). All pairs of bonded particles are assigned in the data.initial file.

```
1 bond_style  harmonic
2 bond_coeff  1 ${kB} ${dB}
3 angle_style none
```

The next section assigns several parameters that determine how the Newton equation of motion is solved numerically. The force fmax is added to all fluid particle in the x-direction, and an artificial viscosity is added for stability reasons.

```
1 fix 2 fluid addforce ${fmax} 0.0 0.0
2 fix 5 fluid meso
3 fix 6 membrane meso
4 fix 8 wall meso/stationary
5 fix 9 all viscous 0.01
```

The last commands determine the value and the number of timesteps used in the simulation plus a variety of computations for output and other purposes that are not discussed here (the reader can refer to the User's Guide).

```
1  compute rho_peratom all meso/rho/atom
2  compute rho_ave all reduce ave c_rho_peratom
3  compute vmax fluid reduce max vx
4  thermo            10000
5  thermo_style custom step c_rho_ave c_vmax
6  thermo_modify        norm no
7  neighbor         ${skin} bin
8  dump   dump_id all custom 10000 dump.lammpstrj id type x y z vx vy
9  timestep          1.e-6
10 run               2500000
```

Appendix B. How to Compile LAMMPS

LAMMPS is build as a library and executable [41] either by using GNU make [61] or a build environment with CMake [62]. In this appendix LAMMPS will be compiled only using make and it is compiled in BlueBEAR. For more details of the compiling process in LAMMPS refer to the user manual [41].

To compile LAMMPS in your own directory you can follow those steps

1. Download the file from here. Select the code you want, click the "Download Now" button, and your browser should download a gzipped tar file. Save the file in your directory on BlueBEAR
2. Unpack the file with the following command line command prompt:

Listing A1: Command to open the tar file on BlueBEAR

```
1  tar -xvf lammps-stable.tar.gz
```

3. Before compiling is important to set up the environment, with BlueBEAR

Listing A2: Commands to set the environment for compile LAMMPS on BlueBEAR

```
1  module purge
2
3  module load bluebear
4
5  module load Eigen/3.3.4-foss-2019a
```

4. Enter in the /src directory in your new LAMMPS directory. The src directory directory contains the C++ source and header files for LAMMPS. It also contains a top-level Makefile and a MAKE sub-directory with low-level Makefile.* files for many systems and machines.
5. Type the following command to compile a serial version of LAMMPS:

Listing A3: Command to compile LAMMPS on BlueBEAR

```
1  make serial
```

or a multi-threaded (parallel) version of LAMMPS:

Listing A4: Command to compile LAMMPS on BlueBEAR

```
1  make mpi
```

If you get no errors and an executable file lmp_mpi is produced.

6. Depending on the features you need, you will have to install same packages in your compiled LAMMPS. Is possible to check which packages is installed in your compiled LAMMPS by typing

Listing A5: Command to check the list of installed packages (you must be inside the /src directory)

```
1  make ps
```

It is possible to install the packages you need with the command line

Listing A6: Command to install a specific package

```
1  make yes-NAMEPACK
```

or un-install them with

Listing A7: Command to un-install a specific package

```
1  make no-NAMEPACK
```

More make commands are explained in LAMMPS user manual [41]. After the installation of the desired packages you need to compile it again (step 5).

References

1. Plimpton, S. *Fast Parallel Algorithms for Short-Range Molecular Dynamics*; Technical Report; Sandia National Labs.: Albuquerque, NM, USA, 1993.
2. Plimpton, S.; Pollock, R.; Stevens, M. Particle-Mesh Ewald and rRESPA for Parallel Molecular Dynamics Simulations. In Proceedings of the Eighth SIAM Conference on Parallel Processing for Scientific Computing, Minneapolis, MN, USA, 14–17 March 1997.
3. Auhl, R.; Everaers, R.; Grest, G.S.; Kremer, K.; Plimpton, S.J. Equilibration of long chain polymer melts in computer simulations. *J. Chem. Phys.* **2003**, *119*, 12718–12728.
4. Parks, M.L.; Lehoucq, R.B.; Plimpton, S.J.; Silling, S.A. Implementing peridynamics within a molecular dynamics code. *Comput. Phys. Commun.* **2008**, *179*, 777–783.
5. Petersen, M.K.; Lechman, J.B.; Plimpton, S.J.; Grest, G.S.; Veld, P.J.; Schunk, P. Mesoscale hydrodynamics via stochastic rotation dynamics: Comparison with Lennard-Jones fluid. *J. Chem. Phys.* **2010**, *132*, 174106.
6. Ganzenmüller, G.C.; Steinhauser, M.O.; Van Liedekerke, P.; Leuven, K.U. The implementation of Smooth Particle Hydrodynamics in LAMMPS. *Paul Van Liedekerke Kathol. Univ. Leuven* **2011**, *1*, 1–26.
7. Jaramillo-Botero, A.; Su, J.; Qi, A.; Goddard III, W.A. Large-scale, long-term nonadiabatic electron molecular dynamics for describing material properties and phenomena in extreme environments. *J. Comput. Chem.* **2011**, *32*, 497–512.
8. Coleman, S.; Spearot, D.; Capolungo, L. Virtual diffraction analysis of Ni [0 1 0] symmetric tilt grain boundaries. *Model. Simul. Mater. Sci. Eng.* **2013**, *21*, 055020.
9. Singraber, A.; Behler, J.; Dellago, C. Library-based LAMMPS implementation of high-dimensional neural network potentials. *J. Chem. Theory Comput.* **2019**, *15*, 1827–1840.
10. Ng, K.; Alexiadis, A.; Chen, H.; Sheu, T. A coupled Smoothed Particle Hydrodynamics-Volume Compensated Particle Method (SPH-VCPM) for Fluid Structure Interaction (FSI) modelling. *Ocean Eng.* **2020**, *218*, 107923.
11. Daraio, D.; Villoria, J.; Ingram, A.; Alexiadis, A.; Stitt, E.H.; Munnoch, A.L.; Marigo, M. Using Discrete Element method (DEM) simulations to reveal the differences in the γ-Al2O3 to α-Al2O3 mechanically induced phase transformation between a planetary ball mill and an attritor mill. *Miner. Eng.* **2020**, *155*, 106374.
12. Qiao, G.; Lasfargues, M.; Alexiadis, A.; Ding, Y. Simulation and experimental study of the specific heat capacity of molten salt based nanofluids. *Appl. Therm. Eng.* **2017**, *111*, 1517–1522.
13. Qiao, G.; Alexiadis, A.; Ding, Y. Simulation study of anomalous thermal properties of molten nitrate salt. *Powder Technol.* **2017**, *314*, 660–664.
14. Anagnostopoulos, A.; Navarro, H.; Alexiadis, A.; Ding, Y. Wettability of $NaNO_3$ and KNO_3 on MgO and Carbon Surfaces—Understanding the Substrate and the Length Scale Effects. *J. Phys. Chem. C* **2020**, *124*, 8140–8152.
15. Sahputra, I.H.; Alexiadis, A.; Adams, M.J. Effects of Moisture on the Mechanical Properties of Microcrystalline Cellulose and the Mobility of the Water Molecules as Studied by the Hybrid Molecular Mechanics–Molecular Dynamics Simulation Method. *J. Polym. Sci. Part B Polym. Phys.* **2019**, *57*, 454–464.
16. Sahputra, I.H.; Alexiadis, A.; Adams, M.J. Temperature dependence of the Young's modulus of polymers calculated using a hybrid molecular mechanics–molecular dynamics method. *J. Phys. Condens. Matter* **2018**, *30*, 355901.
17. Mohammed, A.M.; Ariane, M.; Alexiadis, A. Using Discrete Multiphysics Modelling to Assess the Effect of Calcification on Hemodynamic and Mechanical Deformation of Aortic Valve. *ChemEngineering* **2020**, *4*, 48.
18. Ariane, M.; Vigolo, D.; Brill, A.; Nash, F.; Barigou, M.; Alexiadis, A. Using Discrete Multi-Physics for studying the dynamics of emboli in flexible venous valves. *Comput. Fluids* **2018**, *166*, 57–63.
19. Ariane, M.; Wen, W.; Vigolo, D.; Brill, A.; Nash, F.; Barigou, M.; Alexiadis, A. Modelling and simulation of flow and agglomeration in deep veins valves using discrete multi physics. *Comput. Biol. Med.* **2017**, *89*, 96–103.
20. Ariane, M.; Allouche, M.H.; Bussone, M.; Giacosa, F.; Bernard, F.; Barigou, M.; Alexiadis, A. Discrete multi-physics: A mesh-free model of blood flow in flexible biological valve including solid aggregate formation. *PLoS ONE* **2017**, *12*, e0174795.
21. Schütt, M.; Stamatopoulos, K.; Simmons, M.; Batchelor, H.; Alexiadis, A. Modelling and simulation of the hydrodynamics and mixing profiles in the human proximal colon using Discrete Multiphysics. *Comput. Biol. Med.* **2020**, *121*, 103819.
22. Alexiadis, A.; Stamatopoulos, K.; Wen, W.; Batchelor, H.; Bakalis, S.; Barigou, M.; Simmons, M. Using discrete multi-physics for detailed exploration of hydrodynamics in an in vitro colon system. *Comput. Biol. Med.* **2017**, *81*, 188–198.
23. Ariane, M.; Kassinos, S.; Velaga, S.; Alexiadis, A. Discrete multi-physics simulations of diffusive and convective mass transfer in boundary layers containing motile cilia in lungs. *Comput. Biol. Med.* **2018**, *95*, 34–42.
24. Ariane, M.; Sommerfeld, M.; Alexiadis, A. Wall collision and drug-carrier detachment in dry powder inhalers: Using DEM to devise a sub-scale model for CFD calculations. *Powder Technol.* **2018**, *334*, 65–75.
25. Albano, A.; Alexiadis, A. Interaction of Shock Waves with Discrete Gas Inhomogeneities: A Smoothed Particle Hydrodynamics Approach. *Appl. Sci.* **2019**, *9*, 5435.
26. Albano, A.; Alexiadis, A. A smoothed particle hydrodynamics study of the collapse for a cylindrical cavity. *PLoS ONE* **2020**, *15*, e0239830.
27. Albano, A.; Alexiadis, A. Non-Symmetrical Collapse of an Empty Cylindrical Cavity Studied with Smoothed Particle Hydrodynamics. *Appl. Sci.* **2021**, *11*, 3500.
28. Alexiadis, A. The discrete multi-hybrid system for the simulation of solid-liquid flows. *PLoS ONE* **2015**, *10*, e0124678.

29. Alexiadis, A. A new framework for modelling the dynamics and the breakage of capsules, vesicles and cells in fluid flow. *Procedia IUTAM* **2015**, *16*, 80–88.
30. Alexiadis, A. A smoothed particle hydrodynamics and coarse-grained molecular dynamics hybrid technique for modelling elastic particles and breakable capsules under various flow conditions. *Int. J. Numer. Methods Eng.* **2014**, *100*, 713–719.
31. Rahmat, A.; Barigou, M.; Alexiadis, A. Deformation and rupture of compound cells under shear: A discrete multiphysics study. *Phys. Fluids* **2019**, *31*, 051903.
32. Alexiadis, A.; Ghraybeh, S.; Qiao, G. Natural convection and solidification of phase-change materials in circular pipes: A SPH approach. *Comput. Mater. Sci.* **2018**, *150*, 475–483.
33. Rahmat, A.; Barigou, M.; Alexiadis, A. Numerical simulation of dissolution of solid particles in fluid flow using the SPH method. *Int. J. Numer. Methods Heat Fluid Flow* **2019**, *30*, 290–307.
34. Rahmat, A.; Meng, J.; Emerson, D.; Wu, C.Y.; Barigou, M.; Alexiadis, A. A practical approach for extracting mechanical properties of microcapsules using a hybrid numerical model. *Microfluid. Nanofluidics* **2021**, *25*, 1–17.
35. Ruiz-Riancho, I.N.; Alexiadis, A.; Zhang, Z.; Hernandez, A.G. A Discrete Multi-Physics Model to Simulate Fluid Structure Interaction and Breakage of Capsules Filled with Liquid under Coaxial Load. *Processes* **2021**, *9*, 354.
36. Sanfilippo, D.; Ghiassi, B.; Alexiadis, A.; Hernandez, A.G. Combined Peridynamics and Discrete Multiphysics to Study the Effects of Air Voids and Freeze-Thaw on the Mechanical Properties of Asphalt. *Materials* **2021**, *14*, 1579.
37. Alexiadis, A.; Albano, A.; Rahmat, A.; Yildiz, M.; Kefal, A.; Ozbulut, M.; Bakirci, N.; Garzón-Alvarado, D.; Duque-Daza, C.; Eslava-Schmalbach, J. Simulation of pandemics in real cities: Enhanced and accurate digital laboratories. *Proc. R. Soc. A* **2021**, *477*, 20200653.
38. Alexiadis, A. Deep Multiphysics and Particle–Neuron Duality: A Computational Framework Coupling (Discrete) Multiphysics and Deep Learning. *Appl. Sci.* **2019**, *9*, 5369.
39. Alexiadis, A. Deep multiphysics: Coupling discrete multiphysics with machine learning to attain self-learning in-silico models replicating human physiology. *Artif. Intell. Med.* **2019**, *98*, 27–34.
40. Alexiadis, A.; Simmons, M.; Stamatopoulos, K.; Batchelor, H.; Moulitsas, I. The duality between particle methods and artificial neural networks. *Sci. Rep.* **2020**, *10*, 1–7.
41. Sandia Corporation LAMMPS Users Manual. 2003. Available online: https://lammps.sandia.gov/doc/Developer.pdf (accessed on 11 Jauaury 2021).
42. Plimpton. LAMMPS Developer Guide. Available online: https://lammps.sandia.gov/doc/Developer.pdf (accessed on 11 Jauaury 2021).
43. Plimpton, S.J. *Modifying & Extending LAMMPS*; Technical Report; Sandia National Lab. (SNL-NM): Albuquerque, NM, USA, 2014.
44. Flügge, W. *Viscoelasticity*; Springer Science & Business Media: Berlin/Heidelberg, Germany, 2013.
45. Sahputra, I.H.; Alexiadis, A.; Adams, M.J. A Coarse Grained Model for Viscoelastic Solids in Discrete Multiphysics Simulations. *ChemEngineering* **2020**, *4*, 30.
46. Liu, M.; Liu, G. Smoothed particle hydrodynamics (SPH): An overview and recent developments. *Arch. Comput. Methods Eng.* **2010**, *17*, 25–76.
47. Le Métayer, O.; Saurel, R. The Noble-Abel stiffened-gas equation of state. *Phys. Fluids* **2016**, *28*, 046102.
48. Monaghan, J.J.; Gingold, R.A. Shock simulation by the particle method SPH. *J. Comput. Phys.* **1983**, *52*, 374–389.
49. Lattanzio, J.; Monaghan, J.; Pongracic, H.; Schwarz, M. Controlling penetration. *SIAM J. Sci. Stat. Comput.* **1986**, *7*, 591–598.
50. Morris, J.P.; Fox, P.J.; Zhu, Y. Modeling low Reynolds number incompressible flows using SPH. *J. Comput. Phys.* **1997**, *136*, 214–226.
51. Cornelissen, J.; Waterman, H. The viscosity temperature relationship of liquids. *Chem. Eng. Sci.* **1955**, *4*, 238–246.
52. Seeton, C.J. Viscosity–temperature correlation for liquids. *Tribol. Lett.* **2006**, *22*, 67–78.
53. Stanciu, I. A new viscosity-temperature relationship for vegetable oil. *J. Pet. Technol. Altern. Fuels* **2012**, *3*, 19–23.
54. Gutmann, F.; Simmons, L. The temperature dependence of the viscosity of liquids. *J. Appl. Phys.* **1952**, *23*, 977–978.
55. De Guzman, J. Relation between fluidity and heat of fusion. *Anales Soc. Espan. Fis. Quim* **1913**, *11*, 353–362.
56. Raman, C. A theory of the viscosity of liquids. *Nature* **1923**, *111*, 532–533.
57. Chapman, S.; Cowling, T.G.; Burnett, D. *The Mathematical Theory of Non-Uniform Gases: An Account of the Kinetic Theory of Viscosity, Thermal Conduction and Diffusion in Gases*; Cambridge University Press: Cambridge, UK, 1990.
58. Rathakrishnan, E. *Theoretical Aerodynamics*; John Wiley & Sons: Hoboken, NJ, USA, 2013.
59. Sameen, A.; Govindarajan, R. The effect of wall heating on instability of channel flow. *J. Fluid Mech.* **2007**, *577*, 417–442.
60. Mubin, S.; Li, J. *Extending and Modifying LAMMPS*; Packt Publishing Ltd.: Birmingham, UK, 2021.
61. Project, G. Make-GNU Project-Free Software Fondation. Available online: https://www.gnu.org/software/make/ (accessed on 14 October 2020).
62. Martin, K.; Hoffman, B. *Mastering CMake: A Cross-Platform Build System*; Kitware: Clifton Park, NY, USA, 2010.

chemengineering

MDPI

Article

Numerical Simulations of Red-Blood Cells in Fluid Flow: A Discrete Multiphysics Study

Amin Rahmat [1,*], Philip Kuchel [2], Mostafa Barigou [1] and Alessio Alexiadis [1,*]

[1] School of Chemical Engineering, University of Birmingham, Birmingham B15 2TT, UK; m.barigou@bham.ac.uk
[2] School of Life and Environmental Sciences, University of Sydney, Sydney 2006, Australia; philip.kuchel@sydney.edu.au
* Correspondence: a.rahmat@bham.ac.uk (A.R.); a.alexiadis@bham.ac.uk (A.A.)

Abstract: In this paper, we present a methodological study of modelling red blood cells (RBCs) in shear-induced flows based on the discrete multiphysics (DMP) approach. The DMP is an alternative approach from traditional multiphysics based on meshless particle-based methods. The proposed technique has been successful in modelling multiphysics and multi-phase problems with large interfacial deformations such as those in biological systems. In this study, we present the proposed method and introduce an accurate geometrical representation of the RBC. The results were validated against available data in the literature. We further illustrate that the proposed method is capable of modelling the rupture of the RBC membrane with minimum computational difficulty.

Keywords: discrete multiphysics; the smoothed particle hydrodynamics (SPH) method; fluid–solid interactions (FSI); red-blood cells; numerical modelling; shear flow

check for **updates**

Citation: Rahmat, A.; Kuchel, P.; Barigou, M.; Alexiadis, A. Numerical Simulations of Red-Blood Cells in Fluid Flow: A Discrete Multiphysics Study. *ChemEng* **2021**, *5*, 33. https://doi.org/10.3390/chemengineering5030033

Academic Editor: Evangelos Tsotsas

Received: 4 May 2021
Accepted: 23 June 2021
Published: 30 June 2021

Publisher's Note: MDPI stays neutral with regard to jurisdictional claims in published maps and institutional affiliations.

1. Introduction

The human red blood cell (RBC; erythrocyte) is a homogeneous biconcave disk-like microparticle that is surrounded by a fluidic incompressible lipid bilayer membrane that is underpinned by a thin elastic cytoskeleton [1]. The membrane and its underlying cytoskeleton have been naturally selected to be sufficiently flexible to allow passage of the RBC through capillaries of 4 μm diameter in all tissues; this is approximately half of the cell's main diameter. The main physiological function of the RBC is the transfer of oxygen from the lungs to all tissues and the return of CO_2 to the lungs where it is exhaled. The RBC actively metabolises glucose that provides free energy from its bond cleavage to phosphorylate ADP to make ATP, the 'energy currency' of the cell. In the RBC, an ionic disequilibrium exists across the membrane with a much higher Na^+ concentration outside in the blood plasma, while inside the concentration of K^+ is relatively high. The steady-state of these concentrations is maintained by the membrane protein Na, K-ATPase that catalyses the hydrolysis of one molecule of ATP for three Na^+ ion ejected and two K^+ taken up. The detailed kinetics of RBC metabolism and ATP turnover are encapsulated in a comprehensive computer model [2–4].

Other functions of the RBC are mooted to involve the transfer of ATP and ADP agents inward and outward in the cell to release various molecules in blood flow through cell deformation in shear-induced strain fields [5]. These purines are proposed by some to be key effectors of arteriolar smooth muscle tone and hence affect peripheral resistance to blood flow and thence blood pressure. The dynamics and deformation of the RBC in shear flows are crucial in numerous biomedical applications including the detection of diseases associated with tissue perfusion, the experimental separation of healthy cells from infected parasitised ones, and studying human metabolism in vivo, and using nuclear magnetic resonance spectroscopy (MRS) [6]. The cell might also undergo membrane rupture

159

(haemolysis) under conditions of extreme deformation. To have a deep understanding of these cellular responses, the dynamics of the RBC require investigation.

Furthermore, mechanobiology is concerned with the relationship between cell shape and energy consumption. Experimental methods to study this interrelationship are few, but NMR spectroscopy of RBCs distorted in stretched or compressed hydrogels is one recent example in this field. Specifically, stretching human RBCs by a factor of two enhances their rate of glucose consumption by approximately the same factor [7]. Therefore, being able to model the extent of altered membrane geometry will be key to understanding the role of mechanical or flow induced shape changes of RBCs (and other cells) in whole body energy metabolism.

There are some notable examples of numerical modelling for investigating the complex fluid dynamics of the RBC in shear-induced flows. RBCs in this context have been modelled by many methods including the lattice Boltzmann method (LBM) [8–10] and the finite element (FE) method [11–13] in which the fluid–solid interface is captured by the immersed boundary method (IBM). Particle-based methods such as the smoothed particle hydrodynamics (SPH) [14–16], discrete particle dynamics (DPD) [17], and moving-particle semi-implicit (MPS) [18] methods have also been used for modelling RBC shape in shear induced fluid flow.

The Lagrangian feature of particle-based methods is one of its advantages by which fluid–solid interfaces such as the RBC membrane can be explicitly captured. In all these methods, it is challenging to represent the RBC shape accurately. It is this that undergoes large deformations while leading to no structural damage. The RBC displays in-plane area incompressibility as observed in experimental studies [19,20].

In general, there are two main approaches to representing the RBC membrane, namely shell-based, and spring-based models. In the shell-based model [21], the membrane is represented by a zero-thickness quasi-two-dimensional structure. This view provides a continuum representation of the membrane that can be coupled with several constitutive laws including the neo-Hookean [22–24] and Skalak [13,25,26] laws for evaluating in-plane tension and transverse shear. A detailed comparison and investigation of the effect of the requisite constitutive laws has been conducted by Lac et al. [27] for spherical capsules under shear-induced deformations. On the other hand, spring-based models e.g., the lattice-spring model (LSM) and mass-spring model (MSM) represent the membrane through a (structured) geometrical network of points and particles that are inter-connected by means of virtual springs. In the latter approach, the mechanical response of the membrane depends on the geometry of the structured network (i.e., triangular, tetrahedral, etc.) and it is specified through mathematical correlations that relate the spring constant of the connecting bonds with the overall elastic properties of the membrane. For specifying the in-plane area incompressibility of the membrane in mathematical correlations, we can use linear models that employ Hookean spring characteristics [28], or nonlinear correlations that include virtual dampers [29], or angular springs [30,31]. In addition to these two main approaches (LSM and MSM), other models have also been used for capturing the deformation of the RBC membrane; these include an area-difference-elasticity-theory [32,33]; linear finite element (LFE) method [34], and high fidelity simulations of RBCs [35,36].

As mentioned above, the modelling of RBC behaviour under conditions of flow might also include the sporadic occurrence of haemolysis (membrane rupture) under specific conditions. Taking into account that haemolysis might occur in extreme conditions, a comprehensive model should be able to accommodate all these possible scenarios. On the other hand, and to the best of our knowledge, all models thus far in the literature are not fully comprehensive in this regard. Therefore, this motivated our use of hybrid multiphysics numerical modelling, which is fully capable of capturing all possible scenarios of the RBCs under a shear field.

Therefore, we present a methodological study of modelling RBCs in shear-induced flows based on using what we call the discrete multiphysics (DMP) approach [37–39]. DMP is an alternative approach from traditional multiphysics by combining particle-based methods, e.g., the SPH for solving fluid flow and the MSM for capturing the elastic deformation of flexible solid materials, and it has been particularly successful in modelling biological systems [40–42]. The SPH method is a Lagrangian particle-based technique, which was developed for astrophysical studies [43,44]. The method was later extended for modelling fluid flow [45,46]. In SPH modelling the fluid domain is discretised into Lagrangian computational particles that possess fluid properties viz. density, viscosity, and internal energy, while interacting with their neighbouring particles in a way that is described mathematically by a kernel function [47]. The Lagrangian nature of the SPH method makes it suitable for simulating systems that undergo large deformations, such as occur in free surface flows [46,48], droplet and bubble dynamics [49–51], and fluid–solid interactions [52,53]. Within the DMP framework, the MSM was used to generate the RBC membrane through a linear mathematical correlation. In order to accommodate haemolysis in our model, we extended the DMP by setting a limited strain threshold beyond which the virtual bonds, that connect the membrane particles are dispelled, thus representing membrane rupture.

2. Methodology

2.1. General Equations

An incompressible Newtonian fluid in laminar flow is governed by:

$$\frac{D\rho}{Dt} = -\rho \mathbf{\nabla} \cdot \mathbf{u}, \tag{1}$$

$$\frac{D\mathbf{u}}{Dt} = \frac{-\mathbf{\nabla}p}{\rho} + \frac{1}{\rho}(\mathbf{\nabla} \cdot \mu \mathbf{\nabla}\mathbf{u}) + \mathbf{F}, \tag{2}$$

where $\mathbf{u}$ is the velocity vector, while ρ, μ, and p are the fluid properties density, kinematic viscosity, and pressure, respectively. $\frac{D}{Dt}$ is the material time derivative where t denotes time, and $\mathbf{F}$ denotes denotes the volumetric body force.

We note once again that the present numerical approach was based on the DMP model [38], which combines the SPH method for modelling fluid flow, and the MSM for representing the elastic properties of membrane particles. It is relevant to outline these approaches along with how their coupling has been achieved mathematically.

2.2. The SPH Method

The governing equations of motion i.e., Equations (1) and (2) are solved by the SPH method over computational particles using a kernel function, $W(\mathbf{r}_{ij}, h)$. The kernel function, which can be concisely represented by W_{ij}, relates particle i with its neighbouring particle j [46,54], based on their distance $\mathbf{r}_{ij} = |\mathbf{r}_i - \mathbf{r}_j|$, and the smoothing length, h as the horizon limit for the interactions with neighbouring particles. Figure 1 represents the interactions between two-dimensional SPH particles based on their relative proximity, through a smoothing kernel function. Among several kernel functions in the literature, the Lucy kernel function [44] was used in this study:

$$W_{ij} = \beta \begin{cases} (1 + 3q)(1 - q)^3 & \text{if } 0 \leq q \leq 1 \\ 0 & \text{if } q > 1, \end{cases} \tag{3}$$

where $q = (\mathbf{r}_{ij}/\kappa h)$ and $\beta = (105/16 \, \pi h^3)$ for three-dimensional simulations.

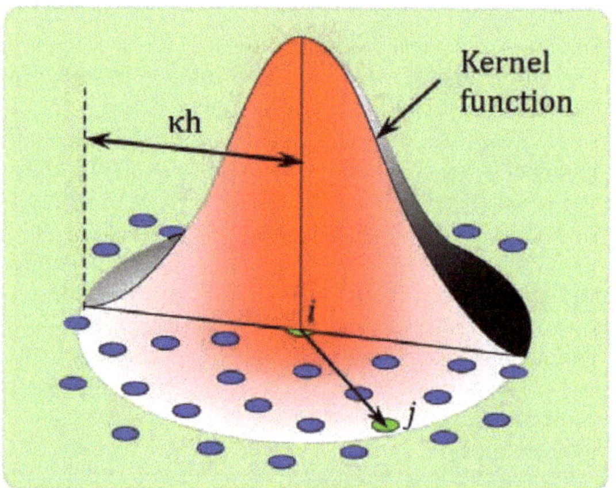

Figure 1. Schematic of the SPH method representing the interaction between particles through the kernel function.

In the SPH method, any field variable f can be approximated using the summation over discrete particles as:

$$f_i \simeq \sum_{j=1}^{J_i} \frac{m_j}{\rho_j} f_j W_{ij},\tag{4}$$

where m_j and J_i denote the mass and the number of computational neighboring particles for particle i. The SPH method discretizes the continuity and momentum equations, Equations (1) and (2) into the following form as:

$$\dot{\rho}_i = -\sum_{j=1}^{J_i} m_j \mathbf{u}_{ij} \frac{\partial W_{ij}}{\partial \mathbf{x}_i},\tag{5}$$

$$\mathbf{f}_i = \quad -\sum_{j=1}^{J_i} m_i m_j \left(\frac{p_i}{\rho_i^2} + \frac{p_j}{\rho_j^2} \right) \frac{\partial W_{ij}}{\partial \mathbf{x}_i} +$$

$$\sum_{j=1}^{J_i} \frac{m_i m_j (\mu_i + \mu_j)}{\rho_i \rho_j} \mathbf{u}_{ij} \frac{\partial W_{ij}}{\partial \mathbf{x}_i} + \mathbf{F}_i.\tag{6}$$

In Equation (6), the first term is the pressure gradient and the second is the dissipation term that conforms to a laminar viscosity model [55]. To evaluate the pressure term in the SPH method, one may solve a Poisson equation to obtain an incompressible solution [50,56,57] or might consider a weakly compressible SPH (WCSPH) approach [48,58,59], in which the pressure is obtained by an equation of state (EOS) using density variations. We used this approach by using the so-called Tait EOS [47,48]:

$$p = \frac{c_0^2 \rho_0}{\gamma} \left[\left(\frac{\rho}{\rho_0} \right)^\gamma - 1 \right].\tag{7}$$

where, c_0 is the reference speed of sound. In order to keep density variations below 1%, it is recommended that the speed of sound should be set at least one order of magnitude larger than the maximum velocity in the domain [45]. ρ_0 is the reference density (1000 kg m^{-3}), and γ is empirically set to be 7 in this power law equation [45,46].

2.3. The MSM

As mentioned, the MSM [60], which is sometimes referred to as coarse grained molecular dynamics (CGMD) [38], was employed to implement elasticity of the RBC membrane. In the MSM [61], a network of harmonic bonds connects computational particles, allowing them to deform and stretch according to the Newtonian equations of motion using linear spring bonds. Figure 2 shows the structure of the MSM for a quasi-two-dimensional RBC membrane. To account for Hookean elasticity, harmonic bonds are used as follows:

$$\mathbf{F}_{i,bond} = k_b\left(\mathbf{r}_{ij} - \mathbf{r}_0\right) \tag{8}$$

where k_b and r_0 are the coefficient and the equilibrium distance of Hookean springs, respectively. In this methodological approach, our emphasis is on the discretization and implementation of the model, so we adopted the linear Hookean model for the sake of simplicity and lower computational costs, while other nonlinear models can be easily replaced without further difficulties. Another advantage of the MSM is inherited in the modelling of membrane rupture, for which we nullified the springs once their length exceed the ultimate threshold set equal to 1.05 of their initial value.

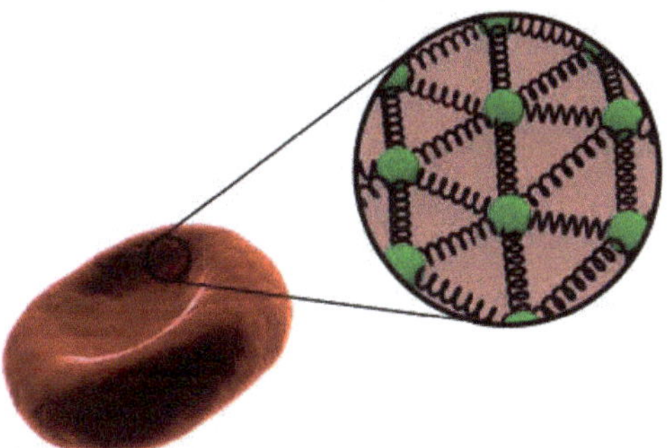

Figure 2. Schematic of the MSM representing the network of interconnected particles via spring network.

2.4. Coupling of SPH and MSM

To model the physical properties at the fluid–solid interface, we considered three types of boundary conditions to be taken into consideration [62], i.e., the no-penetration, the no-slip, and continuity of stresses at the interface. In continuum mechanics these conditions are often represented, respectively, as:

$$\left(\frac{\partial}{\partial t}\mathbf{u}_f - \mathbf{u}_s\right) \cdot \mathbf{n} = 0, \tag{9}$$

$$\left(\frac{\partial}{\partial t}\mathbf{u}_f - \mathbf{u}_s\right) \times \mathbf{n} = 0, \tag{10}$$

and

$$\sigma_s\mathbf{n} = \sigma_f(-\mathbf{n}), \tag{11}$$

where $\mathbf{n}$ represents the unit normal vector. $\mathbf{u}_s$ and $\mathbf{u}_f$ denote the solid displacement and fluid velocity, respectively, while the solid and fluid stresses are represented by σ_s and σ_f. Within the DMP framework, we considered ghost SPH particles, which are assigned to MSM particles at the fluid–solid interface. To satisfy all three boundary conditions, these particles have dual identity; they interact with the SPH fluid particles for solving fluid properties but they interact with other MSM particles as membrane particles.

2.5. Numerical Algorithm

The velocity verlet (VV) algorithm was used to integrate over time by using a first-order Euler approach with a variable time step, specified by the following stability condition, $\Delta t = \zeta h^2 / v$. Here, v represents the dynamic viscosity equal to $v = \rho / \mu$ and $\zeta = 0.125$ provided satisfactory results [38]. Considering that $(*)$ and (n) superscripts denote variables at the intermediate and nth time step, the particle velocities were calculated using the VV algorithm as:

$$\mathbf{u}_i^* = \mathbf{u}_i^{(n)} + \frac{\Delta t}{2m_i}\mathbf{f}_i^{(n)}. \tag{12}$$

Then, particle densities were calculated according to:

$$\rho_i^* = \rho_i^{(n)} + \frac{\Delta t}{2}\dot\rho_i^{(n)}. \tag{13}$$

In the above equation, $\dot\rho$ is the rate of density variations according to Equation (5), in which the density should be updated based on the velocity difference between the particles $\mathbf{u}_{ij}$. However, using $\mathbf{u}_{ij}$ causes numerical instabilities arising from the lag of the velocity in the VV algorithm. So, we use an extrapolated velocity in the calculation of the rate of density variations, which can be represented as:

$$\overline{\mathbf{u}}_i = \mathbf{u}_i^{(n)} + \frac{\Delta t}{m_i}\mathbf{f}_i^{(n)}, \tag{14}$$

where $\mathbf{u}_{ij} = (\overline{\mathbf{u}}_i - \overline{\mathbf{u}}_j)$. Once particle velocities are updated, computational particles are moved to their new positions using:

$$\mathbf{x}_i^{(n+1)} = \mathbf{x}_i^{(n)} + \Delta t\mathbf{u}_i^*. \tag{15}$$

Then, $\dot\rho_i^{(n+1)}$, and $\mathbf{f}_i^{(n+1)}$ were calculated at the $(n+1)$ time step using Equations (5) and (6), respectively. Finally, the velocity and density were calculated at the $(n+1)$ time step, respectively, as:

$$\mathbf{u}_i^{(n+1)} = \mathbf{u}_i^* + \frac{\Delta t}{2m_i}\mathbf{f}_i^{(n+1)}, \tag{16}$$

and,

$$\rho_i^{(n+1)} = \rho_i^* + \frac{\Delta t}{2}\dot\rho_i^{(n+1)}. \tag{17}$$

We have extensively used the above-mentioned numerical algorithm in our previous studies [28,63,64] where we can refer interested readers to for more detailed discussions.

3. Problem Set-Up

The literature contains several models of the geometry of the RBC, including the frequently cited one by Evans et al. [25]. In this study, we used a model which we consider represents the dimples of the biconcave disc more accurately. The model consists of a set of three parametric equations that involve Jacobi elliptic functions [65] as:

$$x = A\frac{\mathcal{CN}(\mu, m)\,\mathcal{CN}(v_1, m_p)}{g}\cos(\psi), \tag{18}$$

$$y = A\frac{\mathcal{CN}(\mu, m)\,\mathcal{CN}(v_1, m_p)}{g}\sin(\psi), \tag{19}$$

and

$$z = A\frac{\mathcal{SN}(\mu, m)\,\mathcal{DN}(\mu', m)\,\mathcal{SN}(v_1, m_p)\,\mathcal{DN}(v_1, m_p)}{g}, \tag{20}$$

where, $\mathcal{SN}$, $\mathcal{CN}$, and $\mathcal{DN}$ are the Jacobi elliptic functions. The coefficient A was defined by using three geometrical distances obtained from experimental electron microscope measurements.

$$A = \frac{1}{2}\sqrt{d^2 - 2h^2 - 2\sqrt{h^2(-b^2 + h^2)}\cos(\psi)}, \qquad (21)$$

This was used to define other intermediate parameters. Also, $g = 1 - (\mathcal{DN}(\mu, m)^2 \mathcal{SN}(v_1, m_p)^2)$ and,

$$m = 1 - m_p = k^2 = 2d\frac{\sqrt{(A^2(-4A^2 - b^2 + d^2))/d^2)}}{\sqrt{(4A^2) + b^2}\sqrt{(-4A^2) + d^2}^2}. \qquad (22)$$

The key specific lengths of the RBC are the cell diameter at the plane splitting the biconcave into two cup-shaped halves $d = 8 \times 10^{-6}$ m, the thickness of the cell at the thickest position (like the width of the tyre on a car wheel) $h = 2.12 \times 10^{-6}$ m, and the thickness at the centre of the dimple, $b = 1 \times 10^{-6}$ m. Considering these parameters, the surface area and volume of the RBC are found to be equal to $S = 85.8$ fL and $V = 128$ μm^2 respectively [65]. Figure 3 shows the shape of the RBC in a two dimensional projection of the three-dimensional body.

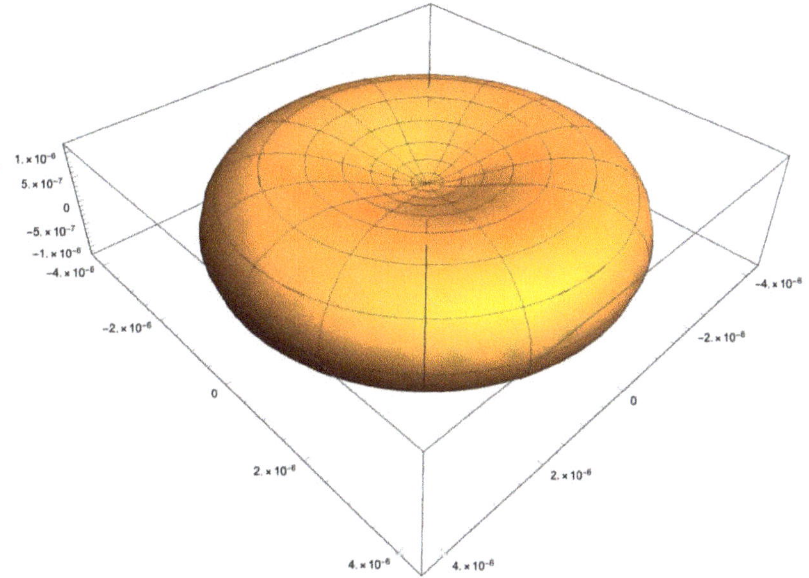

Figure 3. A representative illustration of an RBC.

In the present analysis the RBC was represented by a hybrid model, as described in Section 2 by which the SPH was used for modelling the fluid flow, and the RBC membrane was simulated by the MSM. In order to apply the surface area incompressibility of the membrane (conservation of area), the spring constants were taken to be sufficiently large so that a change in the net surface area was minimized. The spring constant could not be taken to be very large since large spring constants resulted in numerical instabilities. Here, the spring constant was set to $k_b = 0.0005$ N/m. The shear modulus of the RBC was tuned by adjusting the angular spring constant. It has been seen that the angular constant of $k_a = 1 \times 10^{-20}$ Nm/rad led to convincing simulated behaviours in shear fields.

Figure 4 represents the 3D simulation domain with height, width, and depth of H, W, and D, respectively. The top and bottom walls abide the no-slip boundary conditions, while periodic boundary conditions were implemented on the rest of boundaries. To maintain a constant shear rate $\dot{\gamma}$, the top and bottom boundaries were given with velocity U and $-U$, respectively. The RBC was placed at the centre, equidistant from all domain boundaries. A uniformly-spaced Cartesian grid was used to populate the domain with fluid particles. All in all, there were four different types of particles in this model: (1) particles of the top boundary (white); (2) particles of the bottom boundary (red); (3) the RBC particles (blue); (4) the fluid particles (green), as illustrated in Figure 4. In total, the simulation domain contained more than one million computational particles comprising all four particle types. It should be noted that a particle-resolution study was carried out for four different particle resolutions in a shear flow [28] in our previous study, and it was observed that the results and hydrodynamics did not improve significantly beyond the resolution that is equivalent to the one we used in this study. The test-case was simulated on the UK's national super computing facilities (ARCHER with 2.7 GHz, 12-core Intel Xeon E5-2697). The computational cost was 23 h of wall-time on approximately 1500 CPU cores (MPI) for each test-case.

The domain undergoes shear in the x-direction by moving the top and bottom boundary particles in opposite directions. Subsequently, the shear imposed RBC deformation are described and analysed in the next section.

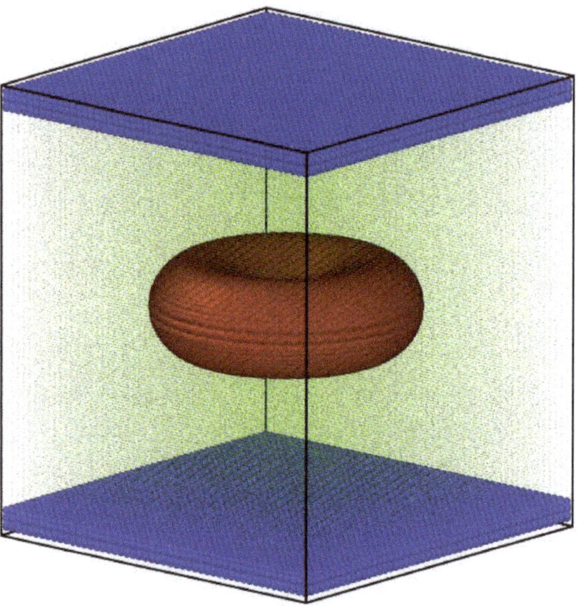

Figure 4. Computational domain for cell under shear rate; the RBC is illustrated by red particles, the top and bottom boundaries are blue, and the fluid particles are green (different particle size, particularly for green and red particles, is a post-processing trick for better illustration and has no further modelling or physical interpretations).

4. Results

In this section, we show that the presented model was capable of capturing the dynamics of the RBC with superior features, relative to other models reported in the literature. Thus, we show how the current model could predict membrane rupture that was induced by extreme flow conditions.

In order to characterise the deformation of the RBC, a quantitative measure was introduced. Specifically, this was the deformation index, $D_{12} = |\mathcal{L} - \mathcal{H}|/(\mathcal{L} + \mathcal{H})$, where $\mathcal{L}$ and $\mathcal{H}$ were the length of the RBC in the plane passing through its origin in the direction of

the main axis, and its transverse direction, respectively. Figure 5 represents the variation of
the deformation index for three capillary numbers with respect to dimensionless simulation
time and compares them with the numerical data in the literature [13]. It was shown that
the present model produced accurate results with respect to the models in the literature
and it captured the frequency of deformation with the sinusoidal variations of flow rate.
Any discrepancy in the results was due to the different geometries used to model a RBC;
these produced a larger amplitude in the sinusoidal variation of the deformation index [13].

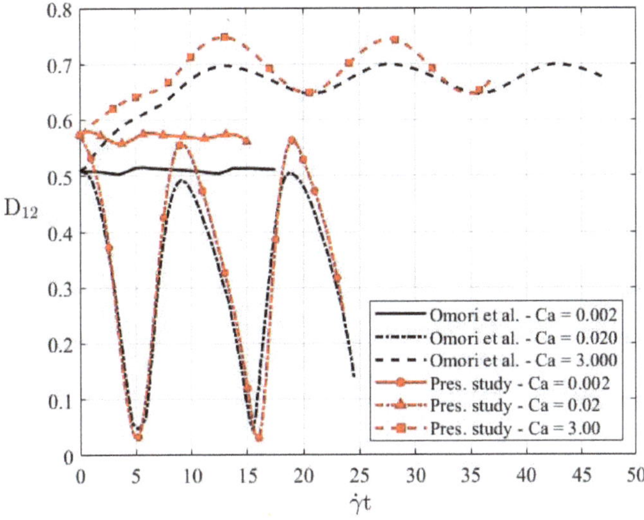

Figure 5. Validation of the present RBC model with those of Omori et al. [13].

Considering the variations of the deformation index shown in Figure 6, it was interesting to note that the RBC underwent a tumbling motion, and it rotated around its centre of
mass when the shear rates were symmetrical. This is the reason why there was a sinusoidal
trend in the deformation index. At small capillary numbers, the RBC did not deform
very much, and accordingly the deformation index exhibited minimal variation over time.
The variation of the deformation index was restricted between $0.56 \leq D_{12} \leq 0.58$ for the
smallest capillary number. At moderate capillary numbers i.e., $Ca = 0.02$, the deformation
index indicated sinusoidal behaviour of large amplitude. This was interpreted to be due to
the resistance of the RBC against the applied shear. When the applied shear met the RBC
when it was oriented such that its main diameter was along the flow direction, the RBC
geometry was similar to its undisturbed shape. This was the point in time when the
deformation index was at its peak value. On the other hand, when the main diameter of the
RBC was transverse to the flow direction, the deformation index decreased considerably.
For large capillary numbers, the shear rate was sufficiently high to stretch the RBC along
the direction of shear. The variation of the deformation index with such high capillary
numbers was also due to the re-orientation of the RBC during its tumbling motion.

Figure 6 represents the RBC shape at different simulation times for three capillary
numbers corresponding to Figure 6. Our numerical model was further compared with
data in the literature [13] and it showed that the shape of the RBC was accurately captured.
The shape of the RBC remained almost undisturbed during its tumbling motion. For moderate values of capillary number, the RBC was similar in shape to its undisturbed form
whereby its major diameter was along the direction of the applied shear. For large capillary
numbers, the RBC was stretched along it direction of shear.

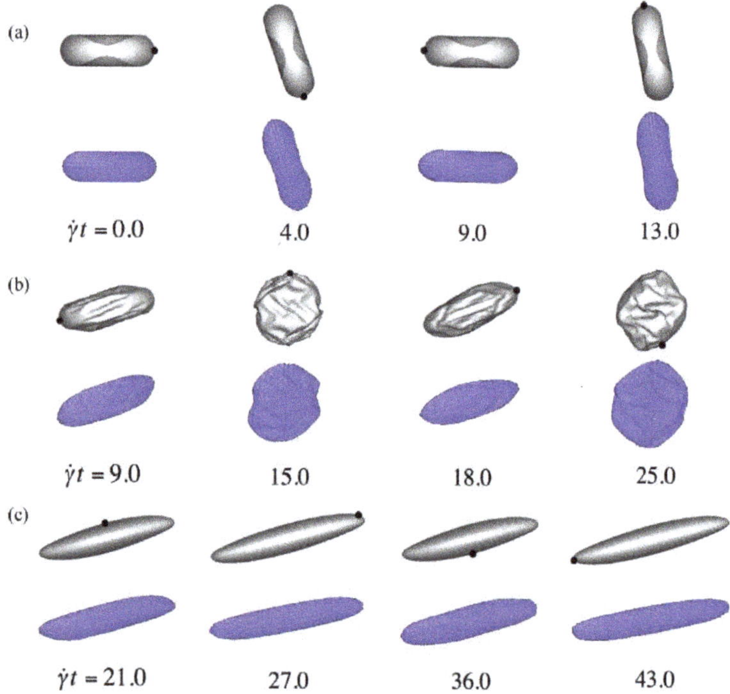

Figure 6. Validation of the present RBC model with those of Omori et al. [13] for three test-cases (**a**) Ca = 0.002, (**b**) Ca = 0.02, and (**c**) Ca = 3.0 at different dimensionless times.

In many applications, once RBCs are exposed to high shear rates, the membranes of unhealthy cells might rupture. To simulate this biomedical eventuality is difficult but our method is one that can model the responses under shear of healthy cells and unhealthy or parasite-infected ones. On the other hand, current numerical techniques generally have systemic problems for modelling the membrane rupture of RBCs under shear fields. In summary, we present in Figure 7 membrane rupture of a RBC under flow-induced shear. It was clear that rupture could occur at relatively low shear rates in contrast to when the RBC was exposed to extreme shear such as those in Figure 5.

The modelling is a numerical depiction of RBCs under shear and serves to illustrate its capabilities for applications to particles of other geometries and elastic properties. In other words, in order to simulate realistic membrane rupture, careful experimental analysis and membrane characterisation were required to determine the threshold strain that must be in the numerical model. These characterisations lie in the realm of cellular biophysics of RBCs (and other cell types) that will be developed as we refine the current numerical modelling to help interpret experimental data.

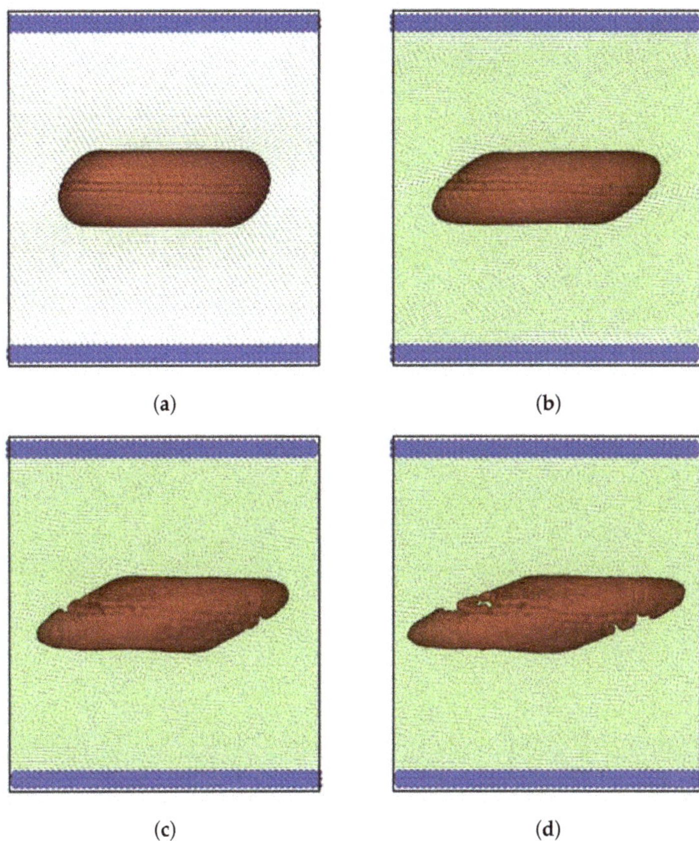

Figure 7. Rupture of the membrane under shear (Ca = 2.0) at different dimensionless times (**a**) $\dot{\gamma}t = 0$, (**b**) $\dot{\gamma}t = 2$, (**c**) $\dot{\gamma}t = 4$, and (**d**) $\dot{\gamma}t = 6$.

5. Conclusions

In this study, we employed a novel numerical technique using the DMP model for simulating the deformation of RBCs in shear-induced flows. This methodological study describes a simple but accurate and flexible numerical technique for simulating a myriad of problems involving RBCs in in vivo and in vitro applications. In this study, we discussed the methodology in details from the mathematical, numerical, and technical perspectives, describing the governing equations, numerical algorithm, and implementation techniques.

The model has a Lagrangian particle-based nature and utilised a complex mathematical formulation derived from the Jacobi elliptic functions, which provided an accurate representation of the RBC geometry. Then, the proposed mathematical formulation was linked to the MSM, which modelled the RBC through a complex network of interconnected particles on its interface. The proposed model is employed for modelling RBC deformation under shear at various capillary numbers. It was observed that RBCs represented a sinusoidal behaviour in their deformation index factor which is accurately comparable with available data in the literature.

We have further extended the model to simulate the rupture of the RBC's membrane under extremely large shear rates, which could be used in many biological and pharmaceutical applications. The membrane rupture was implemented via breaking the spring bonds between particles of which their deformation exceeded a certain strain threshold. Another advantage of the proposed method is that it can be easily extended to other complex physics, and applied to other real-world problems.

Author Contributions: Conceptualization, P.K. and A.A.; methodology, A.R., P.K. and A.A.; software, A.R. and A.A.; validation, A.R.; formal analysis, A.R.; investigation, A.R.; resources, M.B. and A.A.; writing—original draft preparation, A.R.; writing—review and editing, P.K. and A.A.; visualization, A.R.; supervision, A.A.; project administration, A.A.; funding acquisition, M.B. and A.A. All authors have read and agreed to the published version of the manuscript.

Funding: This work was supported by the UK Engineering and Physical Sciences Research Council (EPSRC) under Grant No. EP/N033698/1. Philip Kuchel acknowledges an Australian Research Council Discovery Project Grant, DP190100510. Amin Rahmat acknowledges a financial support from the Birmingham International Engagement Fund (BIEF).

Institutional Review Board Statement: Not applicable.

Informed Consent Statement: Not applicable.

Data Availability Statement: Not applicable.

Conflicts of Interest: The authors declare no conflict of interest.

Abbreviations

CGMD: coarse grained molecular dynamics; DMP: discrete multi-physics; DPD: discrete particle dynamics; EOS: equation of state; FE: finite element; IBM: immersed boundary method; ISPH: incompressible smoothed particle hydrodynamics; LBM: lattice Boltzmann method; LFE: linear finite element; LSM: lattice spring model; MSM: mass spring model; MPS: moving-particle semi-implicit; SPH: smoothed particle hydrodynamics; VV: velocity verlet; WCSPH: weakly compressible smoothed particle hydrodynamics.

References

1. Mohandas, N.; Evans, E. Mechanical properties of the red cell membrane in relation to molecular structure and genetic defects. *Annu. Rev. Biophys. Biomol. Struct.* **1994**, *23*, 787–818. [CrossRef] [PubMed]
2. Mulquiney, P.J.; Bubb, W.A.; Kuchel, P.W. Model of 2, 3-bisphosphoglycerate metabolism in the human erythrocyte based on detailed enzyme kinetic Equations (1): In vivo kinetic characterization of 2, 3-bisphosphoglycerate synthase/phosphatase using 13C and 31P NMR. *Biochem. J.* **1999**, *342*, 567–580. [CrossRef]
3. Mulquiney, P.J.; Kuchel, P.W. Model of 2, 3-bisphosphoglycerate metabolism in the human erythrocyte based on detailed enzyme kinetic equations: Equations and parameter refinement. *Biochem. J.* **1999**, *342*, 581–596. [CrossRef] [PubMed]
4. Mulquiney, P.J.; Kuchel, P.W. Model of 2, 3-bisphosphoglycerate metabolism in the human erythrocyte based on detailed enzyme kinetic equations: Computer simulation and metabolic control analysis. *Biochem. J.* **1999**, *342*, 597–604. [CrossRef]
5. Wan, J.; Ristenpart, W.D.; Stone, H.A. Dynamics of shear-induced ATP release from red blood cells. *Proc. Natl. Acad. Sci. USA* **2008**, *105*, 16432–16437. [CrossRef] [PubMed]
6. Shishmarev, D.; Kuchel, P.W.; Pagès, G.; Wright, A.J.; Hesketh, R.L.; Kreis, F.; Brindle, K.M. Glyoxalase activity in human erythrocytes and mouse lymphoma, liver and brain probed with hyperpolarized 13 C-methylglyoxal. *Commun. Biol.* **2018**, *1*, 1–8. [CrossRef] [PubMed]
7. Kuchel, P.W.; Shishmarev, D. Accelerating metabolism and transmembrane cation flux by distorting red blood cells. *Sci. Adv.* **2017**, *3*, eaao1016. [CrossRef] [PubMed]
8. Dupin, M.; Halliday, I.; Care, C.; Munn, L. Lattice Boltzmann modelling of blood cell dynamics. *Int. J. Comput. Fluid Dyn.* **2008**, *22*, 481–492. [CrossRef]
9. Shi, X.; Lin, G.; Zou, J.; Fedosov, D.A. A lattice Boltzmann fictitious domain method for modeling red blood cell deformation and multiple-cell hydrodynamic interactions in flow. *Int. J. Numer. Methods Fluids* **2013**, *72*, 895–911. [CrossRef]
10. Zhang, J. Lattice Boltzmann method for microfluidics: Models and applications. *Microfluid. Nanofluid.* **2011**, *10*, 1–28. [CrossRef]
11. Chen, M.; Boyle, F.J. Investigation of membrane mechanics using spring networks: Application to red-blood-cell modelling. *Mater. Sci. Eng. C* **2014**, *43*, 506–516. [CrossRef] [PubMed]
12. Eggleton, C.D.; Popel, A.S. Large deformation of red blood cell ghosts in a simple shear flow. *Phys. Fluids* **1998**, *10*, 1834–1845. [CrossRef] [PubMed]
13. Omori, T.; Ishikawa, T.; Barthès-Biesel, D.; Salsac, A.V.; Imai, Y.; Yamaguchi, T. Tension of red blood cell membrane in simple shear flow. *Phys. Rev. E* **2012**, *86*, 056321. [CrossRef]
14. Hosseini, S.M.; Feng, J.J. A particle-based model for the transport of erythrocytes in capillaries. *Chem. Eng. Sci.* **2009**, *64*, 4488–4497. [CrossRef]

15. Freund, J.B. Numerical simulation of flowing blood cells. *Annu. Rev. Fluid Mech.* **2014**, *46*, 67–95. [CrossRef]
16. Wu, T.; Feng, J.J. Simulation of malaria-infected red blood cells in microfluidic channels: Passage and blockage. *Biomicrofluidics* **2013**, *7*, 044115. [CrossRef]
17. Boryczko, K.; Dzwinel, W.; Yuen, D.A. Dynamical clustering of red blood cells in capillary vessels. *J. Mol. Model.* **2003**, *9*, 16–33. [CrossRef] [PubMed]
18. Tsubota, K.; Wada, S.; Yamaguchi, T. Particle method for computer simulation of red blood cell motion in blood flow. *Comput. Methods Programs Biomed.* **2006**, *83*, 139–146. [CrossRef]
19. Hochmuth, R.; Mohandas, N. Uniaxial loading of the red-cell membrane. *J. Biomech.* **1972**, *5*, 501–509. [CrossRef]
20. Hochmuth, R.; Wiles, H.; Evans, E.; McCown, J. Extensional flow of erythrocyte membrane from cell body to elastic tether. II. Experiment. *Biophys. J.* **1982**, *39*, 83–89. [CrossRef]
21. Pozrikidis, C. Effect of membrane bending stiffness on the deformation of capsules in simple shear flow. *J. Fluid Mech.* **2001**, *440*, 269–291. [CrossRef]
22. Zhang, J.; Johnson, P.C.; Popel, A.S. An immersed boundary lattice Boltzmann approach to simulate deformable liquid capsules and its application to microscopic blood flows. *Phys. Biol.* **2007**, *4*, 285. [CrossRef] [PubMed]
23. Bagchi, P. Mesoscale simulation of blood flow in small vessels. *Biophys. J.* **2007**, *92*, 1858–1877. [CrossRef] [PubMed]
24. Barthes-Biesel, D.; Diaz, A.; Dhenin, E. Effect of constitutive laws for two-dimensional membranes on flow-induced capsule deformation. *J. Fluid Mech.* **2002**, *460*, 211–222. [CrossRef]
25. Evans, E.A.; Skalak, R. *Mechanics and Thermodynamics of Biomembranes*; CRC Press: Boca Raton, FL, USA, 1980.
26. Sui, Y.; Chew, Y.; Roy, P.; Cheng, Y.; Low, H. Dynamic motion of red blood cells in simple shear flow. *Phys. Fluids* **2008**, *20*, 112106. [CrossRef]
27. Lac, E.; Barthes-Biesel, D.; Pelekasis, N.; Tsamopoulos, J. Spherical capsules in three-dimensional unbounded Stokes flows: Effect of the membrane constitutive law and onset of buckling. *J. Fluid Mech.* **2004**, *516*, 303–334. [CrossRef]
28. Rahmat, A.; Barigou, M.; Alexiadis, A. Deformation and rupture of compound cells under shear: A discrete multiphysics study. *Phys. Fluids* **2019**, *31*, 051903. [CrossRef]
29. Secomb, T.W.; Styp-Rekowska, B.; Pries, A.R. Two-dimensional simulation of red blood cell deformation and lateral migration in microvessels. *Ann. Biomed. Eng.* **2007**, *35*, 755–765. [CrossRef]
30. Fedosov, D.A.; Caswell, B.; Karniadakis, G.E. A multiscale red blood cell model with accurate mechanics, rheology, and dynamics. *Biophys. J.* **2010**, *98*, 2215–2225. [CrossRef]
31. Fedosov, D.A.; Peltomäki, M.; Gompper, G. Deformation and dynamics of red blood cells in flow through cylindrical microchannels. *Soft Matter* **2014**, *10*, 4258–4267. [CrossRef] [PubMed]
32. Ziherl, P.; Svetina, S. Nonaxisymmetric phospholipid vesicles: Rackets, boomerangs, and starfish. *EPL Europhys. Lett.* **2005**, *70*, 690. [CrossRef]
33. Svetina, S.; Ziherl, P. Morphology of small aggregates of red blood cells. *Bioelectrochemistry* **2008**, *73*, 84–91. [CrossRef] [PubMed]
34. MacMeccan, R.M.; Clausen, J.; Neitzel, G.; Aidun, C. Simulating deformable particle suspensions using a coupled lattice-Boltzmann and finite-element method. *J. Fluid Mech.* **2009**, *618*, 13. [CrossRef]
35. Janoschek, F.; Toschi, F.; Harting, J. Simplified particulate model for coarse-grained hemodynamics simulations. *Phys. Rev. E* **2010**, *82*, 056710. [CrossRef] [PubMed]
36. Melchionna, S. A Model for Red Blood Cells in Simulations of Large-scale Blood Flows. *Macromol. Theory Simul.* **2011**, *20*, 548–561. [CrossRef]
37. Alexiadis, A. A smoothed particle hydrodynamics and coarse-grained molecular dynamics hybrid technique for modelling elastic particles and breakable capsules under various flow conditions. *Int. J. Numer. Methods Eng.* **2014**, *100*, 713–719. [CrossRef]
38. Alexiadis, A. The discrete multi-hybrid system for the simulation of solid-liquid flows. *PLoS ONE* **2015**, *10*, e0124678. [CrossRef] [PubMed]
39. Alexiadis, A. A new framework for modelling the dynamics and the breakage of capsules, vesicles and cells in fluid flow. In Proceedings of the Iutam Symposium on Dynamics of Capsules, Vesicles and Cells in Flow, Rio de Janeiro, Brazil, 8–12 September 2015; Barthes Biesel, D., Blyth, M.G., Salsac, A.V., Eds.; Elsevier: Amsterdam, The Netherlands, 2015; pp. 80–88.
40. Ariane, M.; Vigolo, D.; Brill, A.; Nash, F.; Barigou, M.; Alexiadis, A. Using Discrete Multi-Physics for studying the dynamics of emboli in flexible venous valves. *Comput. Fluids* **2018**, *166*, 57–63. [CrossRef]
41. Mohammed, A.M.; Ariane, M.; Alexiadis, A. Using Discrete Multiphysics Modelling to Assess the Effect of Calcification on Hemodynamic and Mechanical Deformation of Aortic Valve. *ChemEngineering* **2020**, *4*, 48. [CrossRef]
42. Schütt, M.; Stamatopoulos, K.; Simmons, M.; Batchelor, H.; Alexiadis, A. Modelling and simulation of the hydrodynamics and mixing profiles in the human proximal colon using Discrete Multiphysics. *Comput. Biol. Med.* **2020**, *121*, 103819. [CrossRef]
43. Gingold, R.A.; Monaghan, J.J. Smoothed Particle Hydrodynamics: Theory and application to non-spherical stars. *Mon. Not. R. Astron. Soc.* **1977**, *181*, 375–389. [CrossRef]
44. Lucy, L.B. A numerical approach to the testing of the fission hypothesis. *Astron. J.* **1977**, *82*, 1013–1024. [CrossRef]
45. Monaghan, J.J. Simulating free surface flows with SPH. *J. Comput. Phys.* **1994**, *110*, 399–406. [CrossRef]
46. Monaghan, J.J.; Kocharyan, A. SPH simulation of multi-phase flow. *Comput. Phys. Commun.* **1995**, *87*, 225–235. [CrossRef]
47. Monaghan, J.; Kos, A. Solitary waves on a Cretan beach. *J. Waterw. Port. Coast. Ocean Eng.* **1999**, *125*, 145–155. [CrossRef]

48. Ozbulut, M.; Tofighi, N.; Goren, O.; Yildiz, M. Investigation of Wave Characteristics in Oscillatory Motion of Partially Filled Rectangular Tanks. *J. Fluids Eng.* **2018**, *140*, 041204. [CrossRef]
49. Rahmat, A.; Tofighi, N.; Yildiz, M. Numerical simulation of the electrohydrodynamic effects on bubble rising using the SPH method. *Int. J. Heat Fluid Flow* **2016**, *62*, 313–323. [CrossRef]
50. Rahmat, A.; Tofighi, N.; Shadloo, M.; Yildiz, M. Numerical simulation of wall bounded and electrically excited Rayleigh-Taylor Instability using incompressible Smoothed Particle Hydrodynamics. *Colloids Surf. A Physicochem. Eng. Asp.* **2014**, *460*, 60–70. [CrossRef]
51. Shadloo, M.; Rahmat, A.; Yildiz, M. A Smoothed Particle Hydrodynamics study on the electrohydrodynamic deformation of a droplet suspended in a neutrally buoyant Newtonian fluid. *Comput. Mech.* **2013**, *52*, 693–707. [CrossRef]
52. Rahmat, A.; Barigou, M.; Alexiadis, A. Numerical simulation of dissolution of solid particles in fluid flow using the SPH method. *Int. J. Numer. Methods Heat Fluid Flow* **2019**. [CrossRef]
53. Rahmat, A.; Nasiri, H.; Goodarzi, M.; Heidaryan, E. Numerical investigation of anguilliform locomotion by the SPH method. *Int. J. Numer. Methods Heat Fluid Flow* **2019**. [CrossRef]
54. Monaghan, J.J.; Lattanzio, J.C. A refined particle method for astrophysical problems. *Astron. Astrophys.* **1985**, *149*, 135–143.
55. Morris, J.P.; Fox, P.J.; Zhu, Y. Modeling low Reynolds number incompressible flows using SPH. *J. Comput. Phys.* **1997**, *136*, 214–226. [CrossRef]
56. Tofighi, N.; Ozbulut, M.; Rahmat, A.; Feng, J.; Yildiz, M. An incompressible Smoothed Particle Hydrodynamics method for the motion of rigid bodies in fluids. *J. Comput. Phys.* **2015**, *297*, 207–220. [CrossRef]
57. Morris, J.P. Simulating surface tension with Smoothed Particle Hydrodynamics. *Int. J. Numer. Methods Fluids* **2000**, *33*, 333–353. [CrossRef]
58. Fatehi, R.; Rahmat, A.; Tofighi, N.; Yildiz, M.; Shadloo, M. Density-Based Smoothed Particle Hydrodynamics Methods for Incompressible Flows. *Comput. Fluids* **2019**, *185*, 22–33. [CrossRef]
59. Hopp-Hirschler, M.; Shadloo, M.S.; Nieken, U. A Smoothed Particle Hydrodynamics approach for thermo-capillary flows. *Comput. Fluids* **2018**, *176*, 1–19. [CrossRef]
60. Kilimnik, A.; Mao, W.; Alexeev, A. Inertial migration of deformable capsules in channel flow. *Phys. Fluids* **2011**, *23*, 123302. [CrossRef]
61. Lloyd, B.; Székely, G.; Harders, M. Identification of spring parameters for deformable object simulation. *IEEE Trans. Vis. Comput. Graph.* **2007**, *13*. [CrossRef]
62. Esmon, C.T. Basic mechanisms and pathogenesis of venous thrombosis. *Blood Rev.* **2009**, *23*, 225–229. [CrossRef]
63. Rahmat, A.; Meng, J.; Emerson, D.; Wu, C.Y.; Barigou, M.; Alexiadis, A. A practical approach for extracting mechanical properties of microcapsules using a hybrid numerical model. *Microfluid. Nanofluid.* **2021**, *25*, 1–17. [CrossRef]
64. Rahmat, A.; Weston, D.; Madden, D.; Usher, S.; Barigou, M.; Alexiadis, A. Modeling the agglomeration of settling particles in a dewatering process. *Phys. Fluids* **2020**, *32*, 123314. [CrossRef]
65. Kuchel, P.W.; Fackerell, E.D. Parametric-equation representation of biconcave erythrocytes. *Bull. Math. Biol.* **1999**, *61*, 209–220. [CrossRef] [PubMed]

chemengineering

MDPI

Article

Modelling Particle Agglomeration on through Elastic Valves under Flow

Hosam Alden Baksamawi [1,*], Mostapha Ariane [2], Alexander Brill [3], Daniele Vigolo [1,4,5] and Alessio Alexiadis [1,*]

[1] School of Chemical Engineering, University of Birmingham, Birmingham B15 2TT, UK; d.vigolo@bham.ac.uk
[2] Laboratoire Interdisciplinaire Carnot de Bourgogne, Université Bourgogne, 9 Avenue Alain Savary, BP 47 870, 21078 Dijon, France; Mostapha.Ariane@u-bourgogne.fr
[3] Institute of Cardiovascular Sciences, College of Medical and Dental Sciences, University of Birmingham, Birmingham B15 2TT, UK; A.Brill@bham.ac.uk
[4] School of Biomedical Engineering, The University of Sydney, Sydney, NSW 2006, Australia
[5] The University of Sydney Nano Institute, University of Sydney, Sydney, NSW 2006, Australia
* Correspondence: hxb799@bham.ac.uk (H.A.B.); a.alexiadis@bham.ac.uk (A.A.)

Abstract: This work proposes a model of particle agglomeration in elastic valves replicating the geometry and the fluid dynamics of a venous valve. The fluid dynamics is simulated with Smooth Particle Hydrodynamics, the elastic leaflets of the valve with the Lattice Spring Model, while agglomeration is modelled with a 4-2 Lennard-Jones potential. All the models are combined together within a single Discrete Multiphysics framework. The results show that particle agglomeration occurs near the leaflets, supporting the hypothesis, proposed in previous experimental work, that clot formation in deep venous thrombosis is driven by the fluid dynamics in the valve.

Keywords: Deep Vein Thrombosis (DVT); computer simulation; Discrete Multiphysics (DMP); Smoothed Particle Hydrodynamics (SPH); Lattice Spring Model (LSM); solid-solid interaction; agglomeration and venous valves

check for
updates

Citation: Baksamawi, H.A.; Ariane, M.; Brill, A.; Vigolo, D.; Alexiadis, A. Modelling Particle Agglomeration on through Elastic Valves under Flow. *ChemEng* **2021**, *5*, 40. https://doi.org/10.3390/chemengineering5030040

Academic Editor: Timothy Hunter

Received: 31 May 2021
Accepted: 19 July 2021
Published: 26 July 2021

Publisher's Note: MDPI stays neutral with regard to jurisdictional claims in published maps and institutional affiliations.

1. Introduction

Various 'non-return' valves are found in our leg veins [1]. These valves consist of two elastic leaflets that open and close in conjunction with the musculoskeletal system. When we are physically active, the muscles in the leg constantly contract and relax, causing the vein valves to open, allowing blood to return to the heart, and close to avoid blood flowing back in the opposite direction [2,3].

Deep Vein Thrombosis (DVT) occurs when a thrombus forms in the veins as an aggregation of blood components [4]. One hypothesis suggests that these thrombi initially form in the venous valves [5], and subsequently detach from the veins and travel within the blood flow until they reach the pulmonary vascular system. Here, they cause blockage of the pulmonary artery branches, resulting in death or significant disabilities [6–9]. Moreover, it is recognised that the lack of physical activity or long static position causes poor blood circulation, thus increasing the risk of DVT. This suggests that fluid dynamics in the valve play an important role in causing DVT [10].

To understand the flow in venous valves, we carried out a computer simulation of various valve typologies both without [11] and with the presence of thrombi [12]. However, the role of the flow in the initiation of the venous clot due to the aggregation of blood components is still not clear. For this reason, in a previous study conducted by Schofield et al. [10], we developed an in vitro model of DVT. The model comprises a microchannel fabricated out of polydimethylsiloxane (PDMS) by means of soft lithography. Within the microchannel, we fabricated a flexible valve made of cured polyethylene glycol diacrylate (PEGDA). An aqueous dispersion of polystyrene particles was perfused within the microfluidic device using pulsed flow to simulate rhythmic contractions of the leg

173

muscles. In the in vitro models, polystyrene particles tend to form aggregates due to van der Waals and electrostatic interactions between themselves and the solid surfaces, and this simulated the formation of aggregates. We also compared these results with experiments of perfusing blood flow with fluorescently labelled platelets, where platelets tended to form aggregates when activated [13,14].

The results show that the agglomeration of polystyrene particles and platelets occurs near the valve leaflets. These results support that a thrombus forms in the venous valves at least in part due to altered flow. However, they open a new question that could not be answered in the in vitro experiments. Can hydrodynamic alone explain agglomeration?

In the valve, when the leaflets close, the flow streamlines converge and subsequently the probability of particles colliding enhances agglomeration. Moreover, both polystyrene particles and platelets accrue at the leaflet's surface. Therefore, it is not clear how much of the observed agglomeration is caused by hydrodynamics and how much is simply due to the particle sticking to the leaflets surface.

Therefore, in this study, we further developed our original DVT model [11,12] to include particle agglomeration. Since, in the computer model, we can arbitrarily tune the properties of the particles, we can account for 'fictional' particles that are sticky only with each other and not with the leaflets. By studying this virtual system, we can answer the research question of this paper: when removing particle-leaflets adhesion, can hydrodynamics alone explain at least part of the agglomeration observed by Schofield et al. [10]. If, in the virtual system set up in the computer simulations, we observe agglomeration near the leaflets, we can conclude that hydrodynamics play a role; if not, this means that, in Schofield et al. [10], agglomeration is mostly due to the interaction between the particles and the leaflet surface. Thus, this study proposes a novel model (to the best of our knowledge, this is the first DVT model that accounts for agglomeration) and uses this model to answer a specific research question from a medical-related area.

This paper is organized as follows. Firstly, we introduce the general theory behind Discrete Multiphysics (DMP), the modelling approach used to simulate the system. Then, we show how the theory is applied to the concrete case of flow in flexible valves. Lastly, we use the model to simulate particle agglomeration in the valve and show that larger clusters are formed near the leaflets.

2. Methodology

2.1. The Theory of Discrete Multiphysics

Discrete Multiphysics (DMP) is a computational approach on computational particles rather than computational mesh [15]. It links together different discrete models such as Smoothed Particle Hydrodynamics (SPH) [16], Lattice Spring Model (LSM) [17,18], Discrete Element Method (DEM) [19], and Peridynamics [20], which can be used for a range of applications ranging from biological to energy application [21–24]. In particular, DMP was previously used to simulate the flow in cardiovascular [25] and venous valves [11], including the presence of emboli in the blood flow circulation [18].

In this work, we combine the hydrodynamic venous valve model of Ariane et al. [11] and Ariane et al. [18] with the model of particle agglomeration in shear flow of Rahmat et al. [24]. The DMP model combines SPH for the fluid, LSM for the valve and a pseudo-Lennard-Jones potential for particle agglomeration. This section introduces the theory behind these computational methods and explains how they are combined together.

2.1.1. Smoothed Particle Hydrodynamics (SPH)

Smoothed Particle Hydrodynamics is a computational meshless Lagrangian method independently developed by Lucy [26] and Gingold and Monaghan [27], which is used here to simulate the fluid dynamics. Each particle in the SPH domain represents a set of properties such as positions $\mathbf{r}$, mass m, density ρ, pressure p, velocity $\mathbf{v}$ and viscosity μ, which are updated at each timestep. For a desired group of computational particles, the

SPH equation of motion is achieved from the discrete approximations of the Navier–Stokes equation [28,29]:

$$\mathbf{f}_i \; = \; m_i \frac{d\mathbf{v}_i}{dt} \; = \; -\sum_j m_i m_j \left(\frac{p_i}{\rho_i^2} + \frac{p_j}{\rho_j^2} + \Pi_{ij} \right) \nabla_i W_{ij}, \tag{1}$$

where m_i, m_j are the masses of the particles i, j, respectively, $\mathbf{v}_i$ is the velocity of the particle i, p is the pressure, ρ_i is the density of particle i, and $\mathbf{f}$ is the sum of all external forces applied to the system. W is the smoothing kernel function and W is a bell-shaped function that describes how the interaction between particle i_{th} and j_{th} decays with their distance $|\mathbf{r}_i - \mathbf{r}_j|$. Π_{ij} is the so-called artificial viscosity [30]:

$$\Pi_{ij} \; = \; \frac{(\mu_i + \mu_j)\mathbf{v}_{ij}}{\rho_i \, \rho_j \, r_{ij}} \tag{2}$$

where μ is the dynamic viscosity and $\mathbf{v}_{ij} \; = \; \mathbf{v}_i - \mathbf{v}_j$. In this work, we use the so-called Lucy kernel [29]:

$$W(r < h) \; = \; \frac{1}{s}\left[1 + 3\frac{r}{h}\right]\left[1 - \frac{r}{h}\right]^3 \tag{3}$$

where h is the so-called smoothing length and s is a parameter used to normalise the kernel function.

An equation of state (EOS) is required to link the pressure p with the density ρ. In this study, the Tait's equation is used:

$$p \; = \; \frac{c_0^2 \rho_0}{7}\left[\left(\frac{\rho}{\rho_0}\right)^7 - 1\right] \tag{4}$$

where ρ_0 is a reference density and c_0 is a reference for fluid velocity. This formulation refers to the so-called weakly compressible SPH. To keep the variation of the fluid density in the domain less than 1 per cent, c_0 is normally set as ten times the maximum velocity in the flow [31]. This produces repulsive forces between particles aimed at approximately conserving their distance during the simulation [32]

2.1.2. Lattice Spring Model (LSM)

The elastic leaflets and moving walls of the membrane are simulated with the so-called Lattice Spring Model (LSM) or Mass Spring Models (MSM) [33]. The elastic body is subdivided in computational particles which are linked together by Hookean springs. The force between two particles i and j connected with a Hookean spring is given by

$$F_{i,j} \; = \; k(|\mathbf{r}_0 - \mathbf{r}|)^2 \tag{5}$$

where k is the Hookean elasticity coefficient, $\mathbf{r}_0$ is the equilibrium distance between the particles and $\mathbf{r}$ their instantaneous distance. The spring coefficients is determined by the physical properties (e.g., Young's modulus) of the modelled materials as discussed in Kot, Nagahashi and Szymczak [33] and Pazdniakou and Adler [34].

2.1.3. Coupling SPH and LSM (Fluid–Structure Interaction)

In the model, SPH is used to simulate the fluid and LSM used for the elastic structure (valve leaflets). SPH provides the forces acting between two fluid computational particles, where LSM provides the forces between two solid particles. To model the fluid structure interaction, we need to set the forces between liquid and solid computational particles. These forces must ensure no-penetration, no-slip, and continuity of stresses between the solid–liquid interface. In continuum mechanics, these conditions are often represented as

$$\left(\frac{\partial}{\partial_t}\mathbf{u} - \mathbf{v}\right) \cdot \mathbf{n} = 0 \text{ (no - penration)}, \tag{6}$$

$$\left(\frac{\partial}{\partial_t}\mathbf{u} - \mathbf{v}\right) \times \mathbf{n} = 0 \text{ (no - slip)} \tag{7}$$

$$\sigma_s\mathbf{n} = \sigma_f(-|\mathbf{r}|) \text{ (continuity of stresses)} \tag{8}$$

where $\mathbf{n}$ is the normal to the boundary, $\mathbf{u}$ the displacement of the solid, $\mathbf{v}$ the velocity of the liquid, σ_s the stresses in the solid and σ_f in the fluid [15].

In DMP, these conditions need to be 'translated' in terms of forces $F_{i,j}$ in order to be introduced in the model. Here, we use the same approach employed in other DMP studies such as Schütt et al. [35], M. Ariane et al. [12] and Alexiadis [15]. The no-penetration conditions are implemented by means of a repulsive Lennard-Jones potential between SPH and LSM particles:

$$V(r) = K\left[\left(\frac{r^*}{r}\right)^{12} - \left(\frac{r^*}{r}\right)^{6}\right] \text{ for } r < r_{cut} \tag{9}$$

where $r = |\mathbf{r}|$, $r*$ is a reference distance between particles and K is chosen to guarantee no penetration between SPH and LSM particles.

From the potential $V(r)$, the force between two particles is calculated from the potential:

$$F(r) = -\frac{\partial U}{\partial r} \tag{10}$$

Figure 1a shows the potential $V(r)$ and the force $F(r)$ applied between the SPH and LSM, with only values for $r < 2^{1/6}r^*$ being studied so that only the repulsive potential is considered.

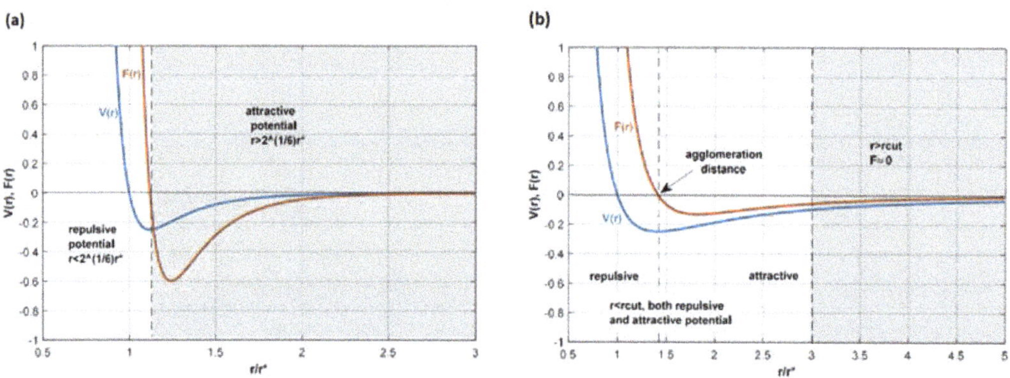

Figure 1. Diagram illustrating (**a**) the 12-6 potential used for the no-penetration conditions and (**b**) the 4-2 potential used for particle agglomeration. In both cases, the cut-off is selected so that only the white area of the diagram is used in the potential. This implies that (**a**) is only repulsive because only the positive part is considered, while in (**b**) the negative part of the force, which is attractive, is considered; particles tend to agglomerate at the location where the force is zero, which is where is the minimum of the potential is located.

No-slip conditions are enforced by imposing SPH-like viscous forces at the solid–liquid interface. Once both the no-penetration and no-slip boundary conditions are enforced, the continuity of stress is automatically satisfied by the fact that particle methods satisfy the Newton equation of motion. The numerical scheme used to solve the resulting equations is reported in Ganzenmüller and Steinhauser [29].

2.1.4. Solid-Solid Interaction (Agglomeration)

Besides fluid–structure interaction, the model also accounts for solid particles moving within the flow. These particles are 'sticky' and prone to agglomeration. To model this phenomenon, we use a similar approach to Rhamat et al. [24] that used soft, pseudo-Lennard-Jones, potentials of the type:

$$U(r) = 4\varepsilon \left[\left(\frac{r^*}{r}\right)^4 - \left(\frac{r^*}{r}\right)^2 \right] \text{ for } r < r_{cut} \tag{11}$$

where ε provides the strength of the agglomeration, to model the interaction between sticky particles. In this case, we consider both repulsive and attractive parts (Figure 1b); this produces a minimum in the potential that represents the equilibrium distance between two agglomerating particles. The value of the cut-off is selected at $3r^*$. For $r > 3r^*$, we assume $F(r) \approx 0$, which simplifies the calculations. Equation (11) is a numerically convenient way to implement agglomeration avoiding the sharp minima of the DLVO theory [36], but it does not represent a very accurate model of agglomeration. As in Rhamat et al. [24], this does not constitute a problem here since we are not interested in a specific type of interaction (the actual potential among particles/platelets in Schofield et al. [10] is unknown anyway); rather, we hope to enable particle–particle agglomeration in our virtual environment. For the same reason, ε is left as a free parameter, and simulations with different values of ε are compared. For a theoretically more accurate approach based on the concept of surface energy, the reader can refer to Ariane et al. [19].

2.2. The Valve Model and Geometry

The DMP methodology discussed previously is applied to a system of flexible valves (Figure 2a), which represents a series of venous valves distributed along a venous vessel located in the leg. The geometry is two-dimensional and adapted from previous work by Ariane [18] and Wijeratne and Hoo [37].

Two consecutive valves are connected by a channel with flexible walls. The model is periodic, meaning that the flow exiting the system from the right boundary re-enters it from the left boundary and vice versa. Therefore, the system is composed of two valves and two flexible sections. These sections are contracted periodically to simulate pulsatile blood flow. During typical daily activities, when the muscles in the leg contract, they squeeze the blood flow which promotes blood circulation [2,3]. The pressure generated by the contraction, opens the valve on the right (Figure 2b) and closes the valve on the left. This mechanism allows flow to circulate in one direction and prevents backflow. The opening and closing rate of the venous valve is around twenty rounds for each minute [38]. To save computational time, in the model, the rhythm is slightly accelerated by considering five cycles in 15 s.

As Figure 2b shows, there are different types of computational particles in the model: SPH particles modelling the blood flow; LSM particles modelling the leaflets and the flexible sections between two valves. The walls that encase the valves are also made of solid particles, but they are fixed and do not change their position during the simulation. Additionally, a certain number of 'sticky' particles are randomly dispersed in the flow at the beginning of the simulation.

In the model the fluid is considered Newtonian and the flow laminar. Table 1 shows all the numerical parameters used in the simulation. The simulations were carried out with the open-source code LAMMPS [39] and the open-source code of OVITO [40] was used for the visualisation and analysis of the data.

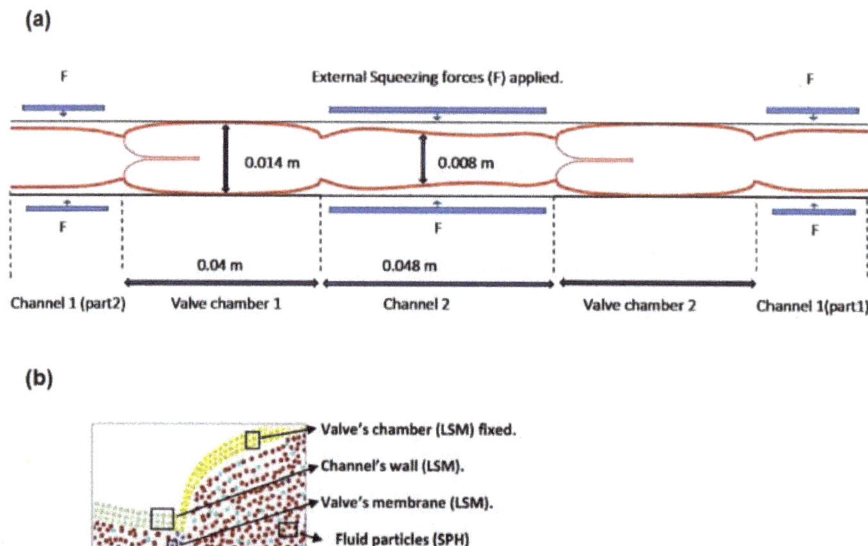

Figure 2. Two-dimensional geometry and structure of a dual venous valve: (**a**) geometry showing the location of the contraction forces and (**b**) geometry showing the computational particles and their location in the model.

Table 1. Model's numerical parameters.

Parameter		Values and Units
SPH		
Number of all particles that created our model domain.		168,676
Number of wall stationary particles (SPH particles), three layers.		4972
Number of wall flexible particles (LSM particles), three layers.		5750
Number of the valve's particles (LSM particles), two layers.	Valve particles	1404
	Each leaflet	351
Number of SPH fluid particles		141,030
Number of SPH agglomerating particles		15,520
Mass of each particle (Fluid)		1.056×10^{-5} kg
Mass of each particle (Solid)		2×10^{-5} kg
Initial distance between particles Δr		10^{-4} m
Density ρ_0		1056 kg m^{-3}
Smoothing length h		2.5×10^{-4} m
Dynamic viscosity μ_0		0.0035 Pa s

Table 1. *Cont.*

Parameter		Values and Units
Virtual sound speed c_0		10 m s^{-1}
Contraction Forces F		0.008 N
Max velocity in the valve		0.04 m s^{-1}
Time step Δt		10^{-6} s
LSM		
Hookian coefficient k_b	Flexible wall	$1 \times 10^5 \text{ J m}^{-2}$
	Valve's membrane	$5 \times 10^6 \text{ J m}^{-2}$
Viscous damping coefficient k_v	Flexible wall	1 kg s^{-1}
	Valve's membrane	0.1 kg s^{-1}
Equilibrium distance r_0		10^{-4} m
Boundaries		
Repulsive radius r^*		1×10^{-4} m
Constant k		1×10^{-4} J
Attractive forces potential		
Mass of solid particles		1.056×10^{-5} kg
Solid diameter		10^{-4} m
Particle density		1056 kg m^{-3}
Pair potential ε		$2 \times 10^{-5} \text{ J} - 1 \times 10^{-16}$ J

3. Results and Discussion:

3.1. Hydrodynamics

The blood flow moves from left to right (Figure 3) under the pressure generated by applying the force F to the flexible sections. As mentioned, F simulates the effect of muscles in the leg contracting around the vein. The higher pressure generated by F opens the valve on the left and simultaneously closes the valve on the right of the contracted section, as is illustrated in Figure 3a, where periodic boundary conditions are applied to the system. Subsequently, the contracted section is released, and F is applied to the other section, causing the open valve to close and the closed valve to open. This prevents backflow and produces a unidirectional flow in Figure 3b from left to right.

In our model, the force F is calibrated to produce a maximal blood velocity of around 0.04 m s^{-1}, which is a reasonable value for blood flow in human veins as it can vary during average physical activity [11]. The velocity magnitude during the closing and opening phases is reported in Figure 3. In all the simulations discussed in the next system, the system is simulated for 15 s, representing five opening and closing cycles.

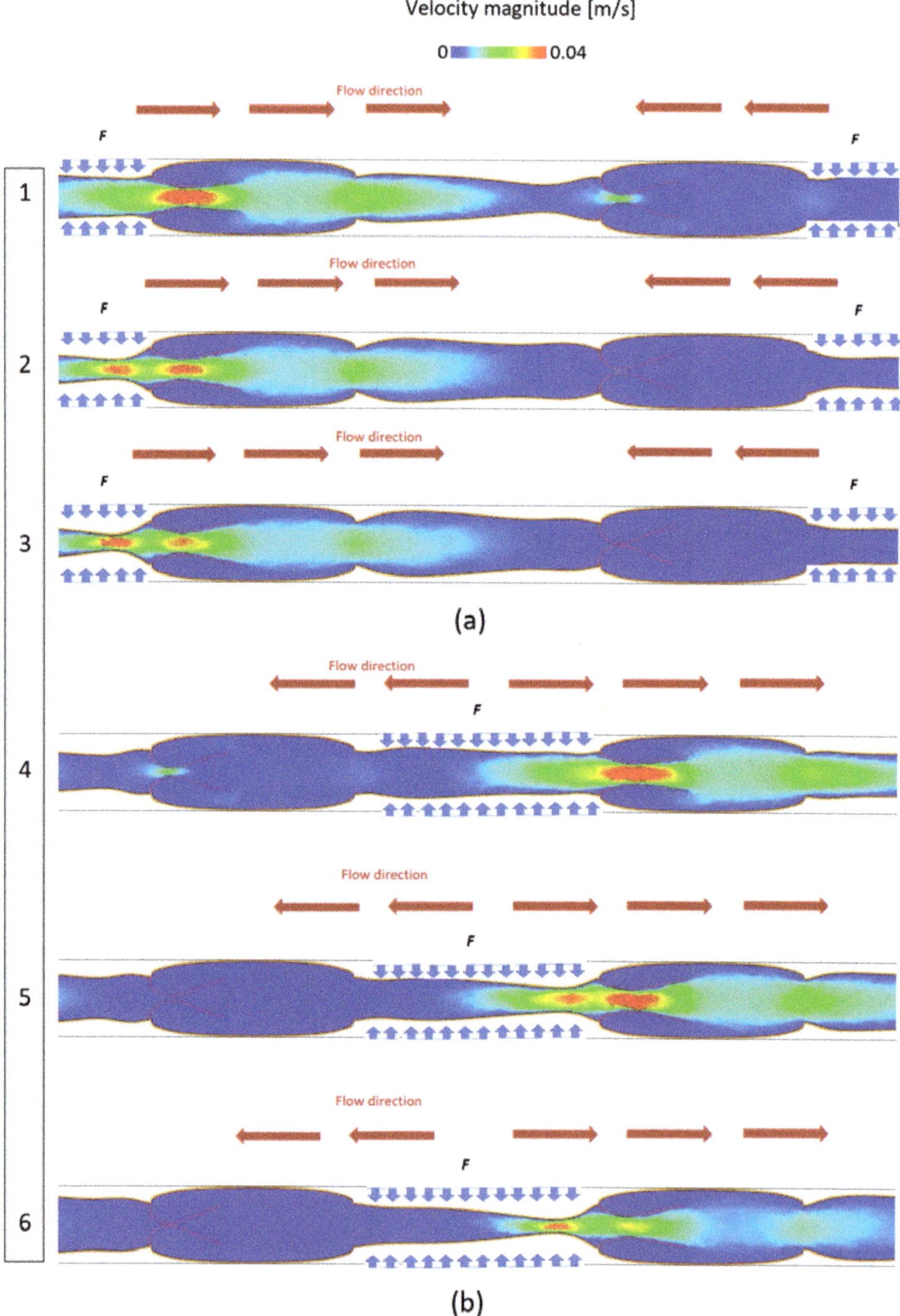

Figure 3. Velocity magnitude in the valve during opening and closing phases: (**a**) *F* is applied to the channel 1 and (**b**) *F* is applied to the channel 2.

3.2. Particle Agglomeration

At the beginning of the simulation, 'sticky' elemental particles are uniformly distributed in the liquid domain with a concentration of ~10% in the flow. When the simulation runs, these sticky particles start to aggregate in larger clusters depending on their 'stickiness', which is controlled by the value of ε in Equation (11). The goal of this study is not to replicate the physiochemical property of actual clots, but to separate the effect of hydrodynamics from the particle–wall interaction in the in vitro experiments by Schofield et al. [10] and verify that agglomeration occurs near the valve even when attractive forces between the leaflets and the particles are arbitrarily removed. The surface energy of the elemental particles is used in the in vitro experiment, and therefore their 'stickiness' is not known. For this reason, in this study, we use the value of ε as a free parameter. Figure 4 shows how the size of the average aggregate changes during the simulation

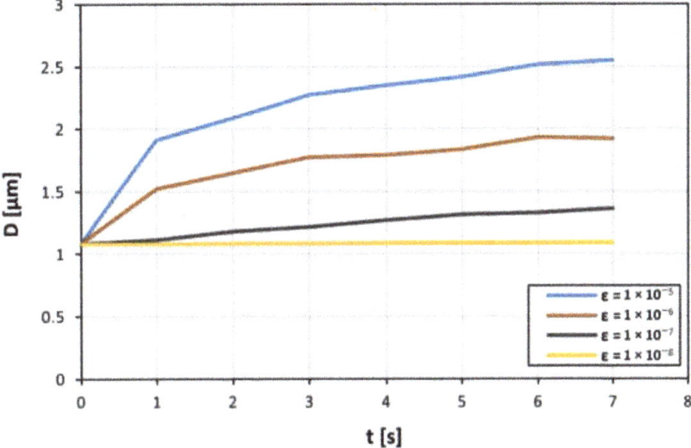

Figure 4. Dependence of the average size of particle agglomerates on the pair potential values between $\varepsilon = 10^{-8}$ J and $\varepsilon = 10^{-5}$ J as the simulation is running.

Initially, all particles are separated, and the average size is equal to one, which represents the size of a solid single particle. As time progresses, particles randomly collide in the flow and form agglomerates. Therefore, the average size of the agglomerate increases. As particles lump into larger and larger agglomerates, the number of agglomerates in the flow decreases, reducing their collision probability. Therefore, the average size tends to plateau as the simulation progresses (Figure 4). An explanation for this is that there are greater attractive forces, as a result of the higher pair potential energy. However, there is a minimum pair potential energy where the attractive forces are not enough for particle agglomeration.

As expected, the higher the value of ε (and, therefore, the 'stickiness' of the particles), the higher the average size of the aggregate. Figure 4 shows that for small values of ε, no agglomeration occurs in the flow. The particles are not sticky enough, and the inertial forces generated in the flow prevent the formation of larger agglomerates. Above the value $\varepsilon = 10^{-8}$ J, agglomeration starts, and the size of the agglomerates increases linearly with $\log(\varepsilon)$ (Figure 5).

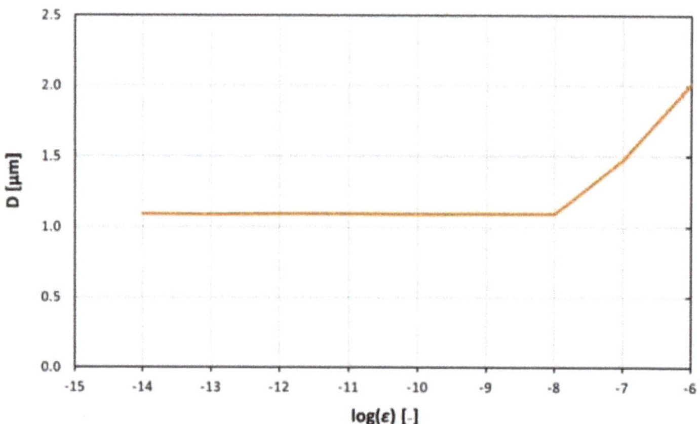

Figure 5. Average size of the agglomerates associated with different pair potential values between $\varepsilon = 10^{-8}$ J and $\varepsilon = 10^{-6}$ J at specific times (t = 12.5 s). Above the value of $\varepsilon = 10^{-8}$ J, agglomeration starts, and the size of the agglomerates increases linearly with $\log(\varepsilon)$.

3.3. Larger Agglomerates

In Figure 5, the average size of the agglomerates is presented. The size distribution, however, is not uniform. Figure 6 shows how the size is distributed in the valve during the simulation.

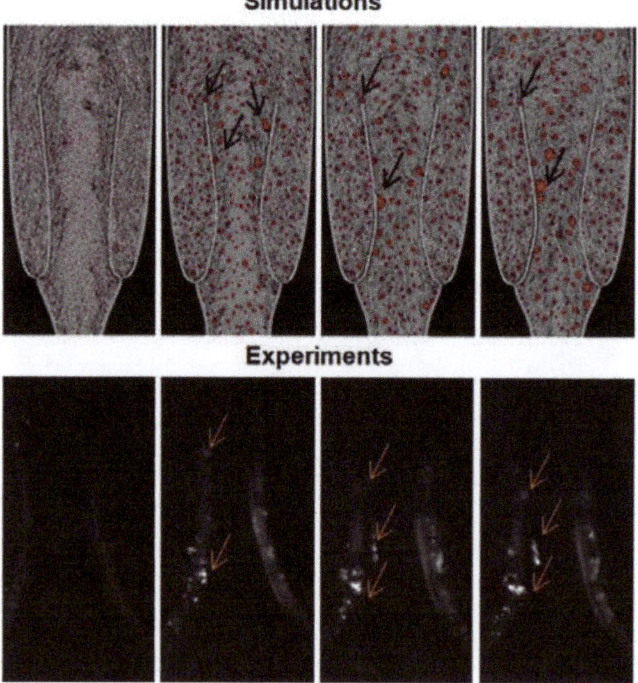

Figure 6. Simulation and experiment (platelets, fluorescently labelled, agglomeration in Schofield et al., 2020) snapshots illustrating the aggregates near the valve's leaflet at different time points. The arrows indicate the aggregates near the valve leaflets.

Moreover, larger agglomerates form near the leaflets. During their motion, the leaflets temporarily reduce the section of the channel available to the flow. This increases the probability of collision between particles forming larger agglomerates. Some of the agglomerates move into the main flow, whereas others remain trapped in the valve district and accumulate in this area. This is very similar to what we observe in the experiments. Therefore, hydrodynamics plays an important role in the in vitro model in a previous study by Schofield et al. (2020) (Figure 6).

4. Conclusions

In this study, we developed a Discrete Multiphysics model combining the fluid–structure interaction model of [11,12] with the agglomeration model of Rhamat et al. [24]. It combines an element of novelty (first DVT model that accounts for agglomeration) with a specific research question concerning the potential role of hydrodynamics in the early stages of agglomeration in DVT.

We investigated agglomeration around the valve leaflets and how this is affected by the hydrodynamics. The results show that larger agglomerates are likely to form near valve leaflets even when the interaction potential between the valve leaflet and the particles is removed. This supports our previous hypothesis [10] that the combination of blood hydrodynamics and the valve's mechanical characteristic is a key factor during agglomeration in venous valves.

Besides its specific results, this study is also a good example of how in vitro and in silico modelling can work together in research areas such as biology and medicine. In vitro models aim at providing a physical replica of a biological system. However, it is sometimes difficult to understand all the interrelated mechanical features of this physical model. At this point, in silico models can offer virtual replicas of the biological system where certain mechanical features can be switched on or off ad libitum. In this way, we can somehow 'dissect' the physics of the system and discern what the most important features are that regulate the system under investigation. In practice, the model can be used to assess which factors can enhance or decrease the tendency of agglomerates to form in the valve.

Author Contributions: Conceptualization, H.A.B. and A.A.; methodology, H.A.B., M.A., A.A. and D.V.; software, H.A.B., A.A. and M.A.; validation, H.A.B., A.A., M.A., A.B. and D.V.; formal analysis, H.A.B., M.A., A.B. and A.A.; investigation, H.A.B., A.B., D.V. and A.A.; resources, H.A.B., M.A., A.B., D.V. and A.A.; data curation, H.A.B., A.A., M.A., A.B. and D.V.; writing—original draft preparation, H.A.B.; writing—review and editing, H.A.B., A.A., M.A., A.B. and D.V.; visualization, H.A.B. and A.A; supervision, M.A., A.B., D.V. and A.A.; project administration, A.B., D.V. and A.A.; funding acquisition, A.B., D.V. and A.A. All authors have read and agreed to the published version of the manuscript.

Funding: This research was funded by NC3Rs and the British Heart Foundation (NC/S001360/1 and FS/18/68/34226).

Acknowledgments: A.B. is supported by the British Heart Foundation Senior Basic Science Research Fellowship (FS/19/30/34173). H.A.B. is supported by NC3Rs and the British Heart Foundation (NC/S001360/1 and FS/18/68/34226).

Conflicts of Interest: The authors declare no conflict of interest.

References

1. Raskob, G.E.; Spyropoulos, A.C.; Cohen, A.T.; Weitz, J.I.; Ageno, W.; De Sanctis, Y.; Lu, W.; Xu, J.; Albanese, J.; Sugarmann, C.; et al. Association between Asymptomatic Proximal Deep Vein Thrombosis and Mortality in Acutely Ill Medical Patients. *J. Am. Hear. Assoc.* **2021**, *10*, e019459.

2. Wu, W.T.; Zhussupbekov, M.; Aubry, N.; Antaki, J.F.; Massoudi, M. Simulation of thrombosis in a stenotic microchannel: The effects of vWF-enhanced shear activation of platelets. *Int. J. Eng. Sci.* **2020**, *147*, 103206. [CrossRef]

3. Lurie, F.; Kistner, R.L.; Eklof, B.; Kessler, D. Mechanism of venous valve closure and role of the valve in circulation: A new concept. *J. Vasc. Surg.* **2003**, *38*, 955–961. [CrossRef]

4. Shen, R.; Gao, M.; Tao, Y.; Chen, Q.; Wu, G.; Guo, X.; Xia, Z.; You, G.; Hong, Z.; Huang, K. Prognostic nomogram for 30-day mortality of deep vein thrombosis patients in intensive care unit. *BMC Cardiovasc. Disord.* **2021**, *21*, 11. [CrossRef]

5.   Bovill, E.G.; van der Vliet, A. Venous valvular stasis–associated hypoxia and thrombosis: What is the link? *Annu. Rev. Physiol.* **2011**, *73*, 527–545. [CrossRef] [PubMed]

6.   Siegal, D.M.; Eikelboom, J.W.; Lee, S.F.; Rangarajan, S.; Bosch, J.; Zhu, J.; Yusuf, S.; the Venous Thromboembolism Collaboration. Variations in incidence of venous thromboembolism in low-, middle-, and high-income countries. *Cardiovasc. Res.* **2021**, *117*, 576–584. [CrossRef]

7.   Das, K.; Biradar, M.S. (Eds.) *Hypoxia and Anoxia*; BoD–Books on Demand; IntechOpen: London, UK, 2018; ISBN 978-1-78984-828-1.

8.   Payne, H.; Brill, A. Stenosis of the Inferior Vena Cava: A Murine Model of Deep Vein Thrombosis. *J. Vis. Exp.* **2017**, *2017*, e56697. [CrossRef]

9.   Cook, D.J.; Crowther, M.A. Thromboprophylaxis in the intensive care unit: Focus on medical–surgical patients. *Critical Care Med.* **2010**, *38*, S76–S82. [CrossRef] [PubMed]

10.   Schofield, Z.; Baksamawi, H.A.; Campos, J.; Alexiadis, A.; Nash, G.B.; Brill, A.; Vigolo, D. The role of valve stiffness in the insurgence of deep vein thrombosis. *Commun. Mater.* **2020**, *1*, 1–10. [CrossRef]

11.   Ariane, M.; Wen, W.; Vigolo, D.; Brill, A.; Nash, F.G.B.; Barigou, M.; Alexiadis, A. Modelling and simulation of flow and agglomeration in deep veins valves using discrete multi physics. *Comput. Biol. Med.* **2017**, *89*, 96–103. [CrossRef]

12.   Ariane, M.; Allouche, H.; Bussone, M.; Giacosa, F.; Bernard, F.; Barigou, M.; Alexiadis, A. Discrete Multiphysics: A mesh-free approach to model biological valves including the formation of solid aggregates at the membrane surface and in the flow. *PLoS ONE* **2017**, *12*, e0174795. [CrossRef]

13.   Bain, B.J. Structure and function of red and white blood cells and platelets. *Medicine* **2021**, *49*, 183–188. [CrossRef]

14.   Cattaneo, M. Light Transmission Aggregometry and ATP Release for the Diagnostic Assessment of Platelet Function. *Semin. Thromb. Hemost.* **2009**, *35*, 158–167. [CrossRef]

15.   Alexiadis, A. The discrete multi-hybrid system for the simulation of solid-liquid flows. *PLoS ONE* **2015**, *10*, e0124678. [CrossRef]

16.   Alexiadis, A.; Stamatopoulos, K.; Wen, W.; Batchelor, H.; Bakalis, S.; Barigou, M.; Simmons, M. Using discrete multi-physics for detailed exploration of hydrodynamics in an in vitro colon system. *Comput. Biol. Med.* **2017**, *81*, 188–198. [CrossRef]

17.   Ariane, M.; Kassinos, S.; Velaga, S.; Alexiadis, A. Discrete multi-physics simulations of diffusive and convective mass transfer in boundary layers containing motile cilia in lungs. *Comput. Biol. Med.* **2018**, *95*, 34–42. [CrossRef] [PubMed]

18.   Ariane, M.; Vigolo, D.; Brill, A.; Nash, F.; Barigou, M.; Alexiadis, A. Using Discrete Multi-Physics for studying the dynamics of emboli in flexible venous valves. *Comput. Fluids* **2018**, *166*, 57–63. [CrossRef]

19.   Ariane, M.; Sommerfeld, M.; Alexiadis, A. Wall collision and drug-carrier detachment in dry powder inhalers: Using DEM to devise a sub-scale model for CFD calculations. *Powder Technol.* **2018**, *334*, 65–75. [CrossRef]

20.   Sanfilippo, D.; Ghiassi, B.; Alexiadis, A.; Hernandez, A.G. Combined peridynamics and discrete multiphysics to study the effects of air voids and freeze-thaw on the mechanical properties of asphalt. *Materials* **2021**, *14*, 1579. [CrossRef] [PubMed]

21.   Alexiadis, A.; Simmons, M.J.H.; Stamatopoulos, K.; Batchelor, H.K.; Moulitsas, I. The virtual physiological human gets nerves! How to account for the action of the nervous system in multiphysics simulations of human organs. *J. R. Soc. Interface* **2021**, *18*, 20201024. [CrossRef] [PubMed]

22.   Ruiz-Riancho, I.N.; Alexiadis, A.; Zhang, Z.; Hernandez, A.G. A discrete multi-physics model to simulate fluid structure interaction and breakage of capsules filled with liquid under coaxial load. *Processes* **2021**, *9*, 354. [CrossRef]

23.   Ng, K.C.; Alexiadis, A.; Chen, H.; Sheu, T.W.H. A coupled Smoothed Particle Hydrodynamics-Volume Compensated Particle Method (SPH-VCPM) for Fluid Structure Interaction (FSI) modelling. *Ocean Eng.* **2020**, *218*, 107923. [CrossRef]

24.   Rahmat, A.; Weston, D.; Madden, D.; Usher, S.; Barigou, M.; Alexiadis, A. Modeling the agglomeration of settling particles in a dewatering process. *Phys. Fluids* **2020**, *32*, 123314. [CrossRef]

25.   Mohammed, A.M.; Ariane, M.; Alexiadis, A. Using discrete multiphysics modelling to assess the effect of calcification on hemodynamic and mechanical deformation of aortic valve. *ChemEngineering* **2020**, *4*, 48. [CrossRef]

26.   Lucy, L.B. A numerical approach to the testing of the fission hypothesis. *Astron. J.* **1977**, *82*, 1013–1024. [CrossRef]

27.   Gingold, R.A.; Monaghan, J.J. Smoothed particle hydrodynamics: Theory and application to non-spherical stars. *Mon. Not. R. Astron. Soc.* **1977**, *181*, 375–389. [CrossRef]

28.   Liu, G.R.; Liu, M.B. *Smoothed Particle Hydrodynamics: A Meshfree Particle Method*; World Scientific: Singapore, 2003.

29.   Ganzenmüller, G.C.; Steinhauser, M.O.; Van Liedekerke, P.; Leuven, K.U. The implementation of Smooth Particle Hydrodynamics in LAMMPS. *Paul Van Liedekerke Kathol. Univ. Leuven* **2011**, *1*, 1–26.

30.   Morris, J.P.; Fox, P.J.; Zhu, Y. Modeling Low Reynolds Number Incompressible Flows Using SPH. *J. Comput. Phys.* **1997**, *136*, 214–226. [CrossRef]

31.   Monaghan, J.J. Simulating Free Surface. *J. Comput. Phys.* **1994**, *110*, 399–406. [CrossRef]

32.   Lee, E.S.; Moulinec, C.; Xu, R.; Violeau, D.; Laurence, D.; Stansby, P. Comparisons of weakly compressible and truly incompressible algorithms for the SPH mesh free particle method. *J. Comput. Phys.* **2008**, *227*, 8417–8436. [CrossRef]

33.   Kot, M.; Nagahashi, H.; Szymczak, P. Elastic moduli of simple mass spring models. *Vis. Comput.* **2015**, *31*, 1339–1350. [CrossRef]

34.   Pazdniakou, A.; Adler, P.M. Lattice Spring Models. *Transp. Porous Media* **2012**, *93*, 243–262. [CrossRef]

35.   Schütt, M.; Stamatopoulos, K.; Simmons, M.; Batchelor, H.; Alexiadis, A. Modelling and simulation of the hydrodynamics and mixing profiles in the human proximal colon using Discrete Multiphysics. *Comput. Biol. Med.* **2020**, *121*, 103819. [CrossRef] [PubMed]

36. Boström, M.; Deniz, V.; Franks, G.; Ninham, B. Extended DLVO theory: Electrostatic and non-electrostatic forces in oxide suspensions. *Adv. Colloid Interface Sci.* **2006**, *123–126*, 5–15. [CrossRef] [PubMed]
37. Wijeratne, N.S.; Hoo, K.A. Numerical studies on the hemodynamics in the human vein and venous valve. In Proceedings of the 2008 American Control Conference, Seattle, DC, USA, 11–13 June 2008; IEEE: New York, NY, USA, 2008; pp. 147–152.
38. Aird, W.C. Vascular bed-specific thrombosis. *J. Thromb. Haemost.* **2007**, *5*, 283–291. [CrossRef]
39. Plimpton, S. Fast parallel algorithms for short-range molecular dynamics. *J. Comput. Phys.* **1995**, *117*, 1–19. [CrossRef]
40. Stukowski, A. Visualization and analysis of atomistic simulation data with OVITO-the Open Visualization Tool. *Model. Simul. Mater. Sci. Eng.* **2010**, *18*, 015012. [CrossRef]

 *chemengineering*

Article

Fluid-Structure Interaction in Coronary Stents: A Discrete Multiphysics Approach

Adamu Musa Mohammed [1,2,*], **Mostapha Ariane** [3] **and Alessio Alexiadis** [1,*]

[1] School of Chemical Engineering, University of Birmingham, Birmingham B15 2TT, UK
[2] Department of Chemical Engineering, Faculty of Engineering and Engineering Technology, Abubakar Tafawa Balewa University, Bauchi 740272, Nigeria
[3] Department of Materials and Engineering, Sayens—University of Burgundy, 21000 Dijon, France; Mostapha.Ariane@u-bourgogne.fr
* Correspondence: amm702@bham.ac.uk or ammohd@atbu.edu.ng (A.M.M.); A.Alexiadis@bham.ac.uk (A.A.); Tel.: +44-(0)-776-717-3356 (A.M.M.); +44-(0)-121-414-5305 (A.A.)

Abstract: Stenting is a common method for treating atherosclerosis. A metal or polymer stent is deployed to open the stenosed artery or vein. After the stent is deployed, the blood flow dynamics influence the mechanics by compressing and expanding the structure. If the stent does not respond properly to the resulting stress, vascular wall injury or re-stenosis can occur. In this work, a Discrete Multiphysics modelling approach is used to study the mechanical deformation of the coronary stent and its relationship with the blood flow dynamics. The major parameters responsible for deforming the stent are sorted in terms of dimensionless numbers and a relationship between the elastic forces in the stent and pressure forces in the fluid is established. The blood flow and the stiffness of the stent material contribute significantly to the stent deformation and affect its rate of deformation. The stress distribution in the stent is not uniform with the higher stresses occurring at the nodes of the structure. From the relationship (correlation) between the elastic force and the pressure force, depending on the type of material used for the stent, the model can be used to predict whether the stent is at risk of fracture or not after deployment.

Keywords: discrete multiphysics; smooth particle hydrodynamics; lattice spring model; fluid-structure interaction; particle-based method; coronary stent; mechanical deformation

 check for
updates

Citation: Mohammed, A.M.; Ariane, M.; Alexiadis, A. Fluid-Structure Interaction in Coronary Stents: A Discrete Multiphysics Approach. *ChemEng* **2021**, *5*, 60. https://doi.org/10.3390/chemengineering 5030060

Academic Editor: Evangelos Tsotsas

Received: 29 June 2021
Accepted: 3 September 2021
Published: 8 September 2021

Publisher's Note: MDPI stays neutral with regard to jurisdictional claims in published maps and institutional affiliations.

1. Introduction

Atherosclerosis is a condition where arteries become clogged with fatty substances called plaque. The plaque is deposited on the arterial wall, and this leads to the narrowing of the artery and subsequently obstruction of the blood flow known as stenosis. This obstruction hinders the smooth transportation of blood through these arteries and consequently poses a serious health problem. When atherosclerosis affects an artery that transports oxygenated blood to the heart, it is called coronary artery disease. Coronary artery disease is the most common heart disease that becomes the leading cause of death globally. Worldwide, it is associated with 17.8 million death annually [1]. The healthcare service for coronary artery disease poses a serious economic burden even on the developed countries, costing about 200 dollars annually in the United States.

The obstruction of the flow line (stenosis) also alters the blood flow regime and causes a deviation from laminar to turbulence, or even transitional flow [2,3], a situation that signifies severely disturbed flow. Studies were carried out on the types of plaque and its morphologies as well as the flow type and its consequence [4–7] that occurred in human arteries. Although the disease is deadly, it is preventable. Therefore, it is paramount to manage or prevent it in order to restore normal blood flow in the affected artery.

One of the ways of managing coronary artery disease is restoring normal blood flow or revascularization in a patient with a severe condition using the percutaneous coronary intervention (with stent) [8]. Coronary stents are tubular scaffolds that are deployed to recover the shrinking size of a diseased (narrowed) arterial segment [9] and stenting is a primary treatment of a stenosed artery that hinders smooth blood flow [10].

The stents used in clinical practice come in differential geometry and design which implies varying stress distribution within the local hemodynamic environment as well as on the plaque and artery [11,12]. The stent structure also induces different levels of Wall Shear Stress (WSS) on the wall of the artery [9,13]. Many cases of stent failure due to unbearable stress were reported and therefore, a careful study on how these stresses are distributed is needed. In fact, stent fracture or failure often occurs after stent implantation, and it can be avoidable if the mechanical property and the performance of the material are predicted. An ideal stent should provide good arterial support after expansion by having high radial strength. It should also cause minimal injury to the artery when expanded and should have high flexibility for easy maneuvering during insertion [14,15].

Studies on the different stent designs and how they affect their mechanical performance were reported [12,13]. Stent deformation and fracture after implantation were also investigated [16–20]. Moreover, numerical modelling and simulations were also used in studying coronary stent and stent implantation. For instance, Di Venuta et al., 2017 carried out a numerical simulation on a failed coronary stent implant on the degree of residual stenosis and discovered that the wall shear stress increases monotonically, but not linearly with the degree of residual stenosis [8]. Simulation of hemodynamics in a stented coronary artery and for in-stent restenosis was performed by [21,22].

With a few exceptions [23–25], the mechanical properties of the stent and the blood fluid dynamics around the stent were studied separately. In this work, we propose a single Fluid-Structure Interaction (FSI) model that calculates the stress on the stent produced by the pulsatile flow around the stent; both the stent mechanics and the blood hydrodynamics are calculated at the same time. The model is based on the Discrete Multiphysics (DMP) framework [26], which has been used in a variety of FSI problems in biological systems such as the intestine [27], aortic valve [28,29], the lungs [28], deep venous valves [30,31]. In this study, therefore, we use the DMP framework to develop an FSI 3D coronary stent model coupled with the blood hydrodynamics and analyse the mechanical deformations produced by the flow hydrodynamics.

2. Methods

2.1. Discrete Multiphysics

Discrete Multiphysics framework combines together particle-based techniques such as Smooth Particle Hydrodynamics (SPH) [32,33], Discrete Element Method (DEM) [34,35], Lattice Spring Model (LSM) [36,37], PeriDynamics (PD) [38], and even Artificial Neural Networks (ANN) [39,40]. In this case, the model couples SPH and LSM. The computational domain is divided into the liquid domain and the solid domain. The liquid domain represents the blood, and it is modelled with SPH particles; the solid domain represents the stent and the arterial walls and it is modelled with LSM particles. Details on SPH theory can be found in [41], and of LSM in [42,43]. Here a brief introduction of the equations used in SPH and LSM is provided.

2.2. Smooth Particle Hydrodynamics (SPH)

This section provides a basic introduction to SPH; additional details can be found in [41,44]. The general idea of SPH is to approximate a partial differential equation over a group of movable computational particles that are not connected over a grid or a mesh [37]. Newton's second law is integrated to give an approximate motion of the particles characterized by their own properties such as mass, velocity, pressure, and density expressed by the fundamental identity:

$$f(r) = f(r')\delta(r - r')dr\prime, \tag{1}$$

where $f(r)$ is a generic function defined over the volume, r is the position where the property is measured, and $\delta(r)$ is the delta function which is approximated by a smoothing

(Kernel) function W over a characteristic with h (smoothing length). In this study, we use the so-called Lucy Kernel [41]. This approximation gives rise to

$$f(r) \approx f(r')W(r - r', h)dr\prime,$$

(2)

which can be discretised over a series of particles of mass $m = \rho(r)dr$ obtaining

$$f(r) \approx \sum_i \frac{m_i}{\rho_i} f(r_i)W(r - r_i, h),$$

(3)

where m_i and ρ_i are the mass and density of the ith particles, and i ranges over all particles within the smoothing Kernel. Using this approximation, the Navier-Stoke equation can be discretised over a series of particles to obtain:

$$m_i \frac{dv_i}{dt} = \sum_j m_i m_j \left(\frac{P_i}{\rho_i^2} + \frac{P_j}{\rho_j^2} + \Pi_{i,j} \right) \nabla_j W_{i,j} + f_i,$$

(4)

where v is the particle velocity, t the time, m is the mass, ρ the density, and P the pressure associated with particles i and j. The term f_i is the volumetric body force acting on the fluid and $\Pi_{i,j}$ introduces the viscous force as defined by [45]. An equation of state is required to relate pressure and density. In this paper, Tait's equation of state is used:

$$P(\rho) = \frac{c_0 \rho_0}{7} \left[\left(\frac{\rho}{\rho_0} \right)^7 - 1 \right],$$

(5)

where c_0 and ρ_0 are a reference sound speed and density. To ensure weak compressibility, c_0 is chosen to be at least 10 times larger than the highest fluid velocity.

2.3. Lattice Spring Model (LSM)

Elastic objects can be simulated using lattice spring models. As already discussed in [29], the main element of this model is composed of a mass point and linear spring which exerts forces at the nodes connected by a linear spring and placed on a lattice. Any material point of the body can be referred to by its position vector $r = (x, y, z)$ [46]; when the body undergoes deformation its position changes and the displacement is related to the applied force as:

$$F = k(l - l_0)$$

(6)

where F is the force, l_0 is the initial distance between two particles, l is the instantaneous distance, and k the spring constant (or Hookean constant).

According to [42], in a regular cubic lattice structure, the spring constant is related to the bulk modulus of the material by

$$K = \frac{5}{3} \frac{k}{l_0}$$

(7)

and

$$E = \frac{3}{2} K$$

(8)

where K is the bulk modulus, E the young modulus, l_0 is the initial particle distance and k the spring constant.

From Equations (7) and (8) the spring constant is then related to the Young modulus of the material by

$$k = \frac{E l_0}{2.5}.$$

(9)

3. Model and Geometry

A three-dimensional stent model including blood flow hydrodynamics and stent mechanics is developed. The model simulates the blood dynamics in a 3D channel similar to a coronary artery with a 1.5×10^{-3} m internal radius, including a PS-shape stent of 4 struts in the x-direction and 4 struts in the z-direction (circumference). The stent has a thickness of 100 μm, and 7.5 mm length, the size that is within the range of the stent used in clinical practice [47]. We choose a PS-Shaped because it performs better compare to most commercially shaped stents as reported by [25].

The geometry was created using CAD. From the geometry, we used MATLAB script to generate the coordinate of the computational particles as the points are created with MATLAB (details can be found in [29]). The script also generates a LAMMPS data file for the simulations and the simulations were run with LAMMPS, an open-source software [48]. The three-dimensional model consists of 1,862,804 particles: 1,609,452 particles for the fluid, 46,336 particles for the stent and 207,016 particles for the arterial wall. The fluid has a density (ρ) of 1056 kg m^{-3} and viscosity (μ) 0.0035 Pa·s. Figure 1 shows the section geometry of the stent within and outside the arterial wall. Local acceleration term g_0 was included to force the fluid to flow at a particular velocity. The inclusion of the local velocity is due to the unsteady or pulsatile flow existing in the cardiovascular system [49]. Womersley parameter α, which is the ratio of unsteady force to viscous force, was used in the model to induce the velocity profile of the flow. The particle spacing is 3.33×10^{-5} m. The optimal spacing value is obtained after several simulations with different particle spacings to make sure the results are independent of the particle resolution. Stress and deformation and other postprocessing calculations were done with the visualization software OVITO [50]. The arterial wall is assumed to be rigid with a no-slip condition [25] whereas the stent is elastic. The no-slip and no penetration boundary condition is imposed at the interface of the solid-liquid interaction as discussed in our previous work [51]. Figure 1a,b shows a section of the solid geometry which includes the stent and the wall and the complete geometry of the stent respectively.

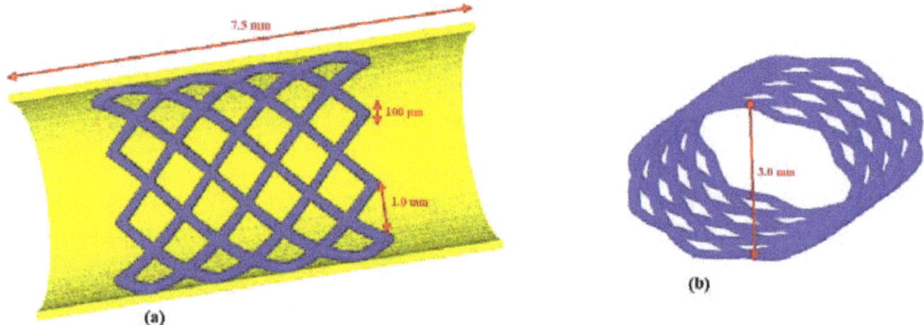

Figure 1. Illustration of the 3D stent geometry at (**a**) section view; and (**b**) front view showing complete stent.

The flow is being driven by a sinusoidal (pulsatile) acceleration (G) of the flow in the axial direction to simulate a heartbeat of 60/min with a period (T) of 1 s; the same approach was used in a previous publication [29], where the reader can find more details.

Eighteen (18) sets of simulations were run with a combination of $v1 - v3$ and $k1 - k6$ (see Section 4 for reasons). The University of Birmingham Bluebear (super-computer) was used for the computation (simulation) where ninety (90) processors were assigned for 72 h to run each simulation. For each simulation, a dump file (result file) of 9–10 GB of memory is obtained.

To better understand the hydrodynamics of the fluid and to be able to access the deformation of the stent, the discussion is carried out in terms of dimensionless numbers. According to the Buckingham π theorem, a physically meaningful equation involving n physical variables can be rewritten in terms of a set of $p = n - k$ dimensionless parameters

$\Pi_1, \Pi_2, \ldots, \Pi_p$, where k is the number of physical dimensions involved. In this case, the physiochemical properties of blood are constant, and the geometry is fixed. Therefore, we want to express the resulting stress on the stent as a function f of the type

$$\sigma = f(P, d, E), \tag{10}$$

where s [kg m^{-1}s^{-2}] is the stress on the stent, P [kg m^{-1}s^{-2}] the dynamic pressure in the fluid (the force exerted by the fluid to the stent depends on P), d [m] a characteristic length of the stent (here, we use the thickness of the stent), and E [kg m^{-1}s^{-2}] the Young Modulus.

In this case, the dynamic pressure can be written in terms of fluid average velocity v and density r,

$$P = \rho v^2, \tag{11}$$

Moreover, k [kg s^{-2}] and E are related in Equation (9). Therefore, assuming that the lattice spacing is fixed, we can replace Equation (10) with

$$\sigma = f(\rho, v, d, k), \tag{12}$$

Since we have 5 variables and 3 units, we can rewrite Equation (12) based on two dimensionless numbers

$$\Pi_2 = \varphi(\Pi_1). \tag{13}$$

The first dimensionless number can be defined as

$$\Pi_1 = \frac{k}{\rho v^2 d} \frac{[\text{elastic forces that contrast deformation (in the solid)}]}{[\text{pressure forces that tend to deform the stent (from the liquid)}]} \tag{14}$$

Knowing the typical ranges of E and d for the stent, and ρ and v for the blood, we can calculate the typical range of Π_1. The second parameter Π_2 can be defined as

$$\Pi_2 = \frac{\sigma \, d}{k}. \tag{15}$$

We can have different types of Π_2 according to the type of stress we use in Equation (15). The stress tensor has 6 independent components that can be composed in different ways to provide different types of information. One possibility is to use the Frobenius norm.

$$\sigma^F = \sqrt{\sigma_{xx}^2 + \sigma_{yy}^2 + \sigma_{zz}^2 + 2\sigma_{xy}^2 + 2\sigma_{xz}^2 + 2\sigma_{yz}^2} \tag{16}$$

In this case, we have a Π_2 based on the Frobenius norm

$$\Pi_2^F = \frac{\sigma^F \, d}{k} \tag{17}$$

that expresses, in dimensionless form, the total stress in the stent. Another possibility is the von Mises stress

$$\sigma^V = \sqrt{\frac{1}{2}\left[(\sigma_{xx} - \sigma_{yy})^2 + (\sigma_{yy} - \sigma_{zz})^2 + (\sigma_{zz} - \sigma_{xx})^2\right] + 3\left(\sigma_{xy}^2 + \sigma_{yz}^2 + \sigma_{zx}^2\right)} \tag{18}$$

which provides another Π_2 number defined as

$$\Pi_2^V = \frac{\sigma^V \, d}{k}. \tag{19}$$

Physically, $\Pi^F{}_2$ and $\Pi^V{}_2$ are dimensionless stresses and can have both a local form Π_2 (x, y, z) (when we calculate them at each x, y, z position), and a global form $<\Pi_2>$ (when we average them over the whole stent). Table 1 shows the parameters used in the simulation.

Table 1. Parameters used in the simulation.

SPH	
Number of SPH fluid particles	1,609,452
Mass of each particle (fluid)	3.41×10^{-12} kg
Length L	7.5×10^{-3} m
Diameter D	3.0×10^{-3} m
Particle spacing l	3.33×10^{-5} m
Smoothing length h	7.5×10^{-5} m
Local acceleration term g_0	0.47138–1.25 m s^{-2}
Fluid Density ρ	1056 kg m^{-3}
Viscosity μ	0.0035 Pa·s
Sound speed c_0	4 m s^{-1}
Alpha α	0.1–0.25 [-]
Time step Δt	1×10^{-7} s

LSM	
Number of SPH stent particles	46,336
Number of SPH wall particles	207,016
Mass of each particle of the stent (Solid)	3.41×10^{-12} kg
Mass of each particle of the wall (Solid)	6.0×10^{-12} kg
Stent thickness d	1.0×10^{-4} m
Elastic constant k	0.5–25 kg s^{-2}

4. Results and Discussion

Three flow velocities of 0.4 ms^{-1}, 0.23 ms^{-1} and 0.16 ms^{-1} were chosen to represent the normal coronary artery and the baseline flow, respectively. The value of k was chosen from 0.5 to 5 to cover materials with the lowest to highest Young modulus. The blood flow velocity observed within the stent ranges from 0.23 ms^{-1} to 0.4 ms^{-1}, whereas the minimum flow velocity which may occur due to stenosis is 0.16 ms^{-1} and taken to be the baseline [52]. The velocity profile at different viewpoints is shown in Figure 2.

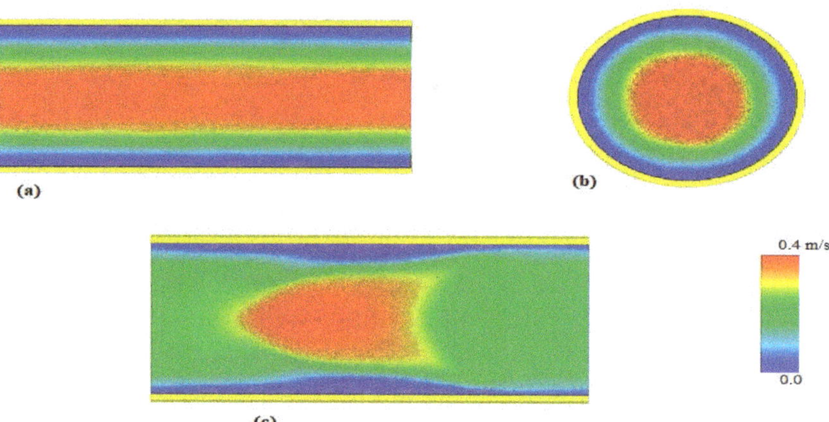

Figure 2. Velocity profile; (**a**) *x-y* view (steady state profile), (**b**) *y-z* view, and (**c**) parabolic profile at the beginning of the flow.

The dimensionless von Mises stress is shown in Figure 3 for two stents at different $\langle \Pi_1 \rangle$. Different values of $\langle \Pi_1 \rangle$ means different k and v. It is shown that the stress is more severe at the nodes (joints) with higher stress at higher $\langle \Pi_1 \rangle$ as clearly shown in Figure 3b. This may lead to potential stent failure (rapture) at the joins or size change of the stent resulting from compression or expansion. The expansion characteristics of a stent are the main causes of vascular wall injuries [53]. Either of these conditions (failure or size change) will cause severe pain and damage to the patient and lead to restenosis and or stent redeployment.

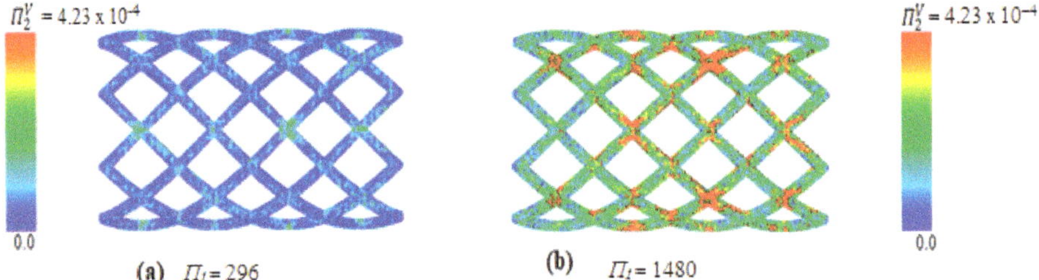

Figure 3. Local Π^V_2 at (a) Π_1 = 296, and (b) Π_1 = 1480.

The result is first presented in Figure 4 and shows how the stress varies with k and v. This is then sorted in dimensionless form and presented in Figures 5 and 6, which shows the average stress <Π^F_2>, and <Π^V_2> versus <Π_1> in dimensionless form. If we use dimensionless numbers the three curves of Figure 4 collapse in only one curve (Figures 5 and 6).

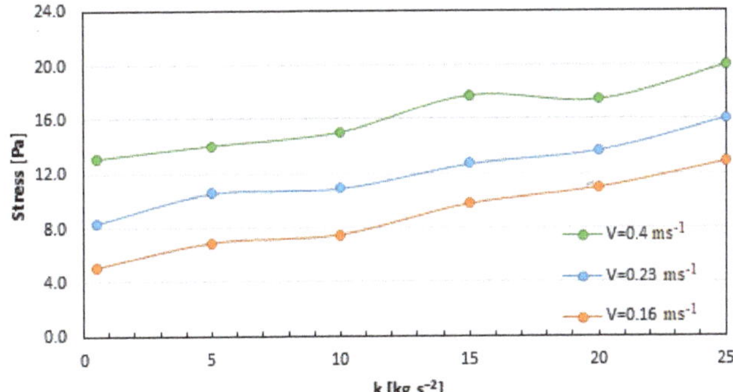

Figure 4. Stress (Frobenius norm of the stress tensor) with respect to k and v.

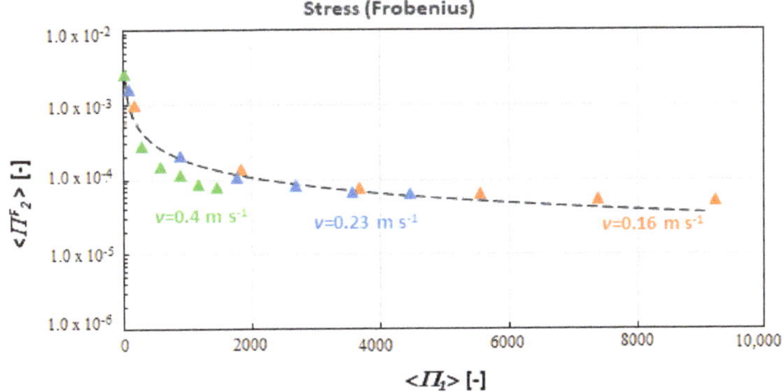

Figure 5. Relationship between average stress <Π^F_2> and <Π_1>.

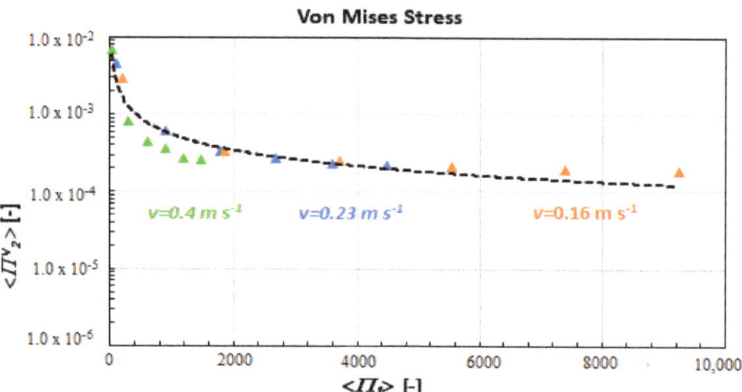

Figure 6. Relationship between average stress $\langle\Pi^V_2\rangle$ and $\langle\Pi_1\rangle$.

The plot confirms that the stress can be effectively sorted out with two dimensionless numbers based on the Buckingham π theorem. The three curves of Figure 4 can be fit by the same function as indicated in Equation (13). This function can be approximated by the following correlation (dotted line in)

$$< \Pi_2^F > = 0.026\langle\Pi_1\rangle^{-0.723}. \tag{20}$$

The same approach can be used for the von Mises stress which also has a correlation

$$< \Pi_2^V > = 0.058\langle\Pi_1\rangle^{-0.6737}. \tag{21}$$

Numerically, we identified that the dimensionless numbers computed can be used as the fundamental group of the system in which the stress can be express in terms of Π_1 and Π_2.

Due to the pulsatile flow, the stent contracts and expands during the simulation. This causes the diameter of the stent to change with the flow. Figure 7 shows that the percentage change in the stent's diameter is fluctuating. This is because the arterial blood flow, which contributed to the stent deformation, is pulsatile in nature [54,55], therefore, it is expected to have a nonlinear change in the diameter. Note that the deformation is not only a function of the pressure forces from the liquid (blood) but also the elastic forces from the solid (stent). For that reason, several oscillation modes occur at the same time and the diameter change is not a simple repetition of the pulsatile flow.

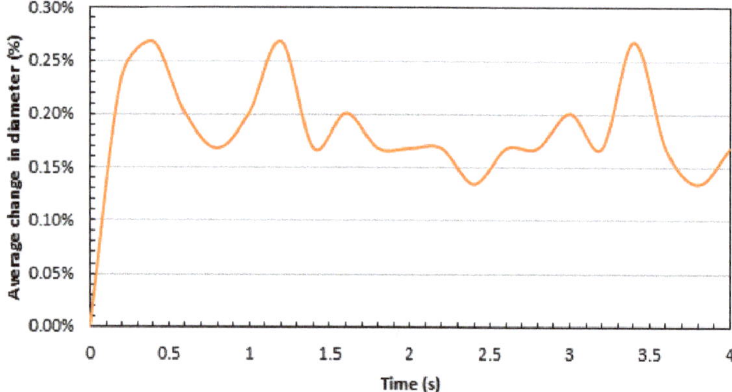

Figure 7. Average diameter change with time.

Another likely incidence of vascular wall injury associated with stent expansion which can be quantified using the model is the so-called dogboning (*DB*) effect/ratio. This phenomenon occurs when the stent expands at the ends, resulting in increased stress and injuries at the arterial wall [53]. This occurs when the diameter expands at both ends of the stent and contracts at the centre. The dogboning ratio is defined as,

$$DB = \frac{D_{max,end} - D_{min,central}}{D_{max,end}} \times 100\% \tag{22}$$

where $D_{max,end}$ is the maximum stent diameter at the end (distal and proximal), and $D_{min,central}$ is the minimum stent diameter at the centre. Our model is capable of analysing the dogboning ratio which could be used to assess and reduce the potential risk of vascular wall injury, and therefore, in this study, DB was calculated to be 4.4%. This is less than the 6.3% reported by [25] for PS-shaped stent.

5. Conclusions

In this paper, a Discrete Multiphysics model is used to simulate a coronary stent accounting for both its hemodynamics and mechanical stress. The model is three-dimensional and includes both the fluid (blood) and the solid structures (arterial wall and stent) and it is used to study the link between the flow dynamics and the mechanical deformation of the stent. The mechanical stress is computed using dimensionless numbers and a relationship between elastic forces and pressure forces was established. The results show that the blood flow contributes significantly to the stent deformation and the stiffness of the stent material affected the rate of deformation. Nonuniform stress distributions are observed. In particular, high stresses are observed at the nodes of the stent.

Given a specific Π_1 and from the corresponding Π_2^V, the maximum stress can be obtained. However, a fracture is not directly accounted for. This will depend on the pressure (stress) and type of material used for the stent as reported by [19]. Different materials have different yield stress (above which the stent fracture occurs). Our model can be used for all types of material by comparing the maximal stress in the structure against the material yield stress. Using the correlation in Equation (21), maximum Π_2^V. can be found when a specific stent material's property (Young modulus) is known. That means, from the value of Π_1, we will be able to predict the Von Mises stress from Π_2^V (which can be compared against the yield stress) as well as the flow velocity. For clinical purposes, one just needs to know the material and the blood flow velocity in the diseased artery to be able to predict whether the stent is at risk of fracture or not. Therefore, the model can be used to predict the deformation of the stent once in place and the conditions that can potentially cause its failure.

Author Contributions: Conceptualization, M.A., A.M.M. and A.A.; simulation and visualization, A.M.M.; numerical calculations, A.M.M., M.A. and A.A.; interpretation and analysis of results, A.M.M. and A.A.; writing—original draft preparation, A.M.M.; supervision, A.A. and M.A.; writing—review and editing, A.A., A.M.M. and M.A.; input script, M.A. and A.M.M. All authors have read and agreed to the published version of the manuscript.

Funding: This research received no external funding.

Institutional Review Board Statement: Not applicable.

Informed Consent Statement: Not applicable.

Acknowledgments: The Nigerian Petroleum Technology Development Fund (PTDF) is acknowledged for the provision of a scholarship to Adamu Musa Mohammed.

Conflicts of Interest: The authors declare no conflict of interest.

References

1. Brown, J.C.; Gerhardt, T.E.; Kwon, E. Risk Factors for Coronary Artery Disease. In *StatPearls*; StatPearls Publishing: Treasure Island, FL, USA, 2021.
2. Kolodgie, F.D.; Nakazawa, G.; Sangiorgi, G.; Ladich, E.; Burke, A.P.; Virmani, R. Pathology of Atherosclerosis and Stenting. *Neuroimaging Clin. N. Am.* **2007**, *17*, 285–301. [CrossRef] [PubMed]
3. Evju, Ø.; Mardal, K.-A. On the Assumption of Laminar Flow in Physiological Flows: Cerebral Aneurysms as an Illustrative Example. In *Modeling the Heart and the Circulatory System*; Quarteroni, A., Ed.; MS&A.; Springer International Publishing: Cham, Germany, 2015; Volume 14, pp. 177–195. ISBN 978-3-319-05229-8.
4. Otsuka, F.; Yasuda, S.; Noguchi, T.; Ishibashi-Ueda, H. Pathology of Coronary Atherosclerosis and Thrombosis. *Cardiovasc. Diagn. Ther.* **2016**, *6*, 396–408. [CrossRef]
5. Griffith, M.D.; Leweke, T.; Thompson, M.C.; Hourigan, K. Effect of Small Asymmetries on Axisymmetric Stenotic Flow. *J. Fluid Mech.* **2013**, *721*, R1. [CrossRef]
6. Jain, K. Transition to Turbulence in an Oscillatory Flow through Stenosis. *Biomech. Model. Mechanobiol.* **2020**, *19*, 113–131. [CrossRef] [PubMed]
7. Ahmed, S.A.; Giddens, D.P. Pulsatile Poststenotic Flow Studies with Laser Doppler Anemometry. *J. Biomech.* **1984**, *17*, 695–705. [CrossRef]
8. Di Venuta, I.; Boghi, A.; Gori, F. Three-Dimensional Numerical Simulation of a Failed Coronary Stent Implant at Different Degrees of Residual Stenosis. Part I: Fluid Dynamics and Shear Stress on the Vascular Wall. *Numer. Heat Transf. Part A Appl.* **2017**, *71*, 638–652. [CrossRef]
9. Pant, S.; Bressloff, N.W.; Limbert, G. Geometry Parameterization and Multidisciplinary Constrained Optimization of Coronary Stents. *Biomech. Model. Mechanobiol.* **2012**, *11*, 61–82. [CrossRef]
10. Hsiao, H.-M.; Lee, K.-H.; Liao, Y.-C.; Cheng, Y.-C. Hemodynamic Simulation of Intra-Stent Blood Flow. *Procedia Eng.* **2012**, *36*, 128–136. [CrossRef]
11. Wei, L.; Chen, Q.; Li, Z. Influences of Plaque Eccentricity and Composition on the Stent–Plaque–Artery Interaction during Stent Implantation. *Biomech. Model. Mechanobiol.* **2019**, *18*, 45–56. [CrossRef]
12. Colombo, A.; Stankovic, G.; Moses, J.W. Selection of Coronary Stents. *J. Am. Coll. Cardiol.* **2002**, *40*, 1021–1033. [CrossRef]
13. Balossino, R.; Gervaso, F.; Migliavacca, F.; Dubini, G. Effects of Different Stent Designs on Local Hemodynamics in Stented Arteries. *J. Biomech.* **2008**, *41*, 1053–1061. [CrossRef]
14. Duraiswamy, N.; Jayachandran, B.; Byrne, J.; Moore, J.E.; Schoephoerster, R.T. Spatial Distribution of Platelet Deposition in Stented Arterial Models under Physiologic Flow. *Ann. Biomed. Eng.* **2005**, *33*, 1767–1777. [CrossRef] [PubMed]
15. Pant, S.; Bressloff, N.W.; Forrester, A.I.J.; Curzen, N. The Influence of Strut-Connectors in Stented Vessels: A Comparison of Pulsatile Flow Through Five Coronary Stents. *Ann. Biomed. Eng.* **2010**, *38*, 1893–1907. [CrossRef] [PubMed]
16. Finet, G.; Rioufol, G. Coronary Stent Longitudinal Deformation by Compression: Is This a New Global Stent Failure, a Specific Failure of a Particular Stent Design or Simply an Angiographic Detection of an Exceptional PCI Complication? *EuroIntervention* **2012**, *8*, 177–181. [CrossRef] [PubMed]
17. Choudhury, T.R.; Al-Saigh, S.; Burley, S.; Li, L.; Shakhshir, N.; Mirhosseini, N.; Wang, T.; Arnous, S.; Khan, M.A.; Mamas, M.A.; et al. Longitudinal Deformation Bench Testing Using a Coronary Artery Model: A New Standard? *Open Heart* **2017**, *4*, e000537. [CrossRef] [PubMed]
18. Ding, H.; Zhang, Y.; Liu, Y.; Shi, C.; Nie, Z.; Liu, H.; Gu, Y. Analysis of Vascular Mechanical Characteristics after Coronary Degradable Stent Implantation. *BioMed Res. Int.* **2019**, *2019*, 8265374. [CrossRef]
19. Chinikar, M.; Sadeghipour, P. Coronary Stent Fracture: A Recently Appreciated Phenomenon with Clinical Relevance. *Curr. Cardiol. Rev.* **2014**, *10*, 349–354. [CrossRef] [PubMed]
20. Alqahtani, A.; Suwaidi, J.; Mohsen, M. Stent Fracture: How Frequently Is It Recognized? *Heart Views* **2013**, *14*, 72. [CrossRef] [PubMed]
21. Faik, I.; Mongrain, R.; Leask, R.L.; Rodes-Cabau, J.; Larose, E.; Bertrand, O. Time-Dependent 3D Simulations of the Hemodynamics in a Stented Coronary Artery. *Biomed. Mater.* **2007**, *2*, S28–S37. [CrossRef]
22. Caiazzo, A.; Evans, D.; Falcone, J.-L.; Hegewald, J.; Lorenz, E.; Stahl, B.; Wang, D.; Bernsdorf, J.; Chopard, B.; Gunn, J.; et al. A Complex Automata Approach for In-Stent Restenosis: Two-Dimensional Multiscale Modelling and Simulations. *J. Comput. Sci.* **2011**, *2*, 9–17. [CrossRef]
23. Beier, S.; Ormiston, J.; Webster, M.; Cater, J.; Norris, S.; Medrano-Gracia, P.; Young, A.; Cowan, B. Hemodynamics in Idealized Stented Coronary Arteries: Important Stent Design Considerations. *Ann. Biomed. Eng.* **2016**, *44*, 315–329. [CrossRef] [PubMed]
24. Xu, J.; Yang, J.; Huang, N.; Uhl, C.; Zhou, Y.; Liu, Y. Mechanical Response of Cardiovascular Stents under Vascular Dynamic Bending. *Biomed. Eng. Online* **2016**, *15*, 21. [CrossRef] [PubMed]
25. Wei, L.; Leo, H.L.; Chen, Q.; Li, Z. Structural and Hemodynamic Analyses of Different Stent Structures in Curved and Stenotic Coronary Artery. *Front. Bioeng. Biotechnol.* **2019**, *7*, 366. [CrossRef] [PubMed]
26. Alexiadis, A. A Smoothed Particle Hydrodynamics and Coarse-Grained Molecular Dynamics Hybrid Technique for Modelling Elastic Particles and Breakable Capsules under Various Flow Conditions: SPH-CGMD HYBRID. *Int. J. Numer. Meth. Eng.* **2014**, *100*, 713–719. [CrossRef]

27. Schütt, M.; Stamatopoulos, K.; Simmons, M.J.H.; Batchelor, H.K.; Alexiadis, A. Modelling and Simulation of the Hydrodynamics and Mixing Profiles in the Human Proximal Colon Using Discrete Multiphysics. *Comput. Biol. Med.* **2020**, *121*, 103819. [CrossRef]
28. Ariane, M.; Kassinos, S.; Velaga, S.; Alexiadis, A. Discrete Multi-Physics Simulations of Diffusive and Convective Mass Transfer in Boundary Layers Containing Motile Cilia in Lungs. *Comput. Biol. Med.* **2018**, *95*, 34–42. [CrossRef] [PubMed]
29. Mohammed, A.M.; Ariane, M.; Alexiadis, A. Using Discrete Multiphysics Modelling to Assess the Effect of Calcification on Hemodynamic and Mechanical Deformation of Aortic Valve. *ChemEngineering* **2020**, *4*, 48. [CrossRef]
30. Ariane, M.; Wen, W.; Vigolo, D.; Brill, A.; Nash, F.G.B.; Barigou, M.; Alexiadis, A. Modelling and Simulation of Flow and Agglomeration in Deep Veins Valves Using Discrete Multi Physics. *Comput. Biol. Med.* **2017**, *89*, 96–103. [CrossRef]
31. Ariane, M.; Vigolo, D.; Brill, A.; Nash, F.G.B.; Barigou, M.; Alexiadis, A. Using Discrete Multi-Physics for Studying the Dynamics of Emboli in Flexible Venous Valves. *Comput. Fluids* **2018**, *166*, 57–63. [CrossRef]
32. Albano, A.; Alexiadis, A. A Smoothed Particle Hydrodynamics Study of the Collapse for a Cylindrical Cavity. *PLoS ONE* **2020**, *15*, e0239830. [CrossRef]
33. Albano, A.; Alexiadis, A. Non-Symmetrical Collapse of an Empty Cylindrical Cavity Studied with Smoothed Particle Hydrodynamics. *Appl. Sci.* **2021**, *11*, 3500. [CrossRef]
34. Liu, W.; Wu, C.-Y. Modelling Complex Particle–Fluid Flow with a Discrete Element Method Coupled with Lattice Boltzmann Methods (DEM-LBM). *ChemEngineering* **2020**, *4*, 55. [CrossRef]
35. Ng, K.C.; Alexiadis, A.; Chen, H.; Sheu, T.W.H. A Coupled Smoothed Particle Hydrodynamics-Volume Compensated Particle Method (SPH-VCPM) for Fluid Structure Interaction (FSI) Modelling. *Ocean Eng.* **2020**, *218*, 107923. [CrossRef]
36. Sahputra, I.H.; Alexiadis, A.; Adams, M.J. A Coarse Grained Model for Viscoelastic Solids in Discrete Multiphysics Simulations. *ChemEngineering* **2020**, *4*, 30. [CrossRef]
37. Ruiz-Riancho, I.N.; Alexiadis, A.; Zhang, Z.; Garcia Hernandez, A. A Discrete Multi-Physics Model to Simulate Fluid Structure Interaction and Breakage of Capsules Filled with Liquid under Coaxial Load. *Processes* **2021**, *9*, 354. [CrossRef]
38. Sanfilippo, D.; Ghiassi, B.; Alexiadis, A.; Hernandez, A.G. Combined Peridynamics and Discrete Multiphysics to Study the Effects of Air Voids and Freeze-Thaw on the Mechanical Properties of Asphalt. *Materials* **2021**, *14*, 1579. [CrossRef] [PubMed]
39. Alexiadis, A. Deep Multiphysics and Particle–Neuron Duality: A Computational Framework Coupling (Discrete) Multiphysics and Deep Learning. *Appl. Sci.* **2019**, *9*, 5369. [CrossRef]
40. Alexiadis, A.; Simmons, M.J.H.; Stamatopoulos, K.; Batchelor, H.K.; Moulitsas, I. The Duality between Particle Methods and Artificial Neural Networks. *Sci. Rep.* **2020**, *10*, 16247. [CrossRef]
41. Liu, G.R.; Liu, M.B. *Smoothed Particle Hydrodynamics: A Meshfree Particle Method*; World Scientific: Singapore, 2003; ISBN 978-981-238-456-0.
42. Kot, M.; Nagahashi, H.; Szymczak, P. Elastic Moduli of Simple Mass Spring Models. *Vis. Comput.* **2015**, *31*, 1339–1350. [CrossRef]
43. Kot, M. Mass Spring Models of Amorphous Solids. *ChemEngineering* **2021**, *5*, 3. [CrossRef]
44. Monaghan, J.J. Smoothed Particle Hydrodynamics. *Annu. Rev. Astron. Astrophys.* **1992**, *30*, 543–574. [CrossRef]
45. Morris, J.P.; Fox, P.J.; Zhu, Y. Modeling Low Reynolds Number Incompressible Flows Using SPH. *J. Comput. Phys.* **1997**, *136*, 214–226. [CrossRef]
46. Pazdniakou, A.; Adler, P.M. Lattice Spring Models. *Transp. Porous Med.* **2012**, *93*, 243–262. [CrossRef]
47. Wall, J.G.; Podbielska, H.; Wawrzyńska, M. (Eds.) *Functionalized Cardiovascular Stents*; Woodhead Publishing Series in Biomaterials; Elsevier: Amsterdam, The Netherlands; Woodhead Publishing: Duxford, UK; Cambridge, MA, USA, 2018; ISBN 978-0-08-100496-8.
48. Plimpton, S. Fast Parallel Algorithms for Short-Range Molecular Dynamics. *J. Comput. Phys.* **1995**, *117*, 1–19. [CrossRef]
49. Ku, D.N. Blood Flow in Arteries. *Annu. Rev. Fluid Mech.* **1997**, *29*, 399–434. [CrossRef]
50. Stukowski, A. Visualization and Analysis of Atomistic Simulation Data with OVITO–the Open Visualization Tool. *Model. Simul. Mater. Sci. Eng.* **2010**, *18*, 015012. [CrossRef]
51. Alexiadis, A. The Discrete Multi-Hybrid System for the Simulation of Solid-Liquid Flows. *PLoS ONE* **2015**, *10*, e0124678. [CrossRef]
52. Vrints, C.J.; Claeys, M.J.; Bosmans, J.; Conraads, V.; Snoeck, J.P. Effect of Stenting on Coronary Flow Velocity Reserve: Comparison of Coil and Tubular Stents. *Heart* **1999**, *82*, 465–470. [CrossRef] [PubMed]
53. Wiesent, L.; Schultheiß, U.; Schmid, C.; Schratzenstaller, T.; Nonn, A. Experimentally Validated Simulation of Coronary Stents Considering Different Dogboning Ratios and Asymmetric Stent Positioning. *PLoS ONE* **2019**, *14*, e0224026. [CrossRef]
54. Huo, Y.; Kassab, G.S. Pulsatile Blood Flow in the Entire Coronary Arterial Tree: Theory and Experiment. *Am. J. Physiol. Heart Circ. Physiol.* **2006**, *291*, H1074–H1087. [CrossRef]
55. Cheung, Y. Systemic Circulation. In *Paediatric Cardiology*; Elsevier: Amsterdam, The Netherlands, 2010; pp. 91–116. ISBN 978-0-7020-3064-2.

chemengineering

MDPI

Article

A Simplified Framework for Modelling Viscoelastic Fluids in Discrete Multiphysics

Carlos Duque-Daza [1,2,*] and Alessio Alexiadis [1,*]

[1] School of Chemical Engineering, University of Birmingham, Birmingham B152TT, UK
[2] GNUM Research Group, Universidad Nacional de Colombia, Bogotá 111321, Colombia
* Correspondence: caduqued@unal.edu.co (C.D.-D.); A.Alexiadis@bham.ac.uk (A.A.)

Abstract: A simplified modelling technique for modelling viscoelastic fluids is proposed from the perspective of Discrete Multiphysics. This approach, based on the concept of linear additive composition of energy potentials, aims to integrate Smooth Particle Hydrodynamics (SPH) with an equivalent elastic potential tailored for fluid flow simulations. The model was implemented using a particle-based software, explored and thoroughly validated with results from numerical experiments on three different flow conditions. The model was able to successfully capture a large extent of viscoelastic responses to external forcing, ranging from pure viscous flows to creep-dominated Bingham type of behaviour. It is concluded that, thanks to the modularity and tunable characteristics of the parameters involved, the proposed modelling approach can be a powerful simulation tool for modelling or mimicking the behaviour of viscoelastic substances.

Keywords: Discrete Multiphysics; viscoelasticity; viscoleastic fluids; Smooth Particle Hydrodynamics; coarse-grained molecular dynamics; viscoelasticity modelling

Citation: Duque-Daza, C.; Alexiadis, A. A Simplified Framework for Modelling Viscoelastic Fluids in Discrete Multiphysics. *ChemEng* **2021**, *5*, 61. https://doi.org/10.3390/chemengineering5030061

Academic Editor: Luca Brandt

Received: 13 July 2021
Accepted: 23 August 2021
Published: 12 September 2021

Publisher's Note: MDPI stays neutral with regard to jurisdictional claims in published maps and institutional affiliations.

1. Introduction

Viscoelastic materials combine mechanical properties typical of solids (i.e., elasticity) and fluids (i.e., viscosity). For instance, many polymer substances exhibit both viscous and elastic properties during deformation of the material. For these substances, instead of a well defined elastic or viscous response to deformation, a time-dependent or shear-dependent strain can be observed. Although polymers can be taken as a clear example of viscoelastic substances, there are many other materials that also exhibit a large range of time-dependent stress–strain behaviour [1–3]. Viscoelasticity, the field devoted to study viscoelastic behaviour, is an important tool to understand a large number of physical processes, including molecular mobility in polymers [4], and analysis of dynamics of defects in crystalline interfaces in solids [5] It is also instrumental in the process of designing new materials and devices employed for a variety of purposes, including vibration abatement, control of mechanical shocks and vibrations, and noise reduction [6]. Viscoelastic flows are important for a great number of industrial applications, as well as present in many common daily situations. For instance, the behaviour ranges from toothpaste flowing by extrusion [7], to metals or molten polymers casting [8], and includes many food industry production processes and analysis [9,10]. The viscoelastic behaviour has been recognised as a dominant and extremely important characteristic of many practical flows, and therefore a key phenomenon that needs to be better understood.

Viscoelasticity is one of the main subjects of rheology, a field of study concerned with the description of the flow of matter and the mechanical properties of substances under various deformation conditions. In rheology experimental methods have usually been the main source of data for the analysis of the flow-type response of matter [11]. Nonetheless, to understand and effectively use available experimental rheological information, it is essential to have a consistent mechanistic framework. Moreover, interpretation of viscoelastic behaviour in terms of clear theoretical concepts should produce guidelines to make a clean sense of observations, to relate behaviour to composition and structure, to predict or estimate physical properties, and to facilitate control of practical applications of viscoelastic substances. In spite of the large number of current practical uses of viscoelastic materials, there are not simple expressions or models that could be considered as a generalised mechanistic framework that fit all experimental data. Although the body of studies on modelling viscoelastic fluids is large, it is still relatively modest compared to the extension of research dedicated to other "more conventional" fluid flow cases.

Modelling is a valid alternative to explore complex viscoelastic materials, to complement any experimental approach and a field to be further explored [12]. For instance, a new spectral modelling approach has been recently proposed where constitutive equations are formulated in terms of spectral invariants (see [13]), and which has been employed to formulate three-dimensional constitutive modelling framework employing a viscoelastic matrix to model residually stressed viscoelastic solids using the finite element method [14]. However, modelling viscoelastic fluids is not an immediate or simple task. The complex nature of the viscoelastic flows has somehow precluded attempts to obtain a general and accurate model. Some of this complexity can be perceived in the number of theoretical models that have been brought about to capture the extended range of possible cases [15–18], which seems to confirm that obtaining a simple universal model is far from easy. Mackay and Phillips [19] highlight that although the experimental contributions to the characterization of polymeric materials in recent decades have been abundant, there seems to be a lack of similar momentum in the development of modelling techniques for those type of fluids. Some examples on viscoelastic behaviour modelling include the use of particle-based viscoelastic fluids modelling [20], modelling based on the finite element method coupled with the generalised bracket method [19], continuum models formulated in terms of tensor diffusion and the Tensor Stokes problem [21], and the use of a formulation based on the Gibbs-potential aiming to obtain thermodynamically consistent modelling of viscoelastic fluids [22]. Besides these rigorous but computationally expensive approaches, viscoelasticity is also of interest for computer graphics [23–25]. In this case, the focus is not so much on the accuracy of the physical model, but rather on obtaining a fast simulation that preserves the overall visual effect of the viscoelastic material.

In the present work, we propose an alternative modelling technique for viscoelastic fluids within the Discrete Multiphysics (DMP) framework (see [26]) that is somehow in between the rigorous approach and computer graphics. DMP is an alternative hybrid approach for modelling multiphysics phenomena in which particle-based methods, e.g., Smooth Particle Hydrodynamics (SPH), are combined with coarse-grained molecular dynamics to capture a wide range of material's behaviour, and which has been particularly successful in modelling multiphysics and multi-phase problems with large interfacial deformations (see [26–30]). In the present work we do not directly modify the equations of motion to account for viscoelasticity, but we build a particle model where force fields typically used for modelling elastic and viscous materials are coupled together. This approach is easy to implement and more accurate than the one used in computer graphics since the viscoelastic properties of the fluid can be verified in detail. However, it loses the ability to directly derive the viscoelastic property of the model from first principles. Thus, it is required to establish these properties ex-post by performing a series of numerical experiments. In [31], a similar approach was used for viscoelastic solids; here, it is extended to viscoelastic fluids.

The paper is organized as follows. In Section 2, we discuss several theoretical concepts related to viscoelasticity. After that, we introduce Smooth particle Hydrodynamics (SPH, Section 3) used for modelling the viscous behaviour and the potentials (Section 4) used for modelling the elastic behaviour. Subsequently in Section 5, three benchmark cases are chosen to validate the results and to explain how the modelling approach works in practical settings.

2. Viscoelastic Behaviour and Standard Models

A compelling feature associated with the deformation of a viscoelastic substance is its simultaneous display of "fluid-like" and "solid-like" characteristics [32]. Arguably, one of the most important features to examine in a fluid-like substance, regardless of its Newtonian or non-Newtonian nature, is the response to shear forcing. In fact, this is the property that has traditionally been taught as the ultimate differentiating element between fluids and solids: if the substance is able to stand shear forcing, without large or permanent deformation, it is generally classified as solid, whereas if the substance is unable to stand shear forcing, thus permanently being deformed, then it is considered a fluid [33]. However, this description is not complete, as there exist a range of materials and substances that clearly show a blended behaviour between pure elastic or pure viscous. This dual response, clearly distinguishable in many aspects of the behaviour of certain substances to external forcing (i.e., shearing), has motivated many of the attempts to describe the viscoelastic behaviour in terms of basic mechanistic models, through which a simple description was always pursued. For instance, in the traditional view of elasticity, the stress found in a substance undergoing deformation is directly proportional to the strain, so the traditional applicable model is the Hooke's law that, in tension, reads as

$$\tau_{ij} = -G\frac{\partial X_j}{\partial x_i} = -G\gamma_{ij} \tag{1}$$

where G is known as Young's modulus, X_j is the shear displacement in any given j-th direction of two elements separated by an infinitesimal gap in the x_i direction, and γ_{ij} is the shear strain. If a deformed substance is able to recover its original shape once the stress is withdrawn, then Equation (1) is an appropriate model for characterizing the elastic mechanical response of the system. Most of the elastic substances might exhibit also a threshold stress beyond which the substance will "flow", and a complete recovery of the shape is not longer possible, a condition known as creep. On the other hand, a Newtonian fluid, the most representative example of a pure viscous substance, will show a shear stress proportional to the shearing rate, and for which a standard simple model can be written as the Newton law of viscosity,

$$\tau_{ij} = \mu\frac{\partial \dot{X}_j}{\partial x_i} = \mu\dot{\gamma}_{ij} \tag{2}$$

where μ is the dynamic viscosity, a proportionality constant in Equation (2), $\dot{X}_j$ is the velocity displacement in any given j-th direction, and $\dot{\gamma}_{ij}$ is the shear strain rate. The fact that the stress in a pure elastic substance is directly defined by the strain by virtue of Equation (1), whereas in a pure viscous substance is proportional to the strain rate (Equation (2)), brings about a phase synchronisation/de-synchronisation between strain and stress when a substance is subject to an oscillating strain. As a simple example, if a material is subject to a periodic shearing strain, for instance a sinusoidal function $\gamma = \gamma_0 \sin(\omega t)$, with angular frequency ω and amplitude γ_0, then the elastic substance will present a stress in phase with the strain. A viscous substance, subject to the same periodic strain, will instead exhibit a 90°-out-of-phase in time stress signal with respect to the same oscillating strain (see Figure 1).

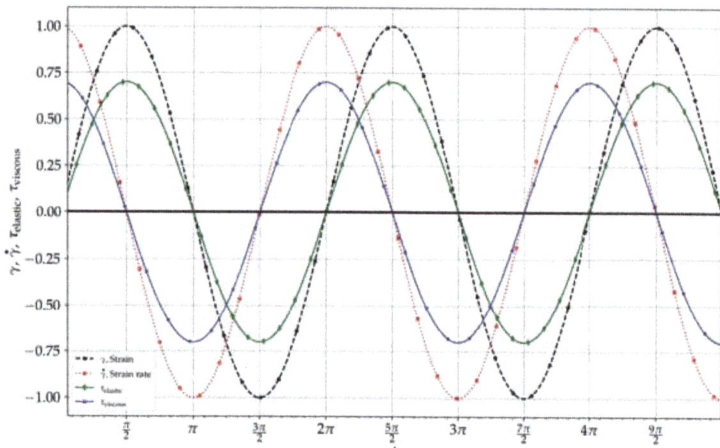

Figure 1. Strain rate and stress of ideal elastic and viscous substances under an oscillating strain $\gamma = \gamma_0 \sin(\omega t)$.

The outlined mechanical response can be somehow captured by using two simple mechanical analogues, i.e., by using a "spring" model to represent an elastic behaviour, and a raw "dashpot" model for representing a viscous response. On the basis of these two simple mechanical models, and aiming to get a quantitative description of the viscoelastic behaviour, the rationale behind many of the models that have been formulated is that by a simple coupling process of the crude mechanical analogues, for instance in an additive way, the more complex response of viscoelastic substances could be attained [34–36]; some of the most known models for modelling viscoelastic substances are Maxwell, Kelvin, Kelvin–Voigt, and Burgers models. Schematic representations of three of these models are presented in Figure 2. Examples of use and applications of these models can be found in [37–39] for the Maxwell model, in [40] for the Kelvin–Voigt and Maxwell fractional models, and in [41] for the Kelvin–Voigt and Burgers models, just to cite a few.

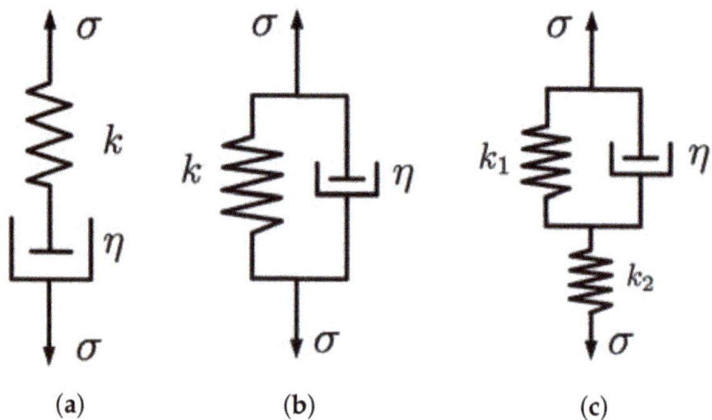

Figure 2. Schematic representation of some standard models based on mechanical analogues. (a) Maxwell model. (b) Kelvin model. (c) Kelvin–Voigt model.

In the Maxwell model the spring and dashpot are arranged in a serial configuration (see Figure 2a), while in the Kelvin model the spring and dashpot are arranged in parallel (see Figure 2b). The simple arrangements of these two models, for example, produce constitutive equations than can be written as [35]

$$\sigma(t) = k\epsilon(t) + \eta\dot{\epsilon} \tag{3}$$

for the Maxwell model, and

$$\sigma(t) + \frac{\eta}{k}\frac{d\sigma(t)}{dt} = \eta\dot{\epsilon} \tag{4}$$

for the Kelvin model, where σ is stress, ϵ is the strain, $\dot{\epsilon}$ is the strain rate, k is the characteristic Hooke elastic spring constant, and η is the equivalent Newtonian viscous dashpot constant. In a general oscillating strain condition, for instance if we prescribe $\epsilon = \epsilon_0 \sin(\omega t)$, and excluding transient periods, it is possible to see that both models, Maxwell and Kelvin, predict similar stress responses in time, although with variations in amplitude and out-of-phase angle with respect to the strain signal. As an illustrative case, if the constant of the viscous component is set as $\eta = k/\omega$, this configuration produces similar stress signals for both models that, in general, are out-of-phase with respect to the driving oscillating strain, although they are in phase to each other, as shown in Figure 3. On the other hand, if the ratio between η and k is changed, the same oscillating strain will produce a variety of responses for the Maxwell and Kelvin models. For instance, if the dashpot constant is set as $\eta = 0.5k/\omega$, it is possible to observe phase difference between the stress response predicted by the Maxwell model and the stress predicted by the Kelvin model, as presented in Figure 4, although both stress responses still exhibit a phase difference with the strain signal. This particular configuration brings about an almost elastic response in the Maxwell model, but a more dissipative behaviour in the Kelvin model. This situation is inverted if the dashpot constant is defined as $\eta = 2.0k/\omega$, as shown in Figure 5.

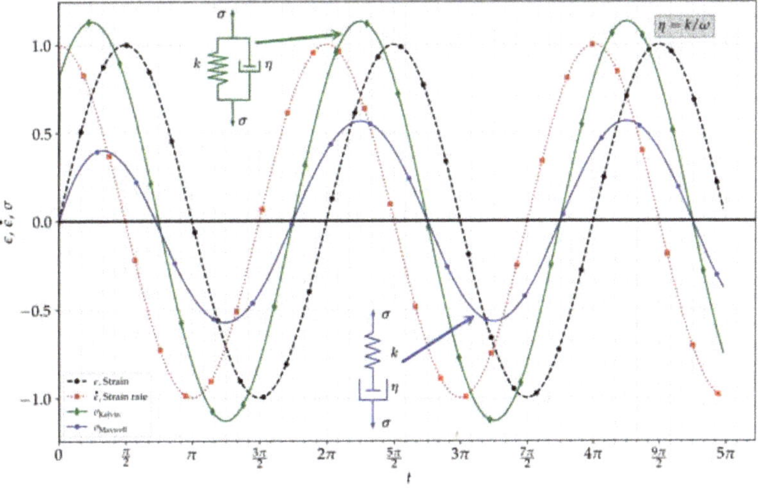

Figure 3. Strain, strain rate, and stress evolution in time for ideal Maxwell and Kelvin models forced with a synthetic strain $\epsilon = \epsilon_0 \sin(\omega t)$. Curves for $\epsilon = 1.0$, $\omega = 1.0$, $k = 0.8$, $\eta = k/\omega$.

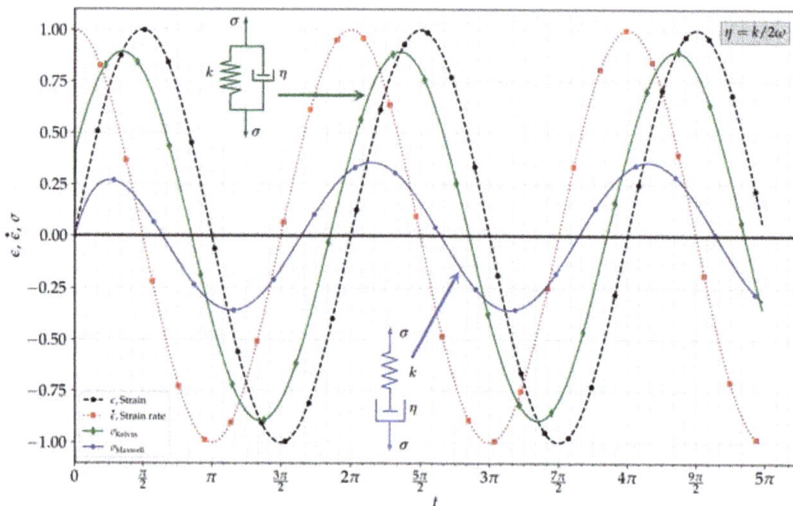

Figure 4. Strain, strain rate, and stress evolution in time for ideal Maxwell and Kelvin models forced with a synthetic strain $\epsilon = \epsilon_0 \sin(\omega t)$. Curves for $\epsilon = 1.0$, $\omega = 1.0$, $k = 0.8$, $\eta = 0.5k/\omega$.

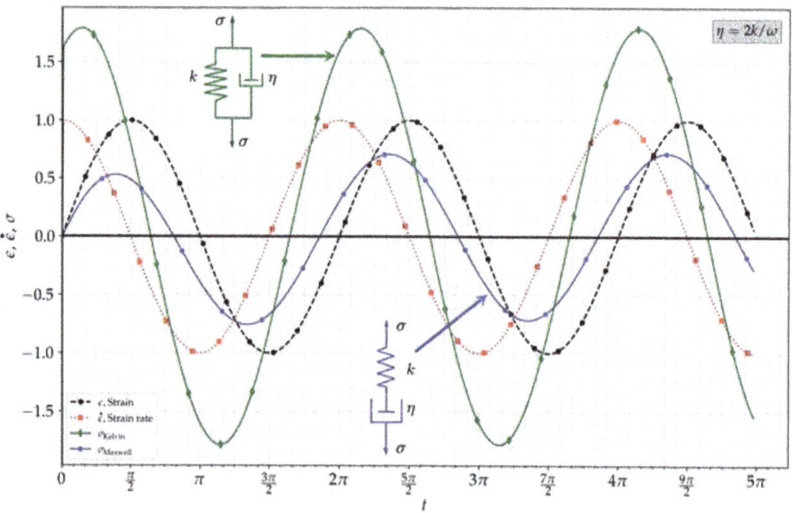

Figure 5. Strain, strain rate, and stress evolution in time for ideal Maxwell and Kelvin models forced with a synthetic strain $\epsilon = \epsilon_0 \sin(\omega t)$. Curves for $\epsilon = 1.0$, $\omega = 1.0$, $k = 0.8$, $\eta = 2.0k/\omega$.

It is important to highlight that the aforementioned models are valid for linear or quasi-linear viscoelastic behaviour, i.e., assuming that the substance is undergoing small deformations. If linear viscoelasticity is adopted, the stress function $\sigma(t)$ and the strain function $\epsilon(t)$ can de considered as linear proportional models, so if the strain function $\epsilon(t)$ is amplified by a constant factor, the outcoming stress would be scaled by the same factor, and if a substance is subject to a linear combination of two (or more) arbitrary strain signals, the stress can be obtained as the linear combination of the two (or more) individual stress responses [35]. The out-of-phase stress response, captured by the linear viscoelastic models, can be gauged through the so-called loss angle (δ), the phase angle between stress and strain during sinusoidal deformation in time, as the example cases presented in Figures 3–5.

The use of the loss angle or its tangent ($\tan \delta$), is extremely useful to produce a measure of damping or internal friction when linear viscous nature is assumed. The loss angle (or $\tan \delta$), depends mostly on the frequency of the external excitation. Although the discussion for the viscoelastic models has been based on a general normal stress σ and longitudinal strain ϵ, it is clear from Equations (1) and (2) that the whole discussion can easily and immediately be extended to a case of a pure oscillating shear applied to a substance. It is also customary to express the oscillatory behaviour of linear viscoelastic substances in complex notation, so a pure oscillating shear condition can be obtained if a shear strain $\gamma(t) = \gamma_0 e^{i\omega t}$ is imposed, which will produce a shear stress response equal to $\tau(t) = G e^{i\omega t}$. The resulting coefficient G will be frequency-dependent and in general a complex number that can be expanded as,

$$G(\omega) = G'(\omega) + iG''(\omega) \tag{5}$$

and from which it is possible to discriminate between the real component $G'(\omega)$, associated to the elastic part of the response, and the complex component $G''(\omega)$, usually considered as a quantification of the viscous part of the response. The first component is called the Storage Modulus, while the complex part is known as the Loss Modulus.

3. SPH Formulation

In this study, we combine different particle methods together. SPH, discussed in this section, is used for viscous behaviour and other specific potentials (see next Section) for the elastic behaviour. This approach of combining different particle methods is called Discrete Multiphysics (e.g., see [26–28]) and it is here extended to account for viscoelastic fluids. The numerical method used in the present work to model the fluid is Smooth Particle Hydrodynamics, or simply SPH, an approximate method to obtain numerical solutions to the equations of fluid dynamics. This is done by replacing the continuum of fluid by a discrete set of particles. It was originally devised for the simulation of stars [42,43], and was later found applicable to molecular dynamics (MD) due to the inherent similarity between SPH and MD. To reproduce the equations of fluid dynamics or continuum mechanics, the statistical concept of kernel interpolation is used to 'smooth' out discrete fields of a quantity of interest (such as density, pressure and velocity). Moreover, in SPH each particle is represented by a kernel function (more generally a window function) $W(\vec{r} - \vec{r'}, \vec{h})$. The local average of the desired property f, in a domain of interest Ω, is the convolution of the quantity f with the chosen smoothing function $W(\vec{r} - \vec{r'}, \vec{h})$,

$$\langle f(\vec{r}) \rangle = \int_\Omega f(\vec{r'}) W(\vec{r} - \vec{r'}, \vec{h}) \, d^3\vec{r'} \tag{6}$$

The kernel functions considered in SPH are generally radially symmetric (spherical) functions centred at $\vec{r'}$ and must decrease monotonically in the outward radial direction. These kernel functions must be normalised to 1 so that constants are interpolated exactly,

$$\int_\Omega W(\vec{r} - \vec{r'}, \vec{h}) \, d^3\vec{r'} \to 1 \quad \text{as} \quad h \to 0. \tag{7}$$

and they should tend to a Dirac delta function $\delta(\vec{r} - \vec{r'})$ as the smoothing length tends to zero, $h \to 0$, in order to recover the original function f in the limit.

$$\langle f(\vec{r}) \rangle \to f(\vec{r}) \quad \text{as} \quad h \to 0 \tag{8}$$

The chosen smoothing length h is often a characteristic length of the target domain Ω. In general, it is advisable for this smoothing length to be a constant value for the prescribed problem so that its spatial and temporal derivatives are identically zero, although variable smoothing lengths in time and space can provide simulation resolutions that adapt to local conditions. It is important to note that the shape of the constructed continuous field will be fully determined by the choice of the kernel function $W(\vec{r} - \vec{r'}, \vec{h})$, the smoothing length

h, and the position of each particle in the domain of interest i.e., the particle distribution, as schematically represented in Figure 6. In SPH the interest is focused on the cases when neighbouring particles are each encircled by the radius of the smoothing length h, as any other particles outside this radius would not be considered by the compact support of the kernel function, and so would generate a discontinuous field. This can be easily enforced by setting the smoothing length to be an appropriate characteristic length.

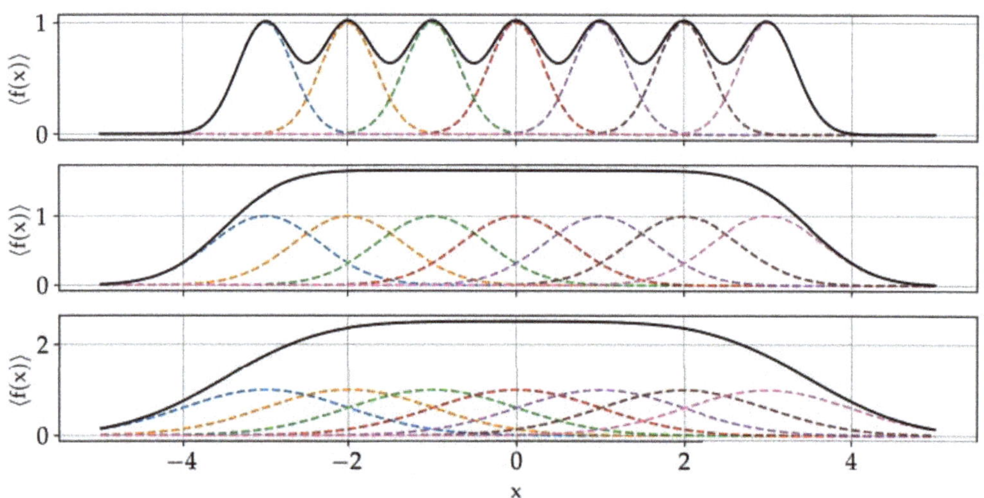

Figure 6. Conceptual example of the effect that the smoothing length has on $\langle f(x) \rangle$ for multiple point-particles, separated one length unit, and with Gaussian function as kernel. The smoothing lengths employed in the examples are: $h = 1$, upper plot; $h = 2$, middle plot; and $h = 3$, bottom plot.

For computational interest, we can discretise Equation (6) as follows,

$$\langle f(\vec{r}) \rangle = \sum_{i=1}^{N} f(\vec{r}_i) W(\vec{r} - \vec{r}_i, \vec{h}) V_i \tag{9}$$

where V_i is the volume of the particle of interest and N is the number of particles within the domain. An equivalent way of this discretisation, common in fluid dynamics applications, can be expressed as

$$\langle f(\vec{r}) \rangle = \sum_{i=1}^{N} \frac{m_i}{\rho_i} f(\vec{r}_i) W(\vec{r} - \vec{r}_i, \vec{h}) \tag{10}$$

in which ρ_i is the mass density of the i-th particle (considered a local density) and m_i its mass. By using index notation this can be simplified further as,

$$\langle f_i \rangle = \sum_{j=1}^{N} m_j \frac{f_j}{\rho_j} W_{ij} \tag{11}$$

with $W_{ij} = W(\vec{r}_i - \vec{r}_j)$ and $f_i = f(\vec{r}_i)$. For fluid dynamics applications it is also highly relevant to define appropriate discrete forms of the gradients present in the governing equations of motion. It is noteworthy that SPH is rather convenient in that neither m_i or f_i are affected by the ∇ operator since they are particle properties. The kernel function is usually a polynomial so we can generate an expression for the gradient by simply taking the gradient of W_{ij},

$$\nabla f_i = \sum_{j=1}^{N} m_j \frac{f_j}{\rho_j} \nabla W_{ij} \tag{12}$$

As a simple example, we demonstrate the SPH approximation of the continuity equation,

$$\frac{d\rho}{dt} + \rho \nabla \cdot \vec{v} = 0 \tag{13}$$

which, by using the identity $\nabla(\vec{v}\rho) = \rho \nabla \vec{v} + \vec{v}\,\nabla \rho$, can be rewritten as,

$$\frac{d\rho}{dt} = \nabla(\rho\vec{v}) - \vec{v}\,\nabla \rho \tag{14}$$

and so using the gradient approximation we have:

$$\frac{d\rho_i}{dt} = \sum_{j=1}^{N} m_j \vec{v}_j \nabla_j W_{ij} - \vec{v}_i \sum_{j=1}^{N} m_j \nabla_j W_{ij} = -\sum_{j=1}^{N} m_j \vec{v}_{ij} \nabla_j W_{ij} \tag{15}$$

Discretised SPH forms of the other relevant conservation equations are widely available in the literature (see [44–46]).

The numerical implementation of our model, including the SPH-viscous component, was performed using LAMMPS [47,48]. In LAMMPS, the SPH method is achieved through the use of pair styles defining the interaction between neighbouring particles following the SPH formulation. For liquids, the two most common SPH models are the compressible model, proposed by [44] apt for high speed flows, and the model proposed by [49], usually better suited for low Reynolds number incompressible flows. One of the main differences between the two models lies in the role of the viscosity within the SPH simulation, which appears naturally in the equations of conservation of momentum in flow dynamics as part of the relationship between stress and strain rate. In SPH, the viscosity is incorporated into the numerical formulation in a number of different ways [44,46]. Nevertheless, some physical phenomena have been more challenging for SPH, requiring the use of numerical strategies aiming to minimize problems like numerical spurious oscillations (generally around shock regions) or unphysical particles penetration, and usually through the introduction of energy dissipation strategies that exploit the viscosity as a dissipative tool [50–52]. In the model proposed in [49], the dynamic viscosity is one of the parameters to be provided, together with the smoothing length h, the speed of sound c, and the density ρ. Alternatively, a common artificial viscosity employed in SPH, introduced in [44] originally as an extension of the von Neumann–Richter artificial viscosity, is defined as

$$\Pi_{ij} = -\alpha h \frac{c_i + c_j}{\rho_i + \rho_j} \frac{\mathbf{v}_i \cdot \mathbf{r}_{ij}}{r_{ij}^2 + \epsilon h^2} \tag{16}$$

where c_i and c_j represent the speed of sound of particles i and j, ϵ is a constant to avoid singularities in the definition of the viscous component Π_{ij}, and α is a dissipation factor that serves as a modulating coefficient of the viscous response from the SPH perspective. A widely accepted equivalence between the real dynamic viscosity μ and the dissipation factor α is given as,

$$\mu = \frac{\alpha c h \rho}{2(n+2)} \tag{17}$$

where, as usual, c represents a numerical speed of sound, h the smoothing length, ρ the density of the target fluid and n is the number of spatial dimensions involved. It is noteworthy that in our DMP approach point-particles rather than finite-size particles are employed, although particle volume is defined to correlate particles mass and density.

4. Proposed Modelling Technique

Viscoelasticity can be directly implemented in SPH (e.g., [53]). However, this implies rewriting the equation of motion to account for a specific model of viscoelasticity. In the case of a particle code like LAMMPS, for instance, this would imply rewriting large sections of the code dedicated to SPH every time we introduce a new model of viscoelasticity. The approach proposed in this study models viscoelasticity by combining different particle potentials, which is a standard procedure in particle codes, rather than rewriting the equations of motion. For the viscous part, the standard SPH approach discussed in the previous section is adopted. The elastic part is discussed in this section.

In an analogue manner to the idea supporting the most widely used basic viscoelastic models discussed previously, we propose an alternative modelling technique within a particle-based framework, where inter-particle potentials tackling viscous and elastic interactions separately are blended in an additive way in order to mimic the dual dissipative and restoring response of viscoelastic substances. For this purpose, the SPH viscous support is blended with a potential that resembles a restorative elastic behaviour, aiming to bring about the characteristic dual response of viscoelastic substances. In our approach, the viscous properties are given to the fluid by the SPH Equations (16) and (17). Furthermore, in principle, we would like to use harmonic springs that are used in the Lattice Spring Model (LSM) to add elasticity to the material (see [54]),

$$U_{\mathrm{LSM}}(r) = \frac{1}{2}k(r - r_0)^2 \tag{18}$$

which would provide a force $F(r) = -\nabla U(r) = -k(r - r_0)$ between two particles connected by the spring with a stiffness k, separated a distance r, and featuring an equilibrium distance r_0. The combination of Equations (16) and (17) and LSM will confer viscoelastic properties to the material. A similar approach to model viscoelastic solids was employed in [31]. However, fluids cannot be modelled in the same way because the springs constrain the particles, which cannot flow as it should occur in fluids. The solution we propose is to employ a blending of two potentials to partially imitate the LSM, but with the advantage of featuring a deactivation or cutoff distance beyond which the composed potential not longer acts, allowing the particles to flow freely. An exclusively attractive potential, first proposed by [55], was chosen to provide the attractive portion of our LSM analogue, defined as,

$$U_{\mathrm{CS}}(r) = \begin{cases} -\epsilon_{\mathrm{CS}} & r < \sigma \\ -\epsilon_{\mathrm{CS}} \cos\left(\dfrac{\pi(r - \sigma)}{2(r_{c,\mathrm{CS}} - \sigma)}\right)^2 & \sigma \leq r < r_{c,\mathrm{CS}} \\ 0 & r \geq r_{c,\mathrm{CS}} \end{cases} \tag{19}$$

which, as schematically illustrated in Figure 7a, is a constant value of ϵ_{CS} below an inter-particle distance σ, that increases proportional to r until it vanishes above a cutoff distance $r_{c,\mathrm{CS}}$, according to Equation (19). This allows us to account for the fact that particles may no longer feel an attractive force when separated far apart; thus fluid particles are not constrained into a lattice structure like in Equation (18), but they are free to flow within the domain.

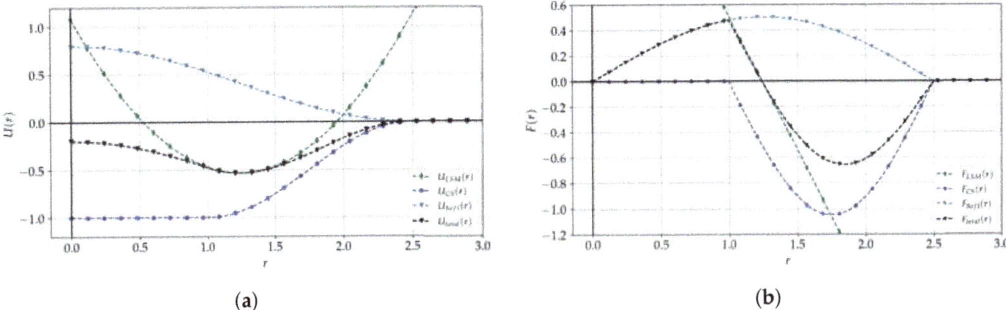

Figure 7. Comparison of energy potentials and forces used for modelling elastic behaviour. *LSM*: Lattice-Spring-Model potential; *CS*: Attractive potential (Cosine/squared); *Soft*: Repulsive potential (Soft); *Total*: Combined Soft-Cosine/Squared potential. Curves obtained for $\epsilon_{CS} = 1$, $\epsilon_{Soft} = 0.4\epsilon_{CS}$, $\sigma = 1$, $r_{c,CS} = r_{c,Soft} = 2.5$. (**a**) Energy potentials. Location of the harmonic LSM potential adjusted for illustration purposes. (**b**) Force field obtained with elastic, attractive, repulsive, and total potentials.

To avoid particle overlap, a repulsive portion in our model is provided by another potential given as,

$$U_{Soft}(r) = \epsilon_{Soft}\left[1 + \cos\left(\frac{\pi r}{r_{c,Soft}}\right)\right] \qquad r < r_{c,Soft} \tag{20}$$

where, ϵ_{Soft} is the magnitude and $r_{c,Soft}$ is the cutoff distance for this repulsive component of our potential therefore, once again, limiting the distance at which repulsion exists for particles far apart. In our proposed model we use a total potential by adding these components aiming to mimic the elastic behaviour of the LSM model, but with the advantage that when the distance between the two particles is above the cutoff value, the force is deactivated and the particle is free to flow. An illustration of the LSM, repulsive, attractive, and total potentials is presented in Figure 7a, while a representation of the forces produced by them is shown in Figure 7b, where the repulsive and attractive components of the elastic equivalent forces are schematically depicted.

As the rationale behind our model is to be able to get a coupling between the two pair styles producing stable repulsion/attraction within a close range of each particle, this coupling requires a sensible choice of values given the possible complex interaction that can be brought about by the number of parameters at play. For instance, an important consideration, usually critical in other mesh-based methods only from the numerical stability perspective, is the spatial resolution. In our case, the spatial resolution is somehow linked to the distribution of particles, so the stability and effectiveness of the model is also dependent of the lattice employed for the initial arrangements of such particles, as well as on the characteristic length in the lattice, or lattice scale Δ_L. Specifically, in order to reduce the number of free parameters, the repulsive potential was defined in terms of the characteristics of the attractive potential, so only a single set of values could define entirely the elastic coupled model. It is important to mention that, as our model is strongly dependent of the inter-particle spacing, the data presented along the present paper as reference values should be taken as a guide for setting up working simulations, rather than as a restrictive range of operating conditions. In any case, some ranges and definitions employed in the characterization of the elastic component of our model, determined through several numerical experiments and found to provide numerical stability and consistency with expected behaviour, are presented in Table 1 for reference purposes. In Table 1 the acronyms *BCC* and *FCC* stand for Body-centred cubic and Face-centred cubic unit cells, respectively, which are some of the most common type of regular lattices for molecular dynamics simulations.

Table 1. Summary of some ranges and relationships between parameters for the elastic potential.

	BCC Lattice	FCC Lattice
Activation distance-attractive potential	$\sigma = \sqrt{3}\,\Delta_L/2$	$\sigma = \sqrt{2}\,\Delta_L/2$
Cutoff distance-attractive potential	$0.95\,\Delta_L \leq r_{c,\text{cs}} \leq 2.1\,\Delta_L$	$0.9\,\Delta_L \leq r_{c,\text{cs}} \leq 1.1\,\Delta_L$
Cutoff distance-repulsive potential	$r_{c,\text{Soft}} \approx 1.05\,r_{c,\text{cs}}$	$r_{c,\text{Soft}} \approx 0.955\,r_{c,\text{cs}}$
Prefactor-repulsive potential	$0.8\,\epsilon_{\text{cs}} \leq \epsilon_{\text{Soft}} \leq 3.0\,\epsilon_{\text{cs}}$	$1.25\,\epsilon_{\text{cs}} \leq \epsilon_{\text{Soft}} \leq 4.0\,\epsilon_{\text{cs}}$

5. Numerical Experiments

The proposed model was characterised and tested using mainly two types of conditions: oscillatory shearing between parallel plates, and constant gradient in a circular pipe flow. Numerical simulations with two types of lattice, namely *Body-centred cubic* or BCC, and *Face-centred cubic* or FCC, were performed for both conditions, although results reported here are mainly based on the numerical experiments using the FCC lattice. Regardless of the lattice employed, the final number of particles in each of our numerical experiments was obtained following a same methodology, i.e., preliminary exploratory simulations were performed first to get time evolution of some representative local measures, like velocity and force ensemble averages, together with an assessment of the level of "mesh independence", and the final selection of the number of particles for each test was decided with basis in those results. Statistical convergence was assumed once the mean and standard deviation of the ensemble averages converged in the preliminary simulations. Furthermore, it is important to mention that although the proposed methodology is general and easy to implement with any SPH and coarse-grained molecular dynamics software, the numerical experiments of the present work were performed using LAMMPS (see [47,48]). The computations described in this paper were completed using the University of Birmingham's BlueBEAR HPC service, and executed in parallel using LAMPPS' MPI capabilities.

5.1. Dynamic Response to Oscillating Shear

The most traditional way to differentiate the viscous from the viscoelastic behaviour is through the determination of the storage and loss moduli presented before. These quantities should be obtained from oscillatory shear tests. To this end, a substance contained between two parallel plates, one of them stationary and the other one oscillating at a fixed frequency, was modelled using our viscoelastic model. The domain for this numerical experiment is a box of dimensions 0.30 m $\times$ 0.14 m $\times$ 0.02 m, including two regions representing the plates. The bottom plate is set as a stationary region, whereas the top plate is set to oscillate at a fixed frequency, and following the simple function,

$$\delta_x = \delta_{x,0}\sin(\omega t) \tag{21}$$

with $\delta_{x,0}$ defining the amplitude of the oscillation, and ω the frequency. The computational domain and lattice structure is presented in Figure 8.

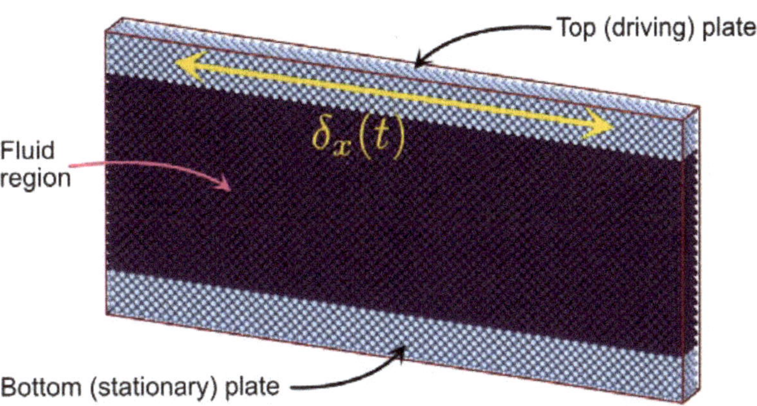

Figure 8. Configuration of oscillatory shear case.

Some preliminary tests were performed in order to gauge the ability of the proposed model to effectively reproduce, separately, the elastic or viscous response. A simple visual example of the test configuration and dynamic response for the pure elastic case is presented in Figure 9. In this figure a group of particles has been selected to be tracked as timeline and three different instants of the oscillatory test, for a pure elastic substance, are presented. Particles not included in the timeline group have been coloured using the velocity in the $x-$direction, which is the direction of the forced oscillating strain. The dynamic relationship between strain and stress for this case is illustrated in Figure 10, where the time evolution of both quantities are plotted and an in-phase evolution can be clearly observed. In Figure 11, are presented the results obtained for the Loss and Storage Moduli from the numerical tests using only the elastic component, for different values of the magnitude of ϵ_{cs}, at different oscillatory shear amplitudes γ, and for three different oscillation frequencies ω. As it can be appreciated, the model performs very well in defined ranges of shear strain amplitude, providing a response that exhibits a clear elastic behaviour. It is noteworthy, however, that some deviations from the expected response can also be observed. For instance, by examining Figure 11, it is always possible to distinguish a range of oscillation amplitudes where the Storage moduli is clearly larger than the Loss moduli, as predicted by theory for elastic materials and, although we did not explore a large number of frequencies, there is a clear frequency dependency, especially observable in the results for $\epsilon_{cs} = 1 \times 10^{-6}$ and $\epsilon_{cs} = 1 \times 10^{-5}$, as expected. Nevertheless, it is important to note that our model exhibits some numerical discrepancies at some of the extreme conditions explored and presented in Figure 11. Clearly, there is an operational range for the model and therefore simulations performed out of those bounds could produce unexpected behaviour, although this can be somehow anticipated. For example, if deformation magnitude is too small, the substance deformation is effectively masked by the simple relaxation of the lattice structure, and therefore not enough to transfer the stress to the neighbouring layers. This can be observed in Figure 11 where some of the explored cases failed to show a clear elastic behaviour at very low values of strain (i.e., at low values of oscillation amplitude). On the other hand, it is important to bear in mind that our proposed modelling approached is based on the concepts and theory of linear viscoelasticity, and therefore applicable mostly to small deformations. This is specially important from the elastic component perspective. It is then expected that the model fails at large values of strain, which can be observed also in Figure 11. In any case, the large range of values explored in the present work is somehow justified because although theory indicates that applicability of the linear viscoelastic behaviour should be expected at small deformations, there is not a clear sense of what "small deformation" might be in a generic substance, like those explored in this work. In fact, the range of parameters was selected only on the basis of exploring the ability of our modelling approach to capture different substances, as well as the reliability and stability of our approach when employed under extreme conditions. Our results indicate that the applicability of our model requires an adequate selection of parameters and operating ranges to avoid problems with very large deformations, as well as with very small deformations, but that it is able to reliably reproduce expected theoretical elastic behaviour within appropriate ranges of shear strain.

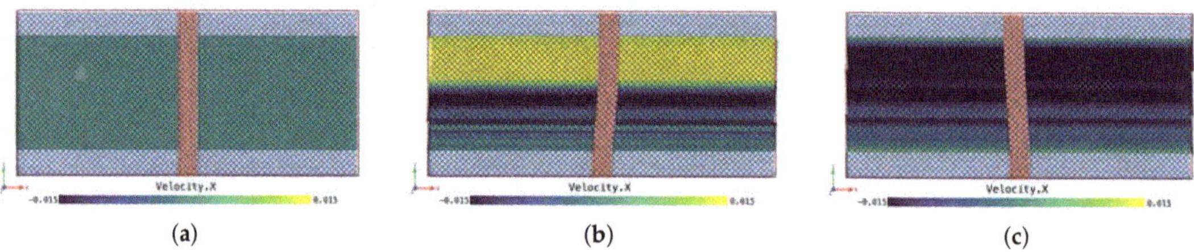

(a)   (b)   (c)

Figure 9. Illustration of set of particles tracked during the oscillatory test in a pure elastic substance. Particles not included in the timelines have been coloured by velocity in x−direction, at three different time instants. (**a**) $t = 0\,$s. (**b**) $t = 5 \times 10^{-5}\,$s. (**c**) $t = 4.5 \times 10^{-4}\,$s.

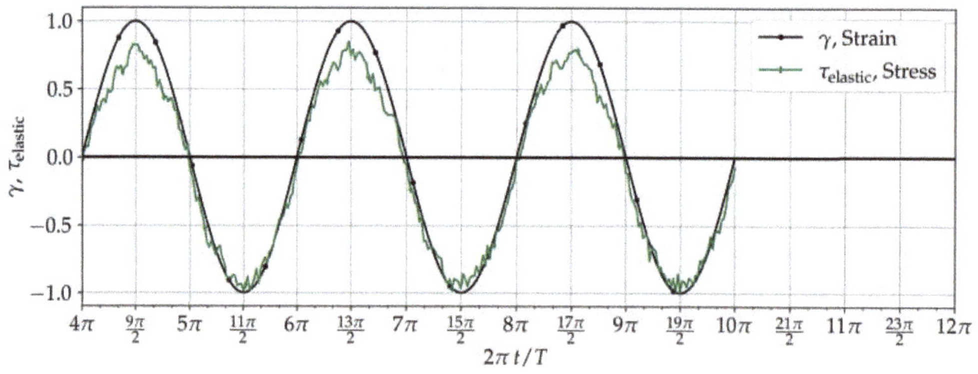

Figure 10. Shear stress τ and strain γ vs. time in the example elastic case.

As our model involves a potential emulating the harmonic LSM model, it is important to asses the impact of oscillatory external forcing or deformation. Then, by performing external forced strain on a pure elastic substance, internal forces and stresses are developed, affecting the movement of each particle in the lattice. This can eventually cause harmonic instabilities to start and grow and, if care is not taken, these internal oscillations and instabilities can render the lattice and model inaccurate. For instance, from the numerical experiments, it was possible to observe that configurations with low level of stiffness caused the generation and propagation of internal longitudinal waves that, given the conditions, might disrupt the structure of the lattice, and therefore rendering the simulation unstable and not physically representative. Clearly, if the elastic potentials are set to low values, but above a given threshold, the behaviour is still elastic but with extremely low stiffness, which might still cause the appearance of some longitudinal waves. The important aspect to consider is then the numerical stability and physical consistency, which will help to determine a good configuration of parameters. Precisely, the parameters assignment for the repulsive component play also an important role in this stability. If the magnitude of repulsive potential, represented by ϵ_{Soft}, is made too small, internal oscillations and waves appear, and they might end up breaking the structure of the lattice. In the limiting stable cases, longitudinal waves appearing and propagating through the whole domain, bring about interference patterns that travel across the modelled substance, but without causing disruption to the particles lattice. Examples of both situations are shown in Figures 12 and 13.

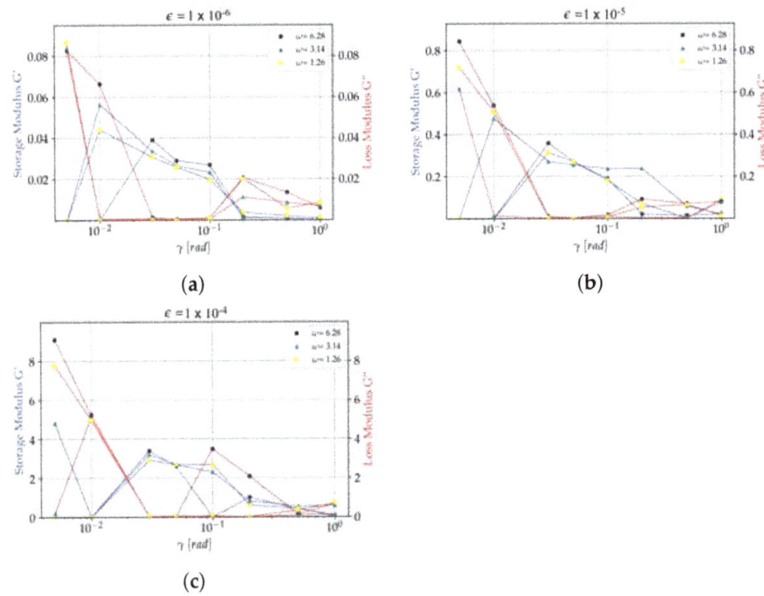

Figure 11. Storage and Loss moduli obtained with an exclusively elastic configuration of the proposed model. Blue: Storage modulus G′, red: Loss modulus G″. Tests obtained for $5 \times 10^{-3} \leq \gamma \leq 1$, and for oscillating frequencies $\omega = 1.26, 3.14, 6.28$ rad s^{-1}. (**a**) Results for $\epsilon_{CS} = 1 \times 10^{-6}$. (**b**) Results for $\epsilon_{CS} = 1 \times 10^{-5}$. (**c**) Results for $\epsilon_{CS} = 1 \times 10^{-4}$.

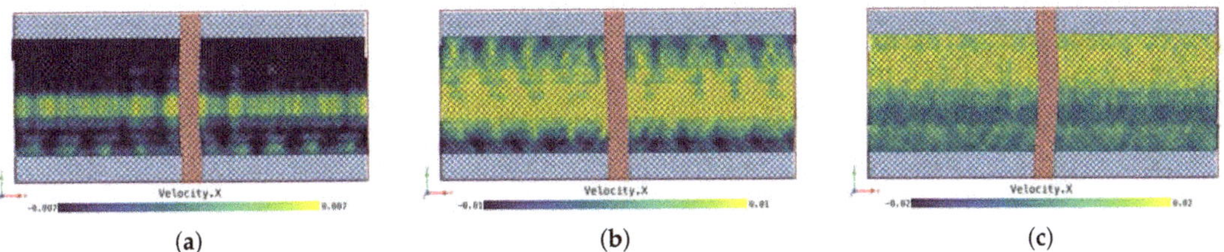

Figure 12. Set of particles tracked during the oscillatory test in a pure elastic substance. Example of a limiting stable case. Longitudinal waves below stability threshold. (**a**) $t = 3.9 \times 10^{-4}$ s. (**b**) $t = 4.3 \times 10^{-4}$ s. (**c**) $t = 7.8 \times 10^{-4}$ s.

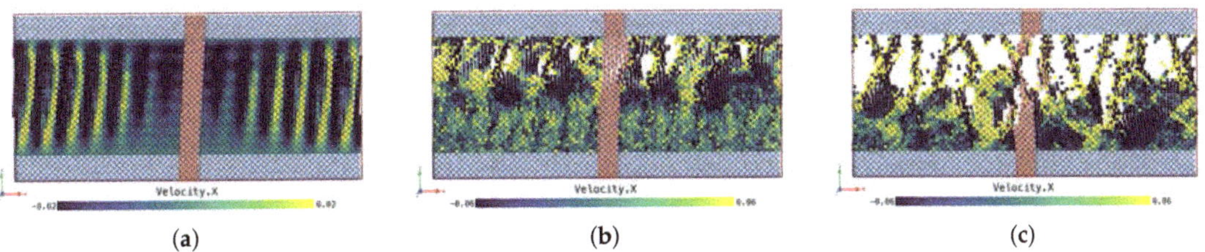

Figure 13. Set of particles tracked during the oscillatory test in a pure elastic substance. Example of a unstable case. Longitudinal waves above stability threshold. (**a**) $t = 5.5 \times 10^{-5}$ s. (**b**) $t = 7.5 \times 10^{-5}$ s. (**c**) $t = 8.5 \times 10^{-5}$ s.

The viscous component in our model is captured by a standard SPH model, as mentioned before. As it can be appreciated from Figures 14 and 15, the viscous component employed in our model is able to capture the intended dissipative nature, exhibiting the expected loss angle $\delta \approx 90°$ between the signals of strain and stress. The consistency of the viscous response was also tested for a number of nominal strain amplitudes and oscillation frequencies with the oscillatory strain test. Numerical results shown in Figure 16 allow us to conclude that the model is able to capture the expected behaviour of higher loss than storage moduli, for all the different frequencies and magnitudes of strain oscillation explored in the numerical experiments.

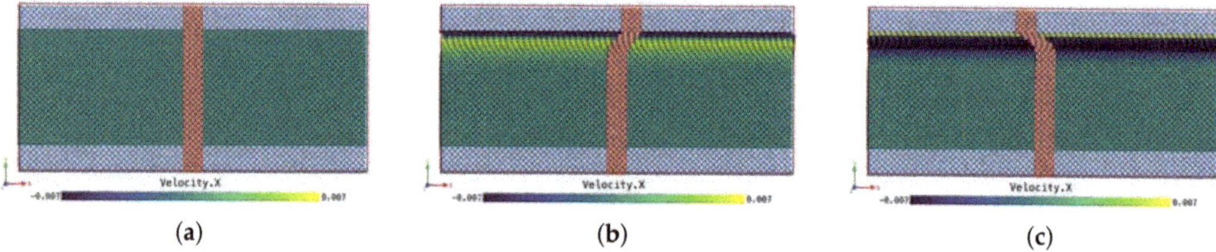

(a)   (b)   (c)

Figure 14. Set of particles tracked during the oscillatory test in a pure viscous substance. (**a**) $t = 0$ s. (**b**) $t = 1 \times 10^{-4}$ s. (**c**) $t = 4.7 \times 10^{-4}$ s.

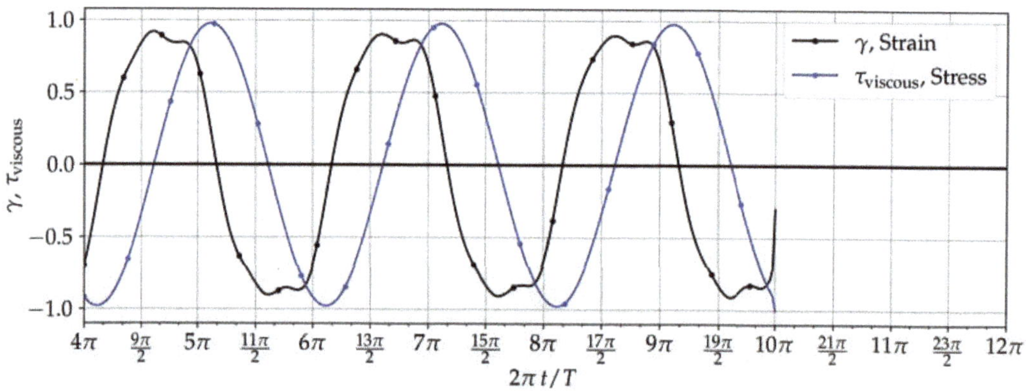

Figure 15. Shear stress τ and strain γ vs. time in the example viscous case.

Results of the oscillatory tests obtained with the proposed viscoelastic methodology are presented graphically in Figure 17 for the timelines of a selected set of particles. Figure 17 shows the presence of elastic internal waves, similar to those longitudinal waves observed for the elastic configuration and presented in Figure 12. However, these internal waves are essentially below any unstable threshold, as they remain stable for more than 4 oscillation periods. The stability is also clear from the consistency of the lattice structure even after several periods of oscillation. In Figure 18, is shown the time evolution of shear stress (τ) and strain (γ) for a selected viscoelastic configuration. It is clear that, compared to the elastic and viscous cases, the viscoelastic fluid shows a loss angle that is neither $\delta = 0°$ or $\delta = 90°$, which is very much in line with the results observed in some real viscoelastic substances that present stress and strain signals out of phase, but not to the degree of a fully viscous substance. This distinctive characteristic was tested for a number of oscillation frequencies and shear strain amplitudes. The results for the different shear conditions are presented in Figure 19. From the graphs it is clear that the response for the different viscoelastic settings are dependent from the oscillation frequency and the amount of strain imposed. This feature has been documented [56–58], and indeed it is one of the main results of many experimental tests showing that substances response to shear exhibit a clear frequency dependency, as well as a strain magnitude dependency.

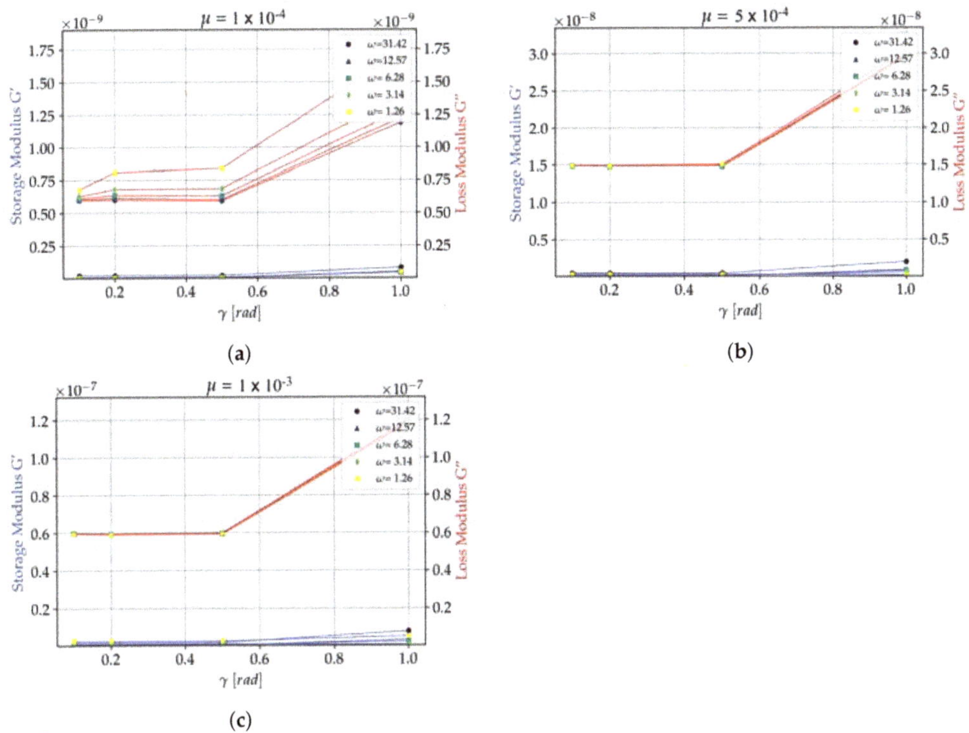

Figure 16. Storage and Loss moduli obtained with an exclusively viscous configuration of the proposed model. Blue: Storage modulus G', red: Loss modulus G''. Tests obtained for $1 \times 10^{-1} \leq \gamma \leq 1$, and for oscillating frequencies $\omega = 1.26$, 3.14, 6.28, 12.57, 31.42 rad s^{-1}. (**a**) Results for $\mu = 1 \times 10^{-4}$. (**b**) Results for $\mu = 5 \times 10^{-4}$. (**c**) Results for $\mu = 1 \times 10^{-3}$.

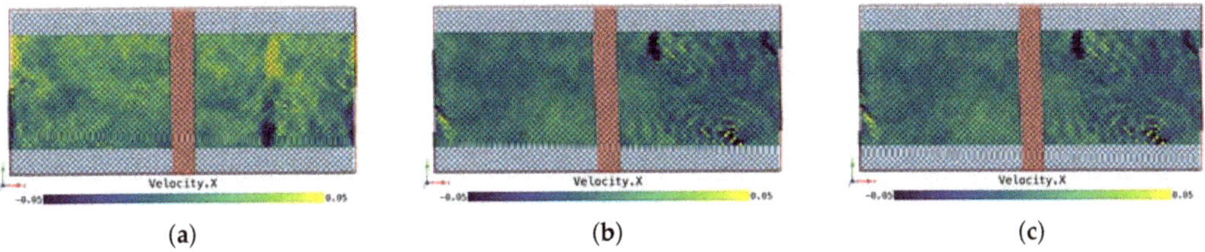

Figure 17. Illustration of set of particles tracked during the oscillatory test in a viscoelastic substance modelled with $\epsilon_{CS} = 1 \times 10^{-5}$ and $\mu = 0.1$. Particles coloured by velocity in $x-$direction, at three different time instants. (**a**) $t = 2.0$ s. (**b**) $t = 2.35$ s. (**c**) $t = 2.6$ s.

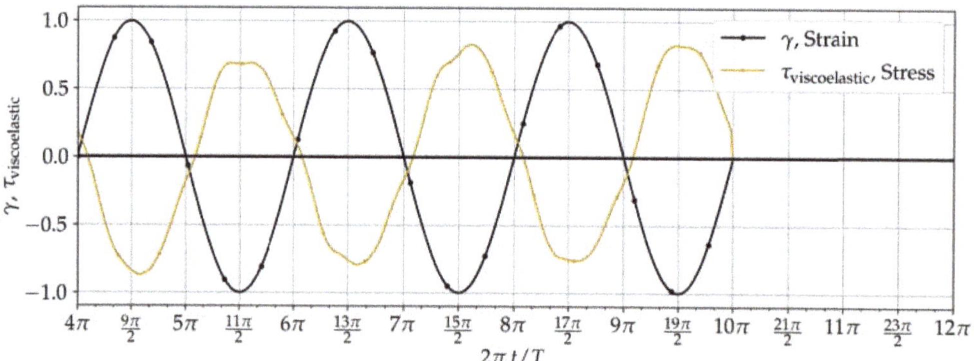

Figure 18. Shear stress τ and strain γ vs. time for a viscoelastic case.

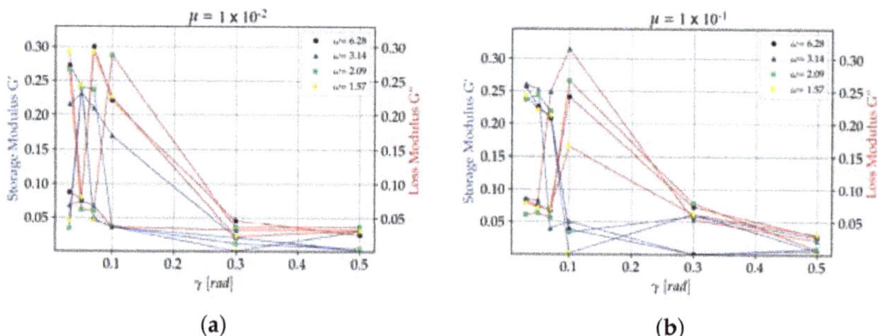

(**a**) (**b**)

Figure 19. Storage and Loss moduli obtained for a combined viscoelastic configuration according to proposed model. Blue: Storage modulus G', red: Loss modulus G''. Tests obtained for $3 \times 10^{-2} \leq \gamma \leq 0.5$, and for oscillating frequencies $\omega = 1.57, 2.09, 3.14, 6.28$ rad s^{-1}. (**a**) Results for $\mu = 1 \times 10^{-2}$. (**b**) Results for $\mu = 1 \times 10^{-1}$.

5.2. Viscoelastic Flows in Cylindrical Pipes

In the previous section, we showed that the model can produce materials whose storage and loss moduli are consistent with those of viscoelastic materials. However, this was the result of very small fluctuations that do not generate a real flow in the material. In this section, we will discuss the case of pipe flow, which produce a fluid-like motion of the particles and it is one of the most common mechanisms of transport of fluid-like substances. This particular flow, from the perspective of a viscoelastic substance, is also extremely important as it represents a standard constant shear condition employed in many viscometers. In this type of flow, a well-known characteristic is the expected velocity profile, specially for laminar low speed flows. In the case of a mostly viscous or Newtonian substance, both experience and theory have demonstrated that a parabolic velocity profile is formed, with maximum or peak velocity present in the centre-line of the pipe. Instead, for a mostly elastic fluid, the viscoelastic nature produces the so-called Bingham velocity profile or Bingham flow, where the velocity remains mostly constant in the cross section within the pipe, except near the pipe walls where there is usually a sharp drop towards the no-slip velocity at the wall. This type of flow is usually also known as "plug flow". The Bingham fluids are characterised by the existence of a yield stress and their ability to also transmit a shear stress without a velocity gradient, unlike Newtonian fluids. Nevertheless, in order to make Bingham fluids flow, the driving shear stress has to be larger than the yield stress. Below this yield stress the fluid will behave almost like a solid body and above as a liquid. Interestingly, although a precise estimation of the nature of the viscoelastic flow will require the estimation of such a yield stress, through the examination of the flow velocity profiles it is possible to ascertain what the respective shear stress is present in a given flow. Specifically, the ability of our model to capture viscoelastic behaviour under constant shear was explored in a constant gradient pipe flow case. Before presenting the simulation setup and results, a brief summary of the models and relationships to estimate some characteristics of a Bingham flow are included next.

5.2.1. Bingham Flows: Velocity Profiles and Yield Stress

As the numerical experiments presented in this section were performed in a cylindrical domain, it is convenient to express the shear stress in cylindrical coordinates and for a general condition (Newtonian or Bingham). For this, we consider an incompressible, laminar flow under the effects of a general pressure gradient, that in general can be decomposed into a gravity component (f_x) and a standard pressure difference ΔP, in a system of length L, which might be at an angle β to the vertical. Neglecting end effects, by assuming the dimension of the system in the radial direction is relatively small compared to that in the axial direction (L), and assuming an axial flow so $v_r = 0$, $v_\theta = 0$, and $v_z \neq 0$, it is possible to obtain a general expression for the momentum conservation equation as,

$$- \mu \nabla^2 v_z = -\nabla p + \rho g_z \tag{22}$$

where μ is the dynamic viscosity, ρ the density of the flow, and v_z the velocity along the axis of the pipe. In this expression it is also assumed that there are small flow rates so that the viscous forces impose strictly uniform flow. With this assumption v_z is independent of z and we may reasonably postulate that the velocity $v_z = v_z(r)$ and pressure $p = p(z)$. In this manner, the only non-vanishing components of the stress tensor are $\tau_{rz} = \tau_{zr}$, which depend only on r, and which can then be expressed as,

$$\tau_{rz} = \frac{\Delta P}{2L} r \tag{23}$$

These equations are derived without making any assumption about the type of fluid and so are applicable to both Newtonian and non-Newtonian fluids. If we use Newton's law of viscosity, $\tau_{rz} = -\mu \frac{d}{dr} v_z$ we can use Equation (23) to generate a differential equation for the velocity, which after integration gives the following flow profile,

$$v_z = -\frac{\Delta p}{4\mu L} r^2 - \frac{C_1}{\mu} \ln(r) + C_2 \tag{24}$$

that, however, it is only applicable to Newtonian fluids due to the use of Newton's law. To construct the flow velocity for Bingham fluids, we must make the following considerations: (i) The velocity profile of Newtonian fluids in pipe flows consists of a velocity gradient which decreases towards the centre of the pipe which in turn causes the shear stress, transmitted by fluid layers, to decline toward the pipe centre. (ii) Since Bingham fluids become solid when the applied shear stress falls below the yield stress we recognize that Bingham-fluids will become solid in the central layers of the pipe. Thus, we will have a solid 'plug' moving within the flow. (iii) In the process of deriving the velocity profile the radius of this solid area has to be additionally determined. Using these considerations, and applying adequate boundary conditions, it is possible to obtain a simple Bingham model where there is no flow until the critical/yield stress τ_0 is reached:

$$\eta \to \infty \quad \text{or} \quad \frac{d}{dr}(v_z) = 0 \qquad \text{when } |\tau_{rz}| \leq \tau_0 \tag{25}$$

$$\eta = \mu_0 + \frac{\tau_0}{\pm \frac{d}{dr}(v_z)} \quad \text{or} \quad \tau_{rz} = -\mu_0 \frac{d}{dr}(v_z) \pm \tau_0 \qquad \text{when } |\tau_{rz}| \geq \tau_0 \tag{26}$$

where η is the non-Newtonian viscosity and μ_0 is a Bingham model parameter with units of viscosity. τ_{rz} is positive when the positive sign is used with τ_0 and the negative sign with $d(v_z)/dr$. We can calculate the yield stress τ_0 by considering that $\tau_{rz} = \tau_0$ at some $r = r_0$. This follows from both Equations (25) and (26) at $|\tau_{rz}| = \tau_0$, which gives the following expression for the yield stress:

$$\tau_0 = \frac{\Delta P}{2L} r_0 \tag{27}$$

It is clear that the velocity profile can be split it into an inner $(r \leq r_0)$ and outer $(r_0 \leq r \leq R)$ region, where appropriate models for these profiles are:

$$v_{z0} = \frac{\Delta P}{4\mu_0 L} R^2 \left(1 - \frac{r^2}{R^2}\right) - \frac{\tau_0}{\mu_0} R \left(1 - \frac{r}{R}\right) \qquad \text{for } r_0 \leq r \leq R \tag{28}$$

for the flow velocity profile in the outer region, and

$$v_{zi} = \frac{\Delta P}{4\mu_0 L} R^2 \left(1 - \frac{r_0}{R}\right)^2 \qquad \text{for } r \leq r_0 \tag{29}$$

for the inner region. This last model gives a constant velocity in the inner region, as expected, for a fluid with plug flow, with $r_0 = 2L\tau_0/\Delta P$ being the radius of the plug-flow region in the central part of the pipe. In this manner, the velocity profile is parabolic in the outer region as given by Equation (28) and is flat in the inner region as given by Equation (29). Finally, the mass flow rate can be obtained by integration by parts of the velocity profile over the cross section of the circular pipe,

$$\dot{m} = \frac{\pi R^3 \rho}{\tau_R^3} \int_0^{\tau_R} \tau_{rz}^2 \left(-\frac{d}{dr}(v_z)\right) d\tau_{rz} \tag{30}$$

where we have used Equation (23) so that $r/R = \tau_{rz}/\tau_R$, with $\tau_R = (\Delta P/(2L))R$ defining the shear stress at the wall. After some manipulation, it is possible to obtain the so-called Buckingham–Reiner equation for the mass flow rate,

$$\dot{m} = \frac{\pi \Delta P R^4 \rho}{8 \mu_0 L} \left(1 - \frac{4}{3} \frac{\tau_0}{\tau_R} + \frac{1}{3} \frac{\tau_0^4}{\tau_R^4}\right) \tag{31}$$

where the mass flow rate is defined in terms of the yield stress τ_0 and the wall shear stress τ_R. Note that no flow occurs below the yield stress, so the equation is valid only for $\tau_R > \tau_0$.

5.2.2. Pipe Flow-Numerical Experiments

In order to test the ability of our model to capture viscoelastic behaviour at different spatial ranges, numerical experiments on small- and medium-scale pipes were performed for a number of parameters adequate to our model. For instance, the small-scale simulations were performed in a pipe with inner radius $r_{inner} = 2 \times 10^{-3}$ m and length $L = 2.4 \times 10^{-2}$ m. The SPH simulation was configured with a FCC lattice cubic of size $\Delta_L = 2.5 \times 10^{-4}$, and smoothing length defined as $h = 1.95 \Delta_L$. A schematic representation of the pipe geometry used is presented in Figure 20. Experiments were performed for a number of values of our base potential elastic model ϵ_{cs}, and using the viscous SPH Taitwater–Morris model with a set of values for the SPH dynamic viscosity. The combination of these parameters allowed us to reproduced a range of viscoelastic behaviours in response to a body force imposed over the set of fluid particles. The virtual fluid was forced to flow by imposing a body force f_x in the $x-$direction, while the domain was defined as periodic along the flow direction.

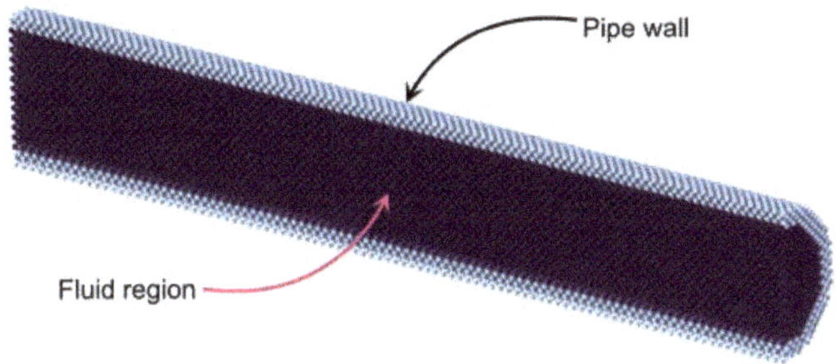

Figure 20. Pipe geometry for the small scale simulations.

In order to visualize the multiple flow regimes obtained, both velocity profiles and snapshots of the particles of our "numerical fluid" at different times have been produced. For the latter, a timeline analysis, a set of particles is preselected to represent a control volume travelling with the particles at different time instants, in the same fashion as in the oscillatory tests. The selection of particles used to construct the timelines is shown in Figure 21 as observed at the initial time, $t = 0$ s.

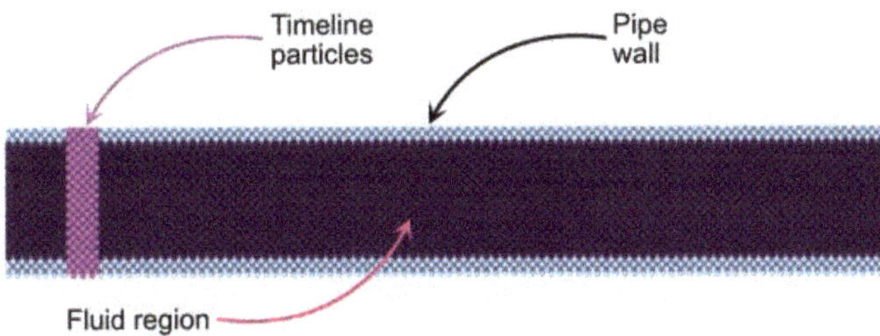

Figure 21. Tracking particles for construction of timelines in pipe simulations.

The model was assessed by a parametric study changing 3 variables: the elastic factor ϵ_{CS}, the driving body force f_x, and the dynamic viscosity μ. It is worth noting that, an additional parameter in our model that is tied to the factor ϵ_{CS}, is the factor of the repulsive potential ϵ_{Soft}. As it was described earlier, this parameter was linked to the ϵ_{CS} by a proportional relation, although for this case a reasonably dependency was found to be given as $\epsilon_{Soft} = 0.4\epsilon_{CS}$. Varying the coefficients in turn means that we are both varying the attractive and the repulsive potentials, therefore changing the ratio between elastic and viscous nature.

From the experiments, the cases exhibiting a Newtonian flow featured the expected predominant parabolic shape. The resolution of the velocity profiles was constrained to to the limit of particle points in the simulation, which caused a few Newtonian cases to lack a well defined parabolic shape and would tend to slightly skew from this pattern when approaching the wall, as it can be seen in Figure 22 below for the Newtonian plot. On the same figure the results for a Bingham flow are also presented. This flow, presented with a blue line, was almost ideal in shape, with the plug flow region containing mainly zero velocity gradient (as expected). As before there are some points that skew from the theoretical shape, specially at the yield stress point, where it can be seen that the flow increases locally before behaving Newtonian as one approaches the wall. Another case presented in Figure 22 is a flow that displays a mixed behaviour. For this generic flow, we see characteristics of Newtonian behaviour as well as features of a Bingham plug flow.

However, the regions in which the plug flow occurs do not follow the theory for Bingham flows, so it cannot be considered properly Bingham. This is because the flow is only plug flow in two regions: at a coaxial annulus and at a smaller radius around the centre, whilst maintaining Newtonian flow features elsewhere.

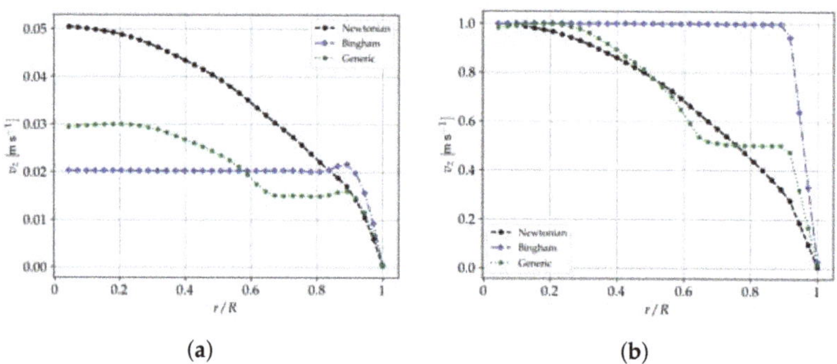

Figure 22. (**a**) Velocity profiles in axial direction in the pipe flow. profiles obtained for Newtonian, Bingham and Generic cases. (**b**) Normalised axial velocity profiles.

For the timelines analysis, we first considered the case when the attractive, and hence the repulsive potentials, were null by assigning $\epsilon_{CS} = 0$. As established earlier, these represent an elastic potential when considered together, and hence we expect a Bingham plug flow when this potential is dominant. On the other hand, for conditions where $\epsilon_{CS} = 0$ we expected flows to be essentially viscous Newtonian in nature. Our model is able to capture this condition, as ratified by the results presented in Figure 23, where the flow can be regarded as laminar, but more importantly Newtonian. Clearly, the flow travels downstream and, as the domain is periodic in the $x-$direction, particles re-enter the domain in the left side of the pipe still maintaining the parabolic profile, as seen in the third time snapshot at $t = 0.4032$ s in this figure.

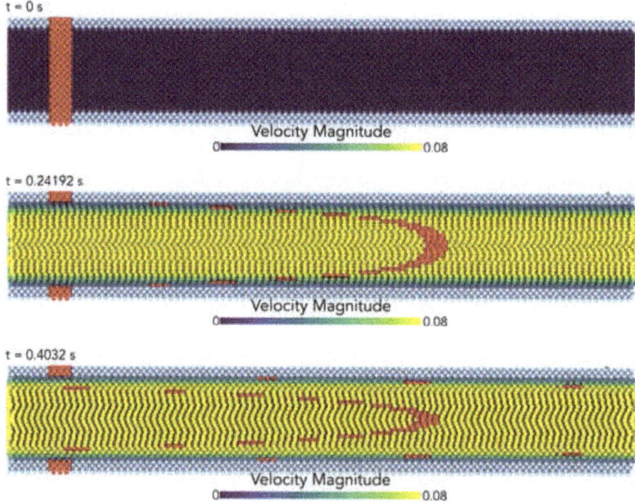

Figure 23. Snapshots of timelines evolution (tracking particles) in a pipe for a Newtonian viscous flow, obtained at three different instants. At third instant particles are going through the domain for the second time, as per periodic configuration. Flow obtained with: $\epsilon_{CS} = 0, \rho = 1 \times 10^3, c = 1 \times 10^{-1}, \mu = 1 \times 10^{-3}, g = 5 \times 10^{-1}$.

By assigning ϵ_{cs} to any non-zero value, it is expected that the model shows a mixed behaviour between complete Newtonian or complete Bingham, being the latter the extreme case of our viscoelastic model. By just setting $\epsilon_{cs} = 1 \times 10^{-14}$, even with a relatively low viscosity of $\mu = 1 \times 10^{-3}$, the model brings about a flow that resembles more a Bingham plug flow, as shown in Figure 24. However, some caution must be advised when assuming only these parameters to be involved in determining the flow type. This can be seen by the next result in which the viscosity was reduced to 3 orders of magnitude below, and we obtained Bingham flow, even with no elastic potential in action Figure 25.

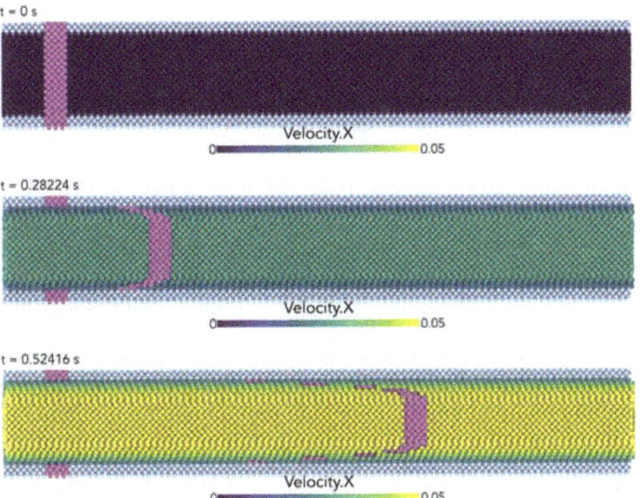

Figure 24. Evolution of timelines (tracking particles) in the pipe for response type "Bingham flow". Flow obtained with: $\epsilon_{cs} = 1 \times 10^{-14}$, $\rho = 1 \times 10^{3}$, $c = 1 \times 10^{-1}$, $\mu = 1 \times 10^{-4}$, $g = 1 \times 10^{-1}$.

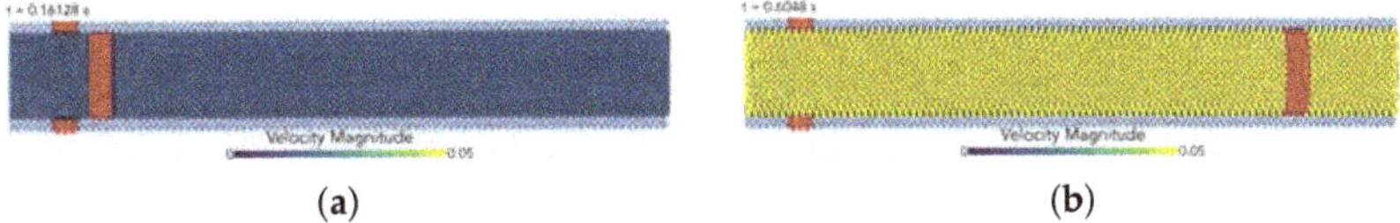

(a)   (b)

Figure 25. Bingham flow obtained with null elastic component. (a) Flow timelines at t = 0.16 s. (b) Flow timelines at t = 0.6 s. Flow obtained with: $\epsilon_{cs} = 0$, $\rho = 1 \times 10^{3}$, $c = 1 \times 10^{-1}$, $\mu = 1 \times 10^{-6}$, $g = 1 \times 10^{-1}$.

For conditions of constant shear as in the pipe flow, the elastic factor ϵ_{cs} is still a variable that plays a role into the possible appearance of some instabilities, as could be expected from the results of the oscillatory shear case. The change of settings in the model, from a case with $\epsilon_{cs} = 0.0$ to a configuration with $\epsilon_{cs} = 1 \times 10^{-12}$ produces a dramatic alteration of the behaviour of the flow, as appreciated in Figure 26. At $t = 0.16$ s the slow speed plug flow that was just beginning to form with $\epsilon_{cs} = 0.0$ (see Figure 25), has already developed an abnormal pattern, with the lattice starting to be disrupted, and some regions of low and high velocity appearing in an alternating fashion in the radial direction. The instability finally produces a completely ill-conditioned lattice, with large scale voids, that are completely clear at $t = 0.4$ s, in contrast with the completely stable plug flow obtained with $\epsilon_{cs} = 0.0$, that even at $t = 0.6$ s keeps a essentially flat velocity profile towards the centre of the pipe. In any case, once a set of parameters is identified as a potential troublesome setting, by a simple re-tuning of the model it is possible to effectively subdue the detected unstable mode. In spite of those few cases developing instabilities, the model behaves extremely well, considering the breadth of potential conditions that can be captured with the proposed methodology.

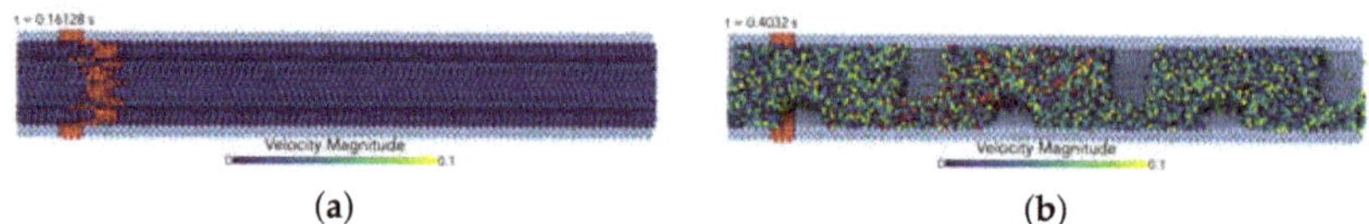

(a)   (b)

Figure 26. Unstable Bingham flow obtained with $\epsilon_{CS} = 1 \times 10^{-12}$. (a) Timelines at t = 0.16 s after the start of the simulation. (b) Timelines at t = 0.40 s. Flow obtained with: $\epsilon_{CS} = 1 \times 10^{-12}$, $\rho = 1 \times 10^3$, $c = 1 \times 10^{-1}$, $\mu = 1 \times 10^{-6}$, $g = 1 \times 10^{-1}$.

To analyse the behaviour of the flow with respect to these variables more rigidly, we propose a dimensionless constant π_1 defined as,

$$\pi_1 = \frac{g\rho^2 h^7}{\mu^2} \tag{32}$$

By constructing graphs of π_1 vs. μ for the different numerical experiments, it is possible to discern an operating region where the flow exhibited a "Bingham flow" response, as presented in Figures 27 and 28. In those plots we have coloured those points that exhibited predominantly plug flow. As can be seen from these figures, at higher values of π_1, or equivalently at lower values of μ, our model brings about a mostly viscoelastic behaviour, as would be expected. From the graphs it seems clear that there is a consistency in the region for which π_1 relates to Bingham flow. As a simple preliminary guide, values of yield stress obtained in the numerical experiments are plotted against π_1 for a number of values of ϵ_{CS} in Figure 29. From this graph, it is clear that yield stress can be obtained at low levels of π_1, but only if the elastic factor ϵ_{CS} is high enough. It is also important to highlight that large values of τ_0 can be obtained with our model even at low levels of stiffness, as there are some points showing high yield stress, even though the elastic factor was relatively reduced in comparison. Noteworthy, although there is an apparent region in the plot where the values of yield stress are clustered, it is clearly necessary to perform additional experiments to corroborate the full validity and application of the region observed in Figure 29.

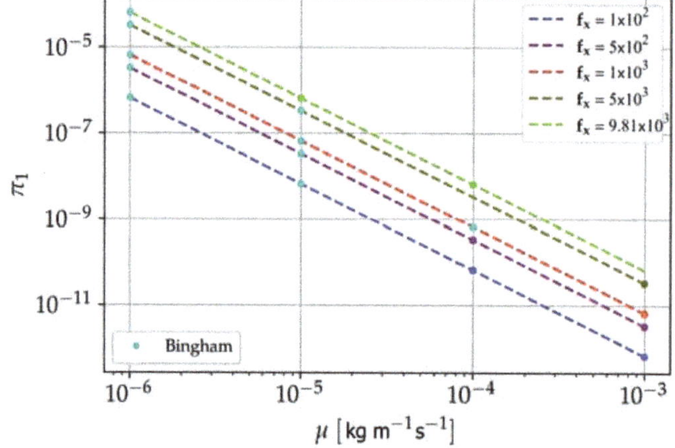

Figure 27. Dimensionless constant π_1 vs. μ at $\epsilon_{CS} = 0$ for different values of f_x.

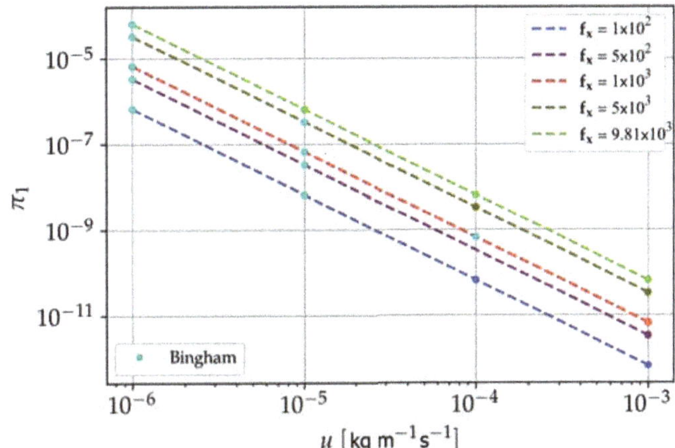

Figure 28. Dimensionless constant π_1 vs. μ at $\epsilon_{CS} = 1e - 14$ for different values of f_x.

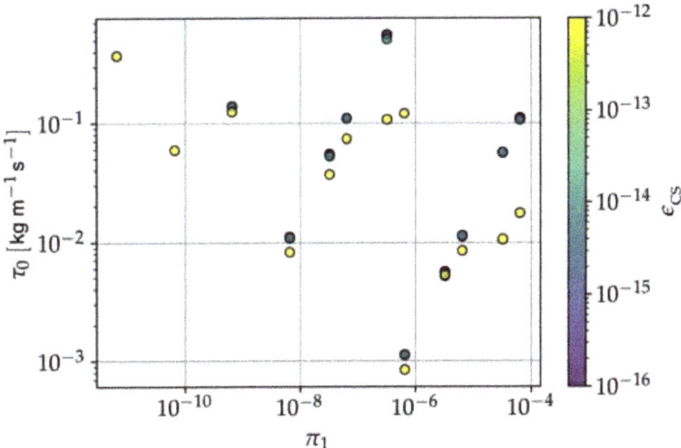

Figure 29. Values of yield shear stress τ_0 obtained in the pipe flow numerical experiments for different values of π_1 and ϵ_{CS}.

5.3. Column Collapse Due to Gravitational Potential

A third set of experiments was devised aiming to get a general overview of the effect of overlaying the potentials as proposed in this work in a more general and everyday situation. A benchmark case employed to show the ability of SPH to capture, among other phenomena, the free surface dynamics of a substance flowing freely is that of a two-dimensional column of water that suddenly collapses due to a gravitational force (see [59–61]). We test the hybrid modelling in two similar alternative cases, i.e., following the same rationale of the liquid column collapse, although with two volumetric domains of substance: a square prismatic column, and a cylindrical column, as shown in Figure 30. In these experiments an initial column of substance is enclosed within a cubic box, indicated partially by the dark grey particles in Figure 30 and used as container with non-permeable walls. The substance is then subjected to a vertical body force (in the form of an acceleration), for instance representing a gravity field in the negative $z - direction$. The box was prescribed as a cube with an internal volume of 1 m $\times$ 1 m $\times$ 1 m. The column of substance was prescribed in one of the experiments as a square prism with dimensions 0.5 m $\times$ 0.5 m $\times$ 0.9 m, in the $x-, y-$ and $z-$directions, respectively. In the second experiment the substance column, also contained within a box of same dimensions as in the previous experiment, was prescribed as a cylinder with a height of 0.9 m, and a base radius of 0.25 m. In both cases gravity was set to 9.81 m/s^2 and density was set to a standard density of 1×10^3 kg/m^3.

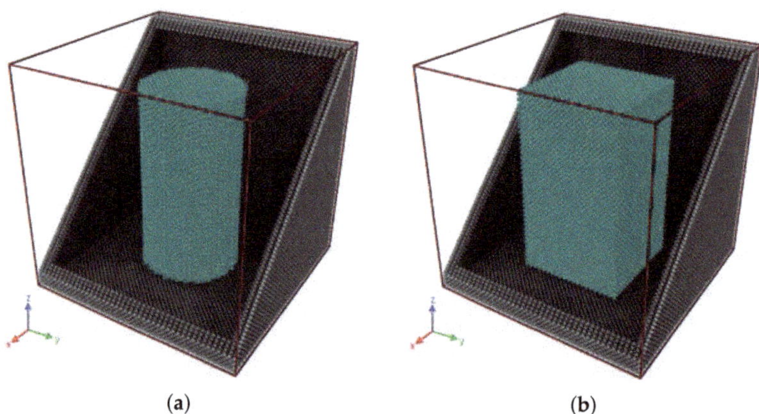

(a)   (b)

Figure 30. General view of particles configuration for the substance column collapse. Wall particles shown in dark-grey; substance particles shown in light blue. Cutting plane only for visualisation. (**a**) Cylindrical column. (**b**) Prismatic column.

Numerical experiments were performed using a similar SPH configuration as previously described, with a smoothing length h = 0.0525 m, and an initial face-centred cubic lattice (fcc). This set of experiments was performed using the SPH model as proposed by [44]. The selected SPH model adopted for these tests aimed at exploiting its ability to simulate flows at relatively high velocities with a consistent numerical stability. As presented previously, and discussed extensively by other authors (e.g., [45,52]), the Monaghan's model uses an artificial viscosity that can easily be converted to the standard absolute viscosity thanks to a widely accepted equivalence between the real dynamic viscosity μ and the dissipation factor α. The relationship between α and μ, presented in Equation (17), was employed here, with a minor modification to account for the effect of the imposed body force. Equally, with the goal of discriminating between the different cases explored numerically, a second non-dimensional relationship was constructed using the main parameters of our model, although tailored to the case of a column of a substance collapsing by its own weight, and given as

$$\pi_2 = \frac{\mu\,h\,c^3}{g\,\epsilon_{cs}} \tag{33}$$

where μ is the dynamic viscosity, c the numerical speed of sound, h the smoothing length and ϵ_{cs} the factor of the attractive component of the equivalent elastic potential. As usual, the value of the factor for the repulsive potential ϵ_{Soft} has been intentionally omitted but defined in terms of ϵ_{cs}. Numerical experiments showed that in this case the model was not strongly affected by variations of ϵ_{Soft} and that changes of its value, in a given range, had minimal influence on the overall performance and prediction capabilities. For instance, at the length and time scales involved in this case, the inclusion of the repulsive potential showed consistent numerical stability for values of potential magnitude in a range between 20% and 60% of the magnitude of the attractive potential. Figure 31, suggests that changing the ratio from 0.2 to 0.6 has a minimal impact in the overall evolution of the collapse of the column of liquid simulated. This appreciation is reinforced when examining the energy budgets presented in Figure 32, where the internal, kinetic, potential, and total energy components are plotted in time. The local maxima and minima for the kinetic and internal potential energies occur essentially at the same instants, and their values are also equivalent, making even difficult to distinguish the energy budgets evolution from each other. The remaining tests presented through the rest of the present section use $\epsilon_{\text{Soft}} = 0.4\,\epsilon_{\text{CS}}$, unless otherwise stated.

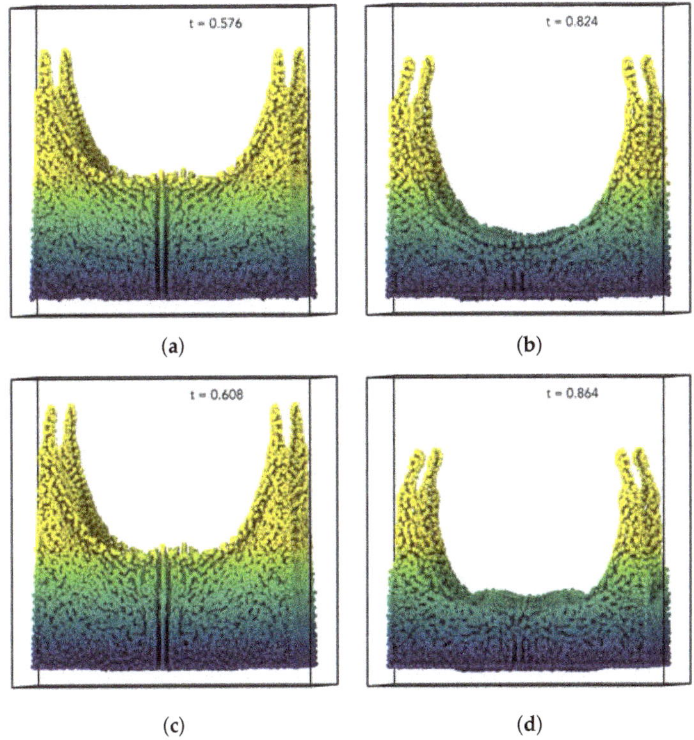

Figure 31. Column collapse at two different instants for different $\epsilon_{\text{Soft}}/\epsilon_{\text{CS}}$ ratios. Experiments performed for $\pi_2 = 9.9 \times 10^2$. (a) $\epsilon_{\text{Soft}} = 0.2\epsilon_{\text{CS}}$, $t \approx 0.4\tau$ ($\tau = 1.50$ s). (b) $\epsilon_{\text{Soft}} = 0.2\epsilon_{\text{CS}}$, $t \approx 0.55\tau$ ($\tau = 1.50$ s). (c) $\epsilon_{\text{Soft}} = 0.6\epsilon_{\text{CS}}$, $t \approx 0.4\tau$ ($\tau = 1.57$ s). (d) $\epsilon_{\text{Soft}} = 0.6\epsilon_{\text{CS}}$, $t \approx 0.55\tau$ ($\tau = 1.57$ s).

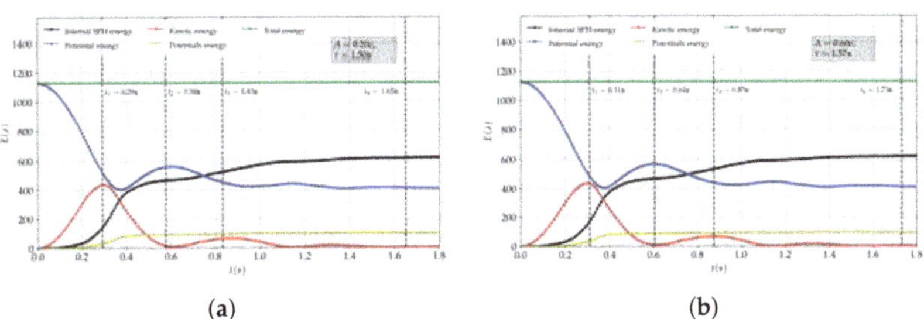

Figure 32. Time evolution of the components of the energy budget for the column collapse case. Viscoelastic response obtained by using $\epsilon_{\text{CS}} = 1 \times 10^{-4}$ and two values for ϵ_{Soft}. Simulation settings: $\mu = 1 \times 10^{-3}$, $c = 10.0$, $\nu_{\text{artf}} = 0.8$, $\rho = 1 \times 10^3$, $h = 5.25 \times 10^{-2}$, $\Delta_L = 2.5 \times 10^{-2}$, $\pi_2 = 9.9 \times 10^2$. (a) $\epsilon_{\text{Soft}} = 0.2\epsilon_{\text{CS}}$. (b) $\epsilon_{\text{Soft}} = 0.6\epsilon_{\text{CS}}$.

An example of the effect of changing or tuning the equivalent quasi-elastic potential, is presented in Figure 33, where a substance with the same viscous component μ, and under the effect of the same vertical downwards acceleration of $g_z = -9.81$, develop three different collapse evolutions in time. For these cases, by examining the evolution of the energy budget, it was possible to assess the impact of modulating the viscoelastic response through several values of the quasi-elastic potential. As in Figure 32, Energy vs. time plots reflecting on the evolution of the main components of the energy budget are used to assess the impact of altering the elastic factor ϵ_{cs}. In the energy budget plots, a simple characteristic time was employed to analyse the results at different instants. This characteristic time was defined as the point in time when the internal SPH energy reaches 99% of its steady-state value,

$$\tau = t_{e_{sph}=0.99e_{ss}}$$

Results for three different elastic factors are presented here: $\epsilon_{cs} = 1 \times 10^{-2}, 1 \times 10^{-3}$, 1×10^{-4}. In the first case, for $\epsilon_{cs} = 1 \times 10^{-2}$, Figure 33a–c show a extremely slow collapse of the substance column, clearly mimicking a gel-like behaviour, where the substance is affected by the downwards acceleration, but not really collapsing completely. Even after $t = 0.464$ s, when the collapse has already stopped and a balanced steady state has been reached, the column of the substance is still distinguishable. The time elapsed until steady state is reached has been estimated as $t \approx 0.43$ s, based on the time-story of the energy budget shown in Figure 34. On the other hand, Figure 33g–i show the evolution of the column collapse for a substance with $\epsilon_{cs} = 1 \times 10^{-4}$, that clearly presents a complete fluid-like behaviour. The collapse lasts $t = 1.76$ s, but during this time there is even a clear formation of a substance bouncing stage, observable at $t = 0.896$ s. The probable formation of ripples, and the occurrence of more than one bounce, can be inferred from the energy budget curve presented in Figure 35, where at least three clear maxima and minima in the potential gravitational energy are observed (blue line in the plot). The case with a relatively moderate elastic intensity of $\epsilon_{cs} = 1 \times 10^{-3}$, also shows a liquid-type of collapse, but it clearly behaves with some solid-like features, observable in the lack of multiple maxima in the energy plot presented in Figure 36.

A simple example of the collapse of a cylindrical column is presented in Figure 37. In this case, the initial particles' lattice has been set in such a way that an intentional slightly unbalanced initial spatial distribution is achieved, allowing the evolution of the collapse to follow a non-symmetrical trajectory. It is important to mention that the simulations with the cylindrical column for the settings used in the prismatic cases offered the same or equivalent type of responses. However, the non-symmetrical evolution captured by this configuration is included here as an example of the flexibility and potential variety of viscoelastic behaviours that can be captured with the methodology proposed in the present work.

Finally, the ability to capture the dynamic behaviour of viscoelastic substances can be better appreciated through animations of the simulations performed using our proposed methodology. Precisely, to get a better grasp of the capabilities of our proposed modelling approach, movies of some of the cases presented in this paper have been produced and made available as supplementary materials. In these videos is possible to appreciate the effect of some of the parameters discussed in this work, and the wide range of possible dynamic response that it is possible to mimic by simply modulating the viscous or elastic components of our model. In Video movie01.mp4 is presented the full evolution represented in Figure 31a,b, whereas in Video movie02.mp4 it is possible to fully appreciate the case presented in Figure 31c,d. The substance column collapse presented in Figure 33a–c can be visualized in Video movie03.mp4, while Videos movie04.mp4 and movie05.mp4 are the animations corresponding to the Figure 33d–i, respectively. The last non-symmetrical column collapse discussed in the previous paragraph, and presented through the snapshots depicted in Figure 37 have been also included in the supplementary material as Video movie06.mp4. From the animations it is clear that our modelling approach is able to represent the wide variety of behaviours and dynamic responses expected from viscoelastic substances.

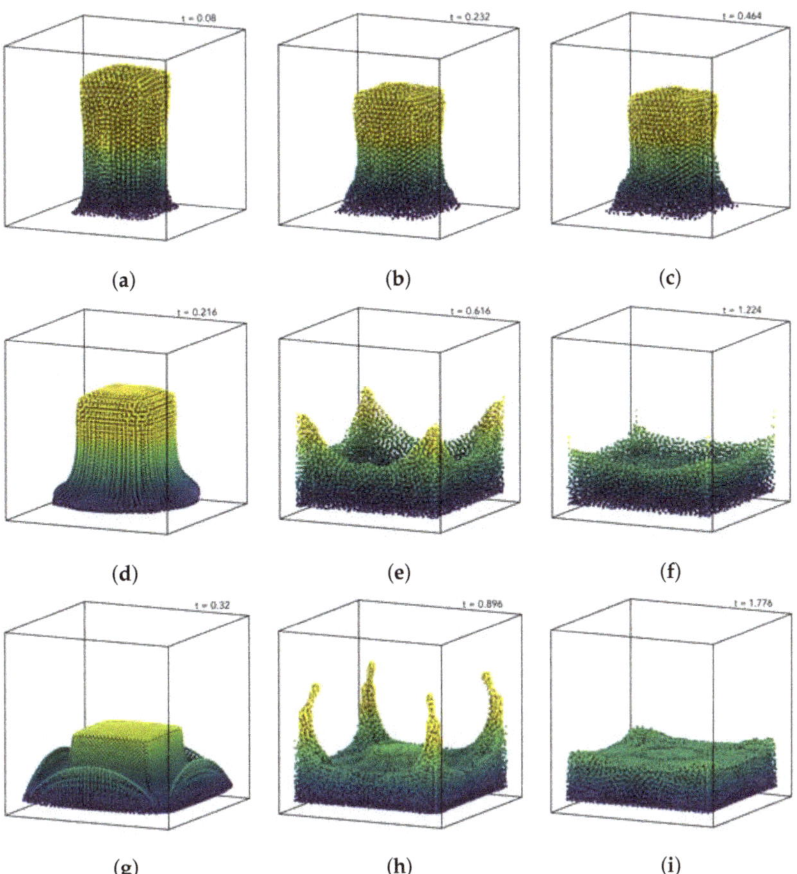

Figure 33. Column collapse at three different instants for different ratios of potentials magnitude.
(a) $\epsilon_{CS} = 1 \times 10^{-2}$, $t \approx 0.20\tau$. (b) $\epsilon_{CS} = 1 \times 10^{-2}$, $t \approx 0.55\tau$. (c) $\epsilon_{CS} = 1 \times 10^{-2}$, $t \approx 1.1\tau$.
(d) $\epsilon_{CS} = 1 \times 10^{-3}$, $t \approx 0.20\tau$. (e) $\epsilon_{CS} = 1 \times 10^{-3}$, $t \approx 0.55\tau$. (f) $\epsilon_{CS} = 1 \times 10^{-3}$, $t \approx 1.1\tau$.
(g) $\epsilon_{CS} = 1 \times 10^{-4}$, $t \approx 0.20\tau$. (h) $\epsilon_{CS} = 1 \times 10^{-4}$, $t \approx 0.55\tau$. (i) $\epsilon_{CS} = 1 \times 10^{-4}$, $t \approx 1.1\tau$.

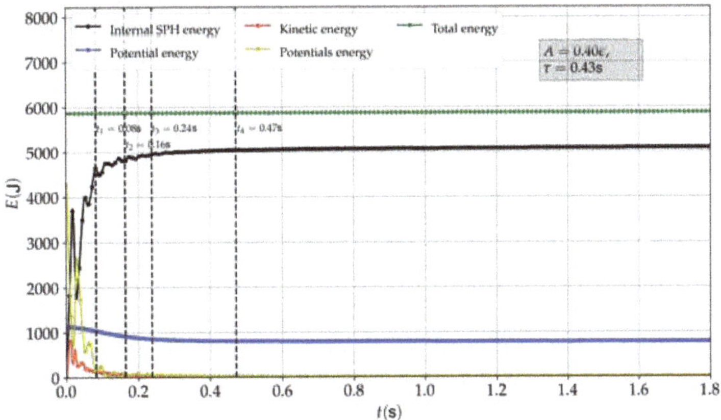

Figure 34. Energy budget vs. time for a "mostly" elastic substance. $\epsilon_{CS} = 1 \times 10^{-2}$, $A = 0.4\epsilon_{CS}$.

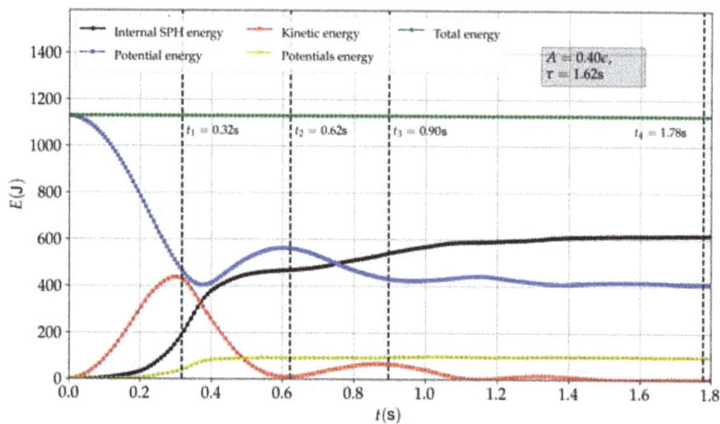

Figure 35. Energy budget vs. time for a "mostly" viscous substance. $\epsilon_{CS} = 1 \times 10^{-4}$, $A = 0.4\epsilon_{CS}$.

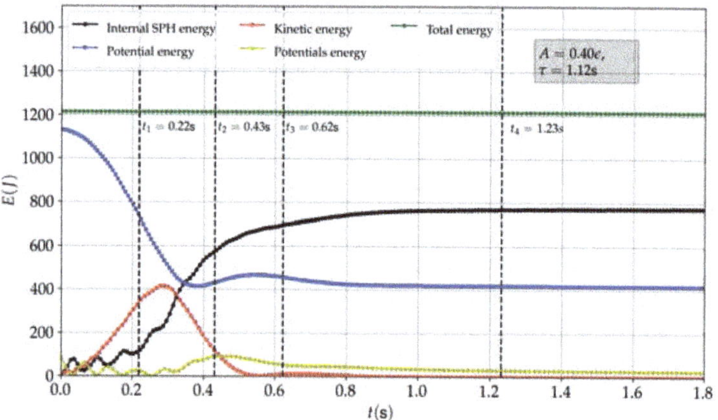

Figure 36. Energy budget vs. time for a viscoelastic substance. $\epsilon_{CS} = 1 \times 10^{-3}$, $A = 0.4\epsilon_{CS}$.

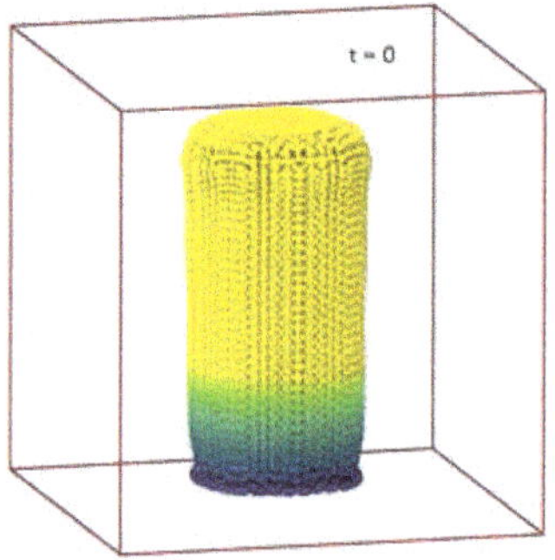
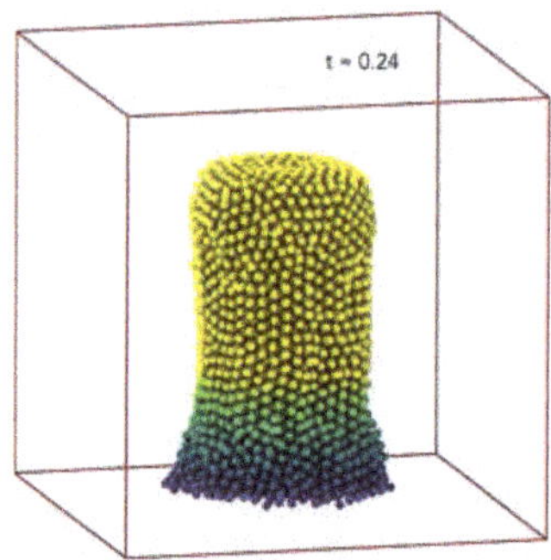

Figure 37. *Cont.*

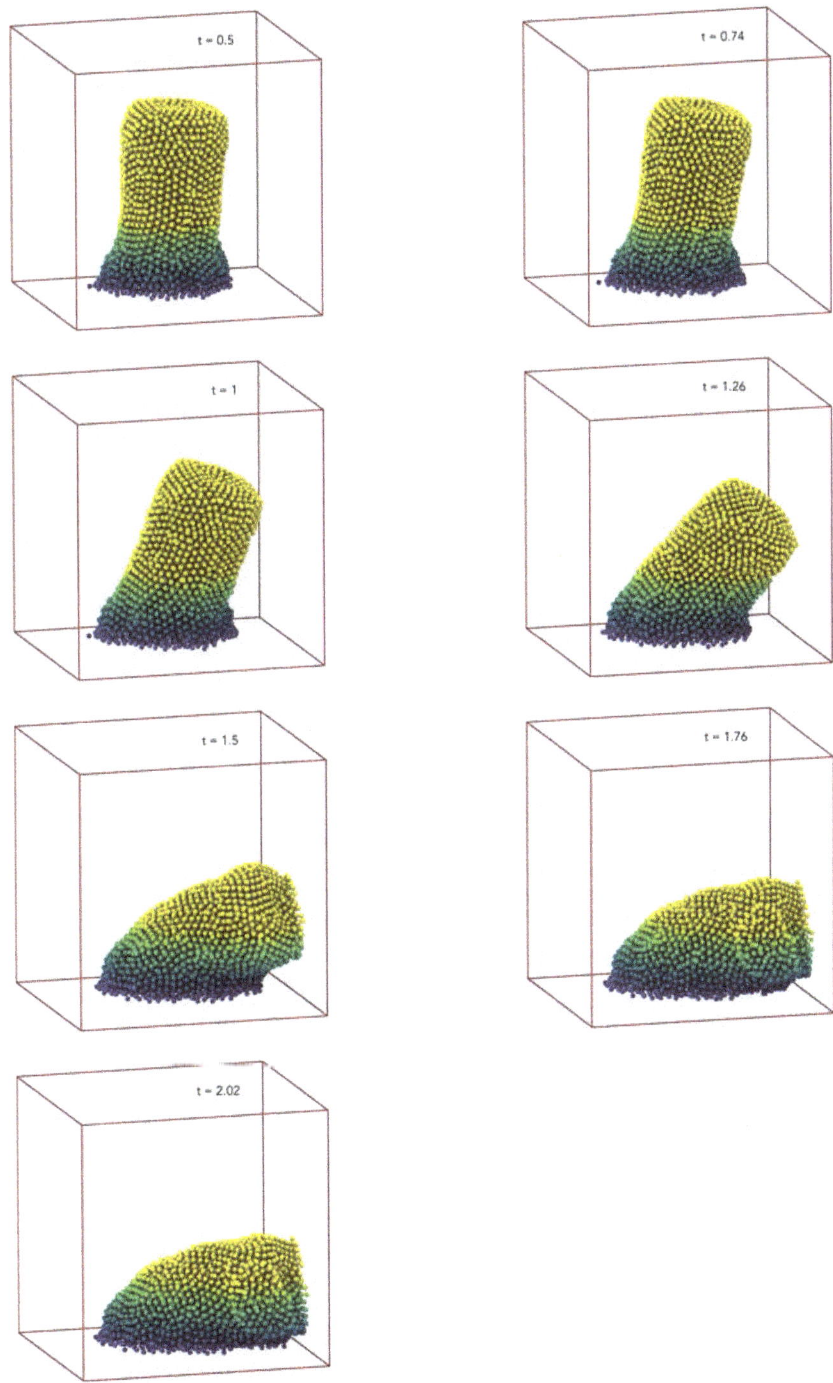

Figure 37. Time evolution of a cylindrical column of a viscoelastic substance in a non-symmetrical fashion. Substance settings: $\epsilon_{CS} = 1 \times 10^{-4}$, $\epsilon_{Soft} = 0.2\epsilon_{CS}$, $\rho = 1000$, $c = 10.0$, $\nu_{artf} = 0.8$.

6. Conclusions

A simplified methodology for simulating and modelling viscoelastic fluids, based on the concept of linear additive composition of energy potentials, has been proposed, validated, and tested in a wide extent of conditions and applications. The model showed a consistent and clear ability to capture classical features of viscoelastic substances from the rheology perspective. Trends obtained for the Loss and Storage moduli in terms of oscillating frequency and strain showed that the model produces consistent results with some well known and observed characteristics of viscoelastic substances. The methodology and its implementation allowed us to capture with relatively ease the elastic, viscous, and viscoelastic behaviours for different frequencies and strain amplitudes. Some instabilities mostly associated with the equivalent or quasi-elastic potential were identified, but it is clear that an early identification of troublesome combinations of parameters is possible, which would facilitate to any modeller or researcher to search for a more convenient or stable setting.

The model was also tested in a more conventional engineering application, i.e., as a model for flow in a circular pipe of a given viscoelastic substance. In this case, again the model was able to obtain and capture the main features, from the classical parabolic profiles of a pure viscous fluid, to the extreme of a substance showing yield stress, and therefore exhibiting plug flow regime. The model proved to be flexible enough to capture both situations, as well as a number of conditions in between.

The simplified modelling approach was also successful in mimicking the behaviour of substances that might be considered as gels, or extremely viscoelastic, as well as the natural free surface evolution of a liquid column collapsing under the effect of gravity. The simplicity and modularity of the modelling framework proposed in this work suggests that it might be used both as a rigorous simulation tool to study phenomena from a rheological/engineering perspective, as well as a practical modelling tool to emulate the flow of substances that exhibit some level of elasticity. Thanks to the modularity, tunable characteristics of the parameters involved, and conceptual simplicity, the proposed modelling approach can be a powerful simulation tool to be used for researchers and visual graphics modellers alike.

This simplified approach is proposed as a sort of "quick and dirty" method for particle simulations involving viscoelastic materials. If the simulation specifically focuses on the mechanical property the viscoelastic material, we suggest a more rigorous approach (see [53]) that requires rewriting the equation of motion to account for the specific viscoelasticity model. However, if the simulation focuses on the effect of the viscoelastic material on a larger computational domain, the proposed method is easier to implement because it only requires combining together different particle potentials, which is a standard procedure in particle simulations. For instance, in [62], we modelled the watery periciliary layer (PCL) located between the respiratory epithelium and a mucus layer. The PCL is a Newtonian fluid, but mucus has a complex viscoelastic response. In this case, it was important to account for the effect of the mucus layer on the PCL, but an easily implementable approximation of the mucus rheology based on the method proposed here would have been sufficient for the scope of that study.

As a future work, amongst some other possibilities, it would be extremely valuable to perform a clear categorization of relations between geometrical lattice properties, and magnitudes of the potentials used to replicate the quasi-elastic interaction. Furthermore, it is necessary to test the model in more intensive applications, to examine performance and computational costs demanded by any implementation. These factors clearly might help to decide on the more extended use of the technique herein proposed.

Supplementary Materials: The following are available online at: https://www.mdpi.com/article/10.3390/chemengineering5030061/s1, Video movie01.mp4: Animation of column collapse simulation using proposed methodology, for a fluid-like substance modelled with $\pi_2 = 9.9 \times 10^2$, $\epsilon_{\text{Soft}} = 0.2\epsilon_{\text{CS}}$; Video movie02.mp4: Animation of column collapse simulation using proposed methodology, for a fluid-like substance modelled with $\pi_2 = 9.9 \times 10^2$, $\epsilon_{\mathrm{Soft}} = 0.6\epsilon_{\mathrm{CS}}$; Video movie03.mp4: Animation of column collapse for a viscoelastic substance modelled with $\epsilon_{\mathrm{CS}} = 1 \times 10^{-2}$; Video movie04.mp4: Animation of column collapse for a viscoelastic substance modelled with $\epsilon_{\mathrm{CS}} = 1 \times 10^{-3}$; Video movie05.mp4: Animation of column collapse for a viscoelastic substance modelled with $\epsilon_{\mathrm{CS}} = 1 \times 10^{-4}$; Video movie06.mp4: Animation of a cylindrical column collapse of a viscoelastic substance modelled with $\epsilon_{\mathrm{CS}} = 1 \times 10^{-4}$, $\epsilon_{\mathrm{Soft}} = 0.2\,\epsilon_{\mathrm{CS}}$, $\rho = 1 \times 10^3$, c = 10, $\nu_{\mathrm{artf}} = 8 \times 10^{-1}$;

Author Contributions: Conceptualization, C.D.-D. and A.A.; methodology, C.D.-D. and A.A.; software, C.D.-D.; validation, C.D.-D.; writing, original draft preparation, C.D.-D.; writing, review and editing, C.D.-D. and A.A.; funding acquisition, C.D.-D. and A.A. All authors read and agreed to the published version of the manuscript.

Funding: This research was funded European Union's Horizon 2020 research and innovation programme under the Marie Sklodowska-Curie grant agreement No 841814.

Institutional Review Board Statement: Not applicable.

Informed Consent Statement: Not applicable.

Data Availability Statement: Metadata for all research data created throughout this project has been recorded in the University of Birmingham's current research information system PURE. These records can be searched within the University's Research Portal, FindtIt@Bham.

Acknowledgments: The computations described in this paper were performed using the University of Birmingham's BlueBEAR HPC service, which provides a High Performance Computing service to the University's research community. See http://www.birmingham.ac.uk/bear (accessed on 9 August 2021) for more details.

Conflicts of Interest: The authors declare no conflict of interest.

References

1. Stamper, R.L.; Lieberman, M.F.; Drake, M.V. Secondary open angle glaucoma. In *Becker-Shaffer's Diagnosis and Therapy of the Glaucomas*; Elsevier: Edinburgh, UK, 2009; pp. 266–293. [CrossRef]
2. Yang, K.H. Material Laws and Properties. In *Basic Finite Element Method as Applied to Injury Biomechanics*; Associated Press: London, UK, 2018; pp. 231–256. [CrossRef]
3. Chandran, N.; Sarathchandran, C.; Thomas, S. Introduction to rheology. In *Rheology of Polymer Blends and Nanocomposites*; Springer: Dordrecht, The Netherlands, 2009; pp. 1–17. [CrossRef]
4. Ferry, J.D. *Viscoelastic Properties of Polymers*; John Wiley & Sons: New York, NY, USA, 1980.
5. Nowick, A. *Anelastic Relaxation in Crystalline Solids*; Academic Press: New York, NY, USA, 1972.
6. Lakes, R.S. Viscoelastic measurement techniques. *Rev. Sci. Instrum.* **2004**, *75*, 797–810. [CrossRef]
7. Ardakani, H.A.; Mitsoulis, E.; Hatzikiriakos, S.G. Thixotropic flow of toothpaste through extrusion dies. *J. Non-Newton. Fluid Mech.* **2011**, *166*, 1262–1271. [CrossRef]
8. Mitsoulis, E.; Khalfalla, Y.; Benyounis, K. Polymer Film Casting: Modeling. In *Reference Module in Materials Science and Materials Engineering*; Elsevier: Amsterdam, The Netherlands, 2016. [CrossRef]
9. Tabilo-Munizaga, G.; Barbosa-Cánovas, G.V. Rheology for the food industry. *J. Food Eng.* **2005**, *67*, 147–156. [CrossRef]
10. Myhan, R.; Białobrzewski, I.; Markowski, M. An approach to modeling the rheological properties of food materials. *J. Food Eng.* **2012**, *111*, 351–359. [CrossRef]
11. Derkach, S.R.; Krägel, J.; Miller, R. Methods of measuring rheological properties of interfacial layers (Experimental methods of 2D rheology). *Colloid J.* **2009**, *71*, 1–17. [CrossRef]
12. Denn, M.M. Issues in Viscoelastic Fluid Mechanics. *Annu. Rev. Fluid Mech.* **1990**, *22*, 13–32. [CrossRef]
13. Shariff, M.H.B.M.; Bustamante, R.; Merodio, J. Rate type constitutive equations for fiber reinforced nonlinearly vicoelastic solids using spectral invariants. *Mech. Res. Commun.* **2017**, *84*, 60–64. [CrossRef]
14. Jha, N.K.; Reinoso, J.; Dehghani, H.; Merodio, J. Constitutive modeling framework for residually stressed viscoelastic solids at finite strains. *Mech. Res. Commun.* **2019**, *95*, 79–84. [CrossRef]
15. Drozdov, A.D.; Kolmanovskii, V.B. Constitutive Models of Viscoelastic Materials. In *Stability in Viscoelasticity*; Academic Press, Elsevier: Amsterdam, The Netherlands, 1994; pp. 1–132. [CrossRef]
16. Wineman, A. Nonlinear Viscoelastic Solids—A Review. *Math. Mech. Solids* **2009**, *14*, 300–366. [CrossRef]
17. Balbi, V.; Shearer, T.; Parnell, W.J. A modified formulation of quasi-linear viscoelasticity for transversely isotropic materials under finite deformation. *Proc. R. Soc. Math. Phys. Eng. Sci.* **2018**, *474*, 20180231. [CrossRef]
18. Zhang, W.; Capilnasiu, A.; Nordsletten, D. Comparative Analysis of Nonlinear Viscoelastic Models across Common Biomechanical Experiments. *J. Elast.* **2021**, 1–36. [CrossRef]

19. Mackay, A.T.; Phillips, T.N. On the derivation of macroscopic models for compressible viscoelastic fluids using the generalized bracket framework. *J. Non-Newton. Fluid Mech.* **2019**, *266*, 59–71. [CrossRef]

20. Clavet, S.; Beaudoin, P.; Poulin, P. Particle-based viscoelastic fluid simulation. In Proceedings of the 2005 ACM SIG-GRAPH/Eurographics symposium on Computer Animation-SCA'05, Los Angeles, CA, USA, 29–31 July 2005; ACM Press: New York, NY, USA, 2005. [CrossRef]

21. Westervoß, P.; Turek, S.; Damanik, H.; Ouazzi, A. The Tensor Diffusion approach for simulating viscoelastic fluids. *J. Non-Newton. Fluid Mech.* **2020**, *286*, 104431. [CrossRef]

22. Rajagopal, K.R.; Srinivasa, A.R. A Gibbs-potential-based formulation for obtaining the response functions for a class of viscoelastic materials. *Proc. R. Soc. Math. Phys. Eng. Sci.* **2011**, *467*, 39–58. [CrossRef]

23. Goktekin, T.G.; Bargteil, A.W.; O'Brien, J.F. A method for animating viscoelastic fluids. *ACM Trans. Graph.* **2004**, *23*, 463–468. [CrossRef]

24. Chang, Y.; Bao, K.; Liu, Y.; Zhu, J.; Wu, E. A particle-based method for viscoelastic fluids animation. In Proceedings of the 16th ACM Symposium on Virtual Reality Software and Technology-VRST'09, Kyoto, Japan, 18–20 November 2009; ACM Press: New York, NY, USA, 2009; pp. 111–117. [CrossRef]

25. Takamatsu, K.; Kanai, T. A fast and practical method for animating particle-based viscoelastic fluids. *Int. J. Virtual Real.* **2011**, *10*, 29–35. [CrossRef]

26. Alexiadis, A. A smoothed particle hydrodynamics and coarse-grained molecular dynamics hybrid technique for modelling elastic particles and breakable capsules under various flow conditions. *Int. J. Numer. Methods Eng.* **2014**, *100*, 713–719. [CrossRef]

27. Alexiadis, A. A new Framework for Modelling the Dynamics and the Breakage of Capsules, Vesicles and Cells in Fluid Flow. *Procedia IUTAM* **2015**, *16*, 80–88. [CrossRef]

28. Alexiadis, A. The Discrete Multi-Hybrid System for the Simulation of Solid-Liquid Flows. *PLoS ONE* **2015**, *10*, e0124678. [CrossRef]

29. Mohammed, A.M.; Ariane, M.; Alexiadis, A. Using Discrete Multiphysics Modelling to Assess the Effect of Calcification on Hemodynamic and Mechanical Deformation of Aortic Valve. *ChemEngineering* **2020**, *4*, 48. [CrossRef]

30. Schütt, M.; Stamatopoulos, K.; Simmons, M.; Batchelor, H.; Alexiadis, A. Modelling and simulation of the hydrodynamics and mixing profiles in the human proximal colon using Discrete Multiphysics. *Comput. Biol. Med.* **2020**, *121*, 103819. [CrossRef]

31. Sahputra, I.H.; Alexiadis, A.; Adams, M.J. A Coarse Grained Model for Viscoelastic Solids in Discrete Multiphysics Simulations. *ChemEngineering* **2020**, *4*, 30. [CrossRef]

32. Chhabra, R.P.; Richardson, J.F. *Non-Newtonian Flow in the Process Industries: Fundamentals and Engineering Applications*; Butterworth-Heinemann: Oxford, MS, USA, 1999.

33. Mai-Duy, N.; Phan-Thien, N. *Understanding Viscoelasticity*; Springer International Publishing AG: Berlin/Heidelberg, Germany, 2017.

34. Flügge, W. *Viscoelasticity*; Springer: Berlin/Heidelberg, Germany, 1975.

35. Bonfanti, A.; Kaplan, J.L.; Charras, G.; Kabla, A. Fractional viscoelastic models for power-law materials. *Soft Matter* **2020**, *16*, 6002–6020. [CrossRef] [PubMed]

36. Wang, S.; Xu, Y.; Li, J. Stationary probability densities of generalized Maxwell-type viscoelastic systems under combined harmonic and Gaussian white noise excitations. *J. Braz. Soc. Mech. Sci. Eng.* **2020**, *42*, 1–9. [CrossRef]

37. Epaarachchi, J.A. The effect of viscoelasticity on fatigue behaviour of polymer matrix composites. In *Creep and Fatigue in Polymer Matrix Composites*; Woodhead Publishing Limited: Cambridge, UK, 2011; pp. 492–513. [CrossRef]

38. Renaud, F.; Dion, J.L.; Chevallier, G.; Tawfiq, I.; Lemaire, R. A new identification method of viscoelastic behavior: Application to the generalized Maxwell model. *Mech. Syst. Signal Process.* **2011**, *25*, 991–1010. [CrossRef]

39. Qi, H.; Xu, M. Unsteady flow of viscoelastic fluid with fractional Maxwell model in a channel. *Mech. Res. Commun.* **2007**, *34*, 210–212. [CrossRef]

40. Lewandowski, R.; Chorążyczewski, B. Identification of the parameters of the Kelvin–Voigt and the Maxwell fractional models, used to modeling of viscoelastic dampers. *Comput. Struct.* **2010**, *88*, 1–17. [CrossRef]

41. Marynowski, K.; Kapitaniak, T. Kelvin–Voigt versus Bürgers internal damping in modeling of axially moving viscoelastic web. *Int. J. Non-Linear Mech.* **2002**, *37*, 1147–1161. [CrossRef]

42. Gingold, R.A.; Monaghan, J.J. Smoothed particle hydrodynamics: Theory and application to non-spherical stars. *Mon. Not. R. Astron. Soc.* **1977**, *181*, 375–389. [CrossRef]

43. Lucy, L.B. A numerical approach to the testing of the fission hypothesis. *Astron. J.* **1977**, *82*, 1013. [CrossRef]

44. Monaghan, J.J.; Gingold, R.A. Shock simulation by the particle method SPH. *J. Comput. Phys.* **1983**, *52*, 374–389. [CrossRef]

45. Ganzenmuller, G.C.; Steinhauser, M.O.; Liedekerke, P.V. The implementation of Smooth Particle Hydrodynamics in LAMMPS. *Liedekerke Kathol. Univ. Leuven* **2011**, *1*, 1–26.

46. Liu, M.B.; Liu, G.R. Smoothed Particle Hydrodynamics (SPH): An Overview and Recent Developments. *Arch. Comput. Methods Eng.* **2010**, *17*, 25–76. [CrossRef]

47. Plimpton, S. Fast Parallel Algorithms for Short-Range Molecular Dynamics. *J. Comput. Phys.* **1995**, *117*, 1–19. [CrossRef]

48. LAMMPS. LAMMPS Molecular Dynamics Simulator. Available online: https://www.lammps.org/index.html (accessed on 1 August 2021).

49. Morris, J.P.; Fox, P.J.; Zhu, Y. Modeling Low Reynolds Number Incompressible Flows Using SPH. *J. Comput. Phys.* **1997**, *136*, 214–226. [CrossRef]

50. Lattanzio, J.C.; Monaghan, J.J.; Monaghan, H.; Schwarz, M.P. Controlling Penetration. *SIAM J. Sci. Stat. Comput.* **1986**, *7*, 591–598. [CrossRef]

51. Albano, A.; Alexiadis, A. Interaction of Shock Waves with Discrete Gas Inhomogeneities: A Smoothed Particle Hydrodynamics Approach. *Appl. Sci.* **2019**, *9*, 5435. [CrossRef]

52. Albano, A.; le Guillou, E.; Danzé, A.; Moulitsas, I.; Sahputra, I.H.; Rahmat, A.; Duque-Daza, C.A.; Shang, X.; Ching Ng, K.; Ariane, M.; et al. How to Modify LAMMPS: From the Prospective of a Particle Method Researcher. *ChemEngineering* **2021**, *5*, 30. [CrossRef]

53. Ellero, M.; Kröger, M.; Hess, S. Viscoelastic flows studied by smoothed particle dynamics. *J. Non–Newton. Fluid Mech.* **2002**, *105*, 35–51. [CrossRef]

54. Pazdniakou, A.; Adler, P.M. Lattice Spring Models. *Transp. Porous Media* **2012**, *93*, 243–262. [CrossRef]

55. Cooke, I.R.; Kremer, K.; Deserno, M. Tunable generic model for fluid bilayer membranes. *Phys. Rev. E* **2005**, *72*, 011506. [CrossRef]

56. Frasca, P.; Harper, R.; Katz, L. Strain and frequency dependence of shear storage modulus for human single osteons and cortical bone microsamples—Size and hydration effects. *J. Biomech.* **1981**, *14*, 679–690. [CrossRef]

57. Mason, T.G.; Weitz, D.A. Optical Measurements of Frequency-Dependent Linear Viscoelastic Moduli of Complex Fluids. *Phys. Rev. Lett.* **1995**, *74*, 1250–1253. [CrossRef]

58. Xu, X.; Gupta, N. Determining elastic modulus from dynamic mechanical analysis: A general model based on loss modulus data *Materialia* **2018**, *4*, 221–226. [CrossRef]

59. Cruchaga, M.A.; Celentano, D.J.; Tezduyar, T.E. Collapse of a liquid column: Numerical simulation and experimental validation. *Comput. Mech.* **2007**, *39*, 453–476. [CrossRef]

60. Cruchaga, M.A.; Celentano, D.J.; Tezduyar, T.E. Computational Modeling of the Collapse of a Liquid Column Over an Obstacle and Experimental Validation. *J. Appl. Mech. Trans. ASME* **2009**, *76*, 021202. [CrossRef]

61. Greaves, D.M. Simulation of viscous water column collapse using adapting hierarchical grids. *Int. J. Numer. Methods Fluids* **2006**, *50*, 693–711. [CrossRef]

62. Ariane, M.; Kassinos, S.; Velaga, S.; Alexiadis, A. Discrete multi-physics simulations of diffusive and convective mass transfer in boundary layers containing motile cilia in lungs. *Comput. Biol. Med.* **2018**, *95*, 34–42. [CrossRef]

chemengineering

MDPI

Article

A 3D Smoothed Particle Hydrodynamics Study of a Non-Symmetrical Rayleigh Collapse for an Empty Cavity

Andrea Albano *,† and **Alessio Alexiadis** *,†

School of Chemical Engineering, University of Birmingham, Birmingham B15 2TT, UK
* Correspondence: AXA1220@student.bham.ac.uk (A.A.); a.alexiadis@bham.ac.uk (A.A.)
† These authors contributed equally to this work.

Abstract: In this work the first 3D Smoothed Particle Hydrodynamics model of a Rayleigh collapse for an empty cavity is proposed with the aim of improving the hydrodynamic analysis of a non-symmetrical collapse. The hydrodynamics of the model is validated against the solution of the Rayleigh-Plesset equation for a symmetrical collapse. The model is then used to simulate a non-symmetrical collapse of an empty cavity attached to a solid surface with $\gamma = 0.6$ induced by an external pressure of 50 [MPa]. The results shows that is possible to identify three regions where the hydrodynamics of the collapsing cavity shows different features. For all the stages of the collapse the simulation shows smooth pressure and velocity fields in the liquid and in the solid phase with the formation of a vortex ring in the final phase of the collapse. Finally, the model is compared to a previous 2D model to highlight strong, weak points and the key differences of both approaches in final phase of the collapse.

Keywords: particle method; smoothed particle hydrodynamics; simulation; cavitation; shock wave

Citation: Albano, A.; Alexiadis, A. A 3D Smoothed Particle Hydrodynamics Study of a Non-Symmetrical Rayleigh Collapse for an Empty Cavity. *ChemEng* **2021**, *5*, 63. https://doi.org/10.3390/chemengineering5030063

Academic Editors: Timothy Hunter and Francesco Di Natale

Received: 31 May 2021
Accepted: 9 September 2021
Published: 14 September 2021

Publisher's Note: MDPI stays neutral with regard to jurisdictional claims in published maps and institutional affiliations.

1. Introduction

The collapse of bubbles, or cavities, has been studied for more than a hundred years [1–11] since the first study by Besant in 1859 [12]. The collapse is the second phase of the cavitation phenomenon, and generally occurs after the growth phase of the so-called cavitation nuclei present in a liquid medium as water [13]. When the driving force of the collapse phase is the pressure difference between the pressure of the liquid and the pressure inside the cavity the collapse is called Rayleigh collapse named after Lord Rayleigh who first derived the analytical expression to describe the dynamics of a collapsing empty cavity [2].

The collapse phase is considered one of the main sources of erosion in applications related to hydraulic machines, propellers and many more [14–16] and, during the decades, has been investigated theoretically [17–21], experimentally [7,8,22–25] and computationally [26–29].

The erosion process induced by the collapse of cavities, called cavitation erosion, is associated to the strong shock waves generated at the collapse [13,15]. The intensity of the interaction between the shock waves and a nearby surface depends on the proximity of the cavity respect to the surface [13]. Moreover, when a spherical cavity is within a specific distance with the surface, of approximately five times the cavity radius, the cavity does not preserve its symmetry generating a high-speed jet in the direction of the surface [30].

When the collapsing cavity is attached to the surface, the jet impacts directly on it generating a strong water hammer shock followed by other post-impact shocks of different intensity induced by the complex hydrodynamic patterns of the collapsing cavity [1,13].

Due to practical difficulties (i.e., short time scale of the collapse, small dimension scale of the cavity) in studying the collapse of a single cavity with an experimental approach, researchers started studying the collapse phase of a single cavity with numerical experiments. Initially, only traditional mesh-based methods were used to address the problem, and lately also mesh-free particle methods [1,31–34] were used.

Mesh-free particle methods, where the domain is discretised using a finite number of particles interacting with each other without a grid, show advantages over mesh-based method in presence of large liquid deformation and break up of solid structures [35]. In particular, Smoothed Particle Hydrodynamics (SPH) has been successfully used for simulating shock waves [36,37], underwater explosion [38,39], multiphase flow and high velocity impact phenomena [40]. For this reasons, it looks promising for simulating the hydrodynamics of the collapse phase of a cavity and its interaction with a solid surface.

However, to the best of our knowledge, all cavitation studies based on mesh-free methods are limited to two-dimensional domains. In this paper, we present the first three-dimensional SPH model of a collapsing cavity near a solid surface with the aim to investigate the hydrodynamics of the last phases of a non-symmetrical collapse. The results show several key differences between two- and three-dimensional simulations regarding the last phase of the collapse.

2. Smoothed Particle Hydrodynamics

The SPH method has been originally developed by Lucy [41] and Gingold & Monaghan [42] in 1977 to simulate astrophysics problems. In the following years it was expanded to be able to simulate a wide range of applications such as shock waves [43,44], Riemann problem [45], explosion [46], non Newtonian fluid flow [47,48], multiphase flow [49,50], thermo-capillary flows [51], nano-fluid flows [52], and thermo-fluid application [53].

The SPH method uses a particle representation of continuous functions to discretise set of equations [35]. To derive a particle expression of a function and of the gradient of a function we start from the integral representation of a function: given a domain delimited within a volume V, is possible to express any continuum function $f(\mathbf{r})$, function of the position $\mathbf{r}$, with the integral representation

$$f(\mathbf{r}) = \iiint f(\mathbf{r}')\delta(|\mathbf{r} - \mathbf{r}'|)d\mathbf{r}', \tag{1}$$

where $\delta(\mathbf{r} - \mathbf{r}')$ is the Dirac delta function. In the SPH framework the Dirac delta function is replaced with a bell-shaped, normalised, symmetric function with compact support function called smoothing function or kernel, W. The kernel only depends on the position $\mathbf{r}$ and on the smoothing length h [35]. With this substitution we obtain the so-called SPH interpolant.

$$f(\mathbf{r}) \approx \iiint f(\mathbf{r}')W(|\mathbf{r} - \mathbf{r}'|, h)d\mathbf{r}'. \tag{2}$$

From the SPH interpolant, we can now derive the particle representation of the function: in the SPH framework, the control volume defined before is represented with a finite number of particles with their own volume and mass, and carrying physical information. The idea of the particle approximation is to assume that a smaller portion of V, dr^3, is occupied by a particle with a finite volume and a mass $m = \rho dr^3$. With this assumption, it is possible to discretise Equation (2) as follows

$$f(\mathbf{r_i}) \approx \sum \frac{m_j}{\rho_j} f(\mathbf{r_j})W(|\mathbf{r}_i - \mathbf{r}_j|, h) = \sum \frac{m_j}{\rho_j} f_j W_{ij}, \tag{3}$$

where r_i is the position of the i-th particle and m_j, ρ_j and r_j are mass, density and position of the j-th neighbour particle, which is the neighbour of the i-th particle. In fact, particles for which $|\mathbf{r}_i - \mathbf{r}_j| < h_i$ are considered neighbouring particles and accounted for in the summation [35].

To discretise a PDE or ODE we also need a particle expression for the gradient operator: since f and m are particle properties the gradient operator is only going to operate on W. We obtain

$$\nabla f(\mathbf{r_i}) \approx \nabla \sum \frac{m_j}{\rho_j} f(\mathbf{r}_j) W(|\mathbf{r}_i - \mathbf{r}_j|, h) = \sum \frac{m_j}{\rho_j} f_j \nabla W_{ij}. \tag{4}$$

Kernel Function

There are several kernels in literature, all of them must satisfy the Unity, Delta, Compact, and Positivity conditions [35]. In this work we choose to use the original Lucy kernel:

$$W(R, h) = \begin{cases} \chi(1 + 3S)(1 - S)^3 & S \leq 1 \\ 0 & S > 1, \end{cases} \tag{5}$$

where $S = |\mathbf{r} - \mathbf{r}'|/h$ and χ is the parameter used to satisfy the unity condition. χ is, for one, two and three dimensions, equal to $5/4h$, $5/\pi h^2$ and $105/16\pi h^3$.

Despite being the first kernel used by Lucy in 1977, it performs well in the model presented in this work. In fact, as shown in the next sections, the simulation does not show any forms of instability and its simple form helps to lower the overall computational cost of the simulation. Elsewhere [34], we compared the effect of different kernels for the case of void collapse and no significant difference was found in the results.

3. Model

3.1. Particle Expression of Governing Equations

The following continuity and momentum equations

$$\begin{cases} \frac{d\rho}{dt} = -\rho \frac{\partial v^\beta}{\partial x^\beta}, \\ \frac{dv^\alpha}{dt} = -\frac{1}{\rho} \frac{\partial P}{\partial x^\alpha}, \end{cases} \tag{6}$$

in the SPH framework have the following particle expressions

$$\begin{cases} \frac{d\rho_i}{dt} = \sum_j m_j v_{ij}^\beta \frac{\partial W_{ij}}{\partial x_i^\beta}, \\ m_i \frac{dv_i^\eta}{dt} = \sum_j m_i m_j \left(\frac{P_i}{\rho_i^2} + \frac{P_j}{\rho_j^2} + \Pi_{ij} \right) \frac{\partial W_{ij}}{\partial x_i^\eta}, \end{cases} \tag{7}$$

where η and β are the Einstein notation indexes, $\mathbf{v}$ is the velocity vector with $\mathbf{v}_{ij} = \mathbf{v_i} - \mathbf{v_j}$, and Π_{ij} is the artificial viscosity introduced by Monaghan [43] to model shock wave:

$$\Pi_{ij} = -\alpha h \frac{c_i + c_j}{\rho_i + \rho_j} \frac{\mathbf{v}_{ij} \cdot \mathbf{r}_{ij}}{r_{ij}^2 + \epsilon h^2}, \tag{8}$$

where α is the dimensionless dissipation factor, c_i and c_j the speed of sound of particle i and j, and $\epsilon = 0.01$ is used to avoid singularities when particles are very close to each other.

To solve the set of Equation (7) an Equation Of State (EOS) linking the pressure P with density ρ is required to solve the set of equations. In SPH a common EOS is the Tait equation [35]

$$P(\rho) = \frac{c_0^2 \rho_0}{7} \left(\left(\frac{\rho}{\rho_0} \right)^7 - 1 \right), \tag{9}$$

where c_0 is the speed of sound of the liquid and ρ_0 is the reference density. In this work $c_0 = 1484$ [ms], that is the real speed of sound of the water, and $\rho_0 = 978.46$ [kg m^{-3}]. The value of the reference density was chosen to set the initial pressure of the system to 50 [MPa].

3.2. Problem Description

The aim of this work is to simulate a non-symmetrical Rayleigh collapse of a wall-attached empty cavity. The term Rayleigh collapse is generally used to describe a collapse whose collapse driving force is the pressure difference between the pressure of the liquid, P_∞, and the pressure inside the cavity, p_b, as described by the Rayleigh-Plesset equation:

$$\frac{p_B - P_\infty(t)}{\rho_L} = R(t)\frac{d^2 R(t)}{dt^2} + \frac{3}{2}\left(\frac{dR(t)}{dt}\right)^2 + \frac{4 v_L}{R(t)}\frac{dR(t)}{dt} + \frac{2S}{\rho_L R(t)}, \tag{10}$$

where v_L is the liquid viscosity, S the surface tension of the cavity, and the $R(t)$ the radius of the cavity.

In Equation (10), the viscosity and the surface tension terms are often neglected since their order of magnitude is smaller than that of the inertial term [15]. Under this assumption is possible to directly determine the Rayleigh collapse time [2] with an analytical expression:

$$t_{TC} = 0.915 R_0 \left(\frac{\rho_L}{p_\infty - p_B}\right)^{\frac{1}{2}}, \tag{11}$$

where R_0 is the initial radius of the cavity.

Another reason to use the term Rayleigh collapse is to differentiate it from another collapse mechanism called shock-induced collapse. In the shock-induced collapse the collapse driving force is the interaction of the cavity with a travelling shock wave [28,33]. Another difference is that the shock-induced collapse is always non-symmetrical since a re-entrant jet is formed in the shock direction. On the contrary, the Rayleigh collapse can be either symmetric in presence of an isotropic pressure field or non-symmetric in presence of anisotropic pressure field. The anisotropic pressure field is generated by the presence of anisotropic drivers such as gravitational field, nearby rigid or free surface, stationary potential flow, liquid interfaces, or inertial boundaries [30].

To quantify the degree of anisotropy in the pressure field induced by a nearby rigid surface anisotropic driver we use the standoff defined as:

$$\gamma = \frac{d}{R_0}, \tag{12}$$

where d is the distance between the centre of the cavity and the surface, see Figure 1.

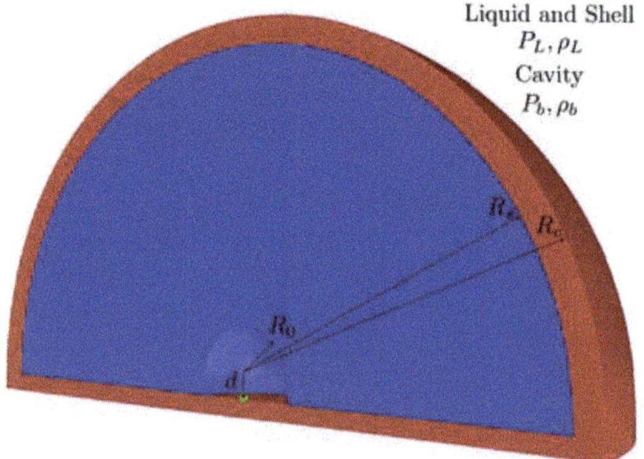

Figure 1. Section of the simulation box obtained with a slicing plan perpendicular to the y-axis and the collapsing cavity at $t = 0$. The radii dimensions are: $R_0 = 1 \times 10^{-4}$ [m], $R_S = 3 \times 10^{-3}$ [m], and $R_C = 3.1 \times 10^{-3}$ [m] .

3.3. Geometry

The 3D domain shown in Figure 1 is divided in three concentric regions, delimited by three different radii. In those regions the particles are distributed using a lattice, a set of point in space, determined by a face-centred cubic (fcc) unit cell with basis atoms, with characteristic size equal to 2.5×10^{-6} [m], that is replicated in all dimensions. In this model, two types of computational particles are used:

- Cavity ($r < R_0$): the particles in this region are removed to model an empty cavity with $p_b = 0$ and $\rho_b = 0$. Modelling the cavity as a void region is a common procedure in computational [1,17,33,34] and theoretical [2,20] studies.
- Liquid ($R_0 < r < R_s$): the particles inside this region are modelled as water following the Tait EOS. The density is set as $\rho = 1000$ [Kg m^{-3}] with an initial pressure $P = 50$ [MPa].
- Shell ($R_s < r < R_c$): In this region the particles are modelled as described in the liquid region. However, the have fixed position and density to keep constant pressure as boundary condition. The lower part of this region also acts as anisotropy driver inducing an anisotropic pressure field during the collapse. Between the shell region and the liquid region a non compenetration condition is used.

A green spot, with a radius of 0.01 mm, is highlighted in Figure 1; this spot will be used to monitor the pressure evolution during the collapse.

In this work, we focus on a single case of non-symmetrical Rayleigh collapse with $\gamma = 0.6$ to investigate the hydrodynamic evolution of the cavity during the collapse and the pressure developed on the surface. The reason for that is that the 3D simulations are considerably more computationally intensive than the 2D simulations. Therefore, a complete parametric study would be impractical. In a previous work [1], we carried out a parametric study in 2D. We identified the case of $\gamma = 0.6$ as particularly interesting because of its complex mechanics. Therefore, we focus the 3D simulations on this value of γ and compare the 2D and 3D results.

3.4. Validation

To validate the hydrodynamics of the model we compare evolution of the dimensionless radius, $R(t)/R_0$, plotted against the dimensionless time, $\tau = t/t_{TC}$ (where t_{TC} is the Rayleigh collapse time for this configuration determined using Equation (11)), of our model to the solution of Equation (10) for a symmetrical collapse

For validation, we use the case of symmetric collapse, where the cavity is not near a solid surface ($\gamma \to \infty$) because it has a theoretical solution (see Equation (10)). The model is in good agreement with the theoretical solution. There is a certain difference in the final phase of the collapse where the particles resolution is no enough to preserve the spherical symmetry (this issue is also been discussed in a previous work [34]). The final dimensionless time is $\tau = 1.08$.

During the validation, we choose the minimal resolution of $dL/R_0 = 40$, with dL initial spacing between particle, identified in previous works [32–34]. Although this resolution was identified for the 2D model, Figure 2 shows that it is also good for the 3D case. However, for the non-symmetrical collapse, a higher resolution is required, see Section 4, to see all the hydrodynamic features [1].

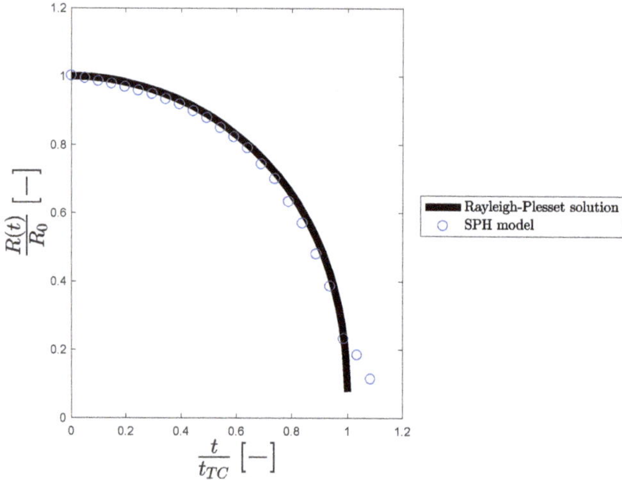

Figure 2. Dimensionless ratio (R/R_0) against dimensionless time (t/t_c) for both SPH (blue circle dot) and the numerical solution of the Rayleigh-Plesset equation (continuum black curve) for the empty cavity collapse ($P_\infty = 50$ [MPa], ρ_L =1000 [Kg m^{-3}] , $P_b = 0$, $\rho_b = 0$).

3.5. Software for Simulation, Visualisation and Post-Process

The simulations were run with the open source code simulator LAMMPS [54–56]. The visualisation and data post-processing were generated with the Open Source code OVITO [57].

4. Results

The Results shown in this section are obtained using the Lucy Kernel and smoothing length of $h = 1.3dL$. The dimensionless dissipation factor is set as $\alpha = 1$ as usually done for shock wave problems [1,34,37,44] while the time step t_s =1 × 10^{-10} has been chosen following the CFL criterion. The resolution, particle numbers, the collapse driving force, the initial radius and the magnitude of the anisotropic diver are respectively, $dL/R_0 = 133$, N_p = 20,292,752, $P = 50$ [MPa], $R_0 = 100$ μm and $\gamma = 0.6$.

Figure 3 shows the pressure history over the green region in Figure 1.

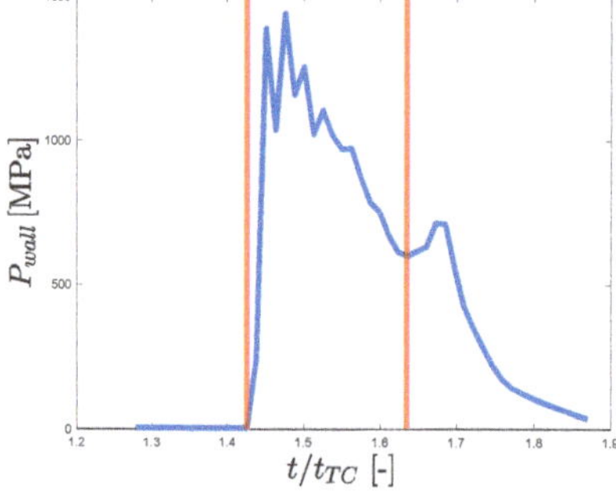

Figure 3. Pressure trend over the green region of Figure 1 for a non-symmetrical wall attached collapse (dL/R_0=133, $\gamma = 0.6$ & $P_\infty = 50$ [MPa]). The profile is divided in three regions: (*I*) Jet formation, (*II*) Jet impact and ring formation, (*III*) Ring expansion.

The pressure trend can be divided in three regions corresponding to different collapsing phase and hydrodynamics features of the cavity that are discussed in the next sections.

4.1. Region I: Jet Formation

With a stand off of $\gamma = 0.6$, the collapse cannot be symmetric and an anisotropic pressure field is expected. Figure 4a shows a high-pressure area above the cavity.

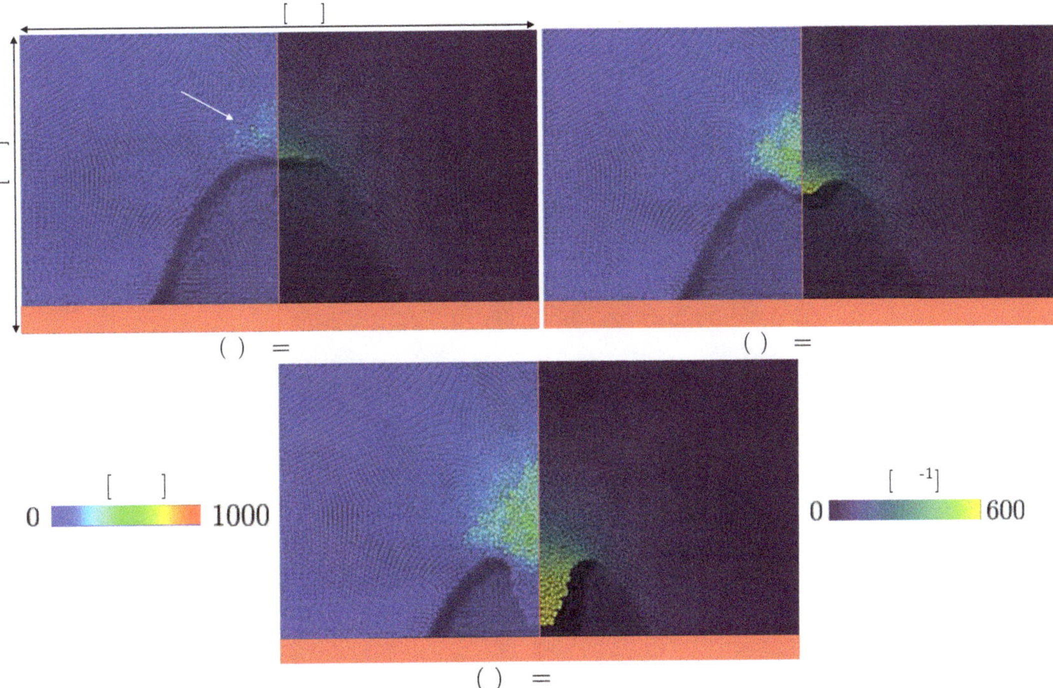

Figure 4. Pressure and velocity field in the domain represented in Figure 1 for different collapse snapshots in Region I ($dL/R_0 = 133$, $\gamma = 0.6$ & $P_\infty = 50$ [MPa]): (**a**) Anisotropic pressure field formation in the liquid phase (**b**) Generation of the jet towards the surface. (**c**) High-speed/Low pressure profile of the fully-developed jet.

As the time passes the pressure of this area raises and accelerate the liquid of the top of the cavity pushing it the direction of the solid surface. This generates the re-entrant jet (Figure 4b).

The jet, before the impact with the surface, has a high speed/ low-pressure profile as can be seen from Figure 4c.

4.2. Region II: Jet Impact and Ring Formation

At the impact, Figure 5a, the pressure at the centre of the surface reaches a maximal peak of around 1500 MPa (see Figure 6a). Because of the impact the jet splits in circular lateral jet (see Figures 5b and 6b) that will eventually hit the side of the collapsing cavity generating side shocks, see Figures 5c and 6c. When the lateral jets impact with the cavity side the cavity assumes the shape of a ring.

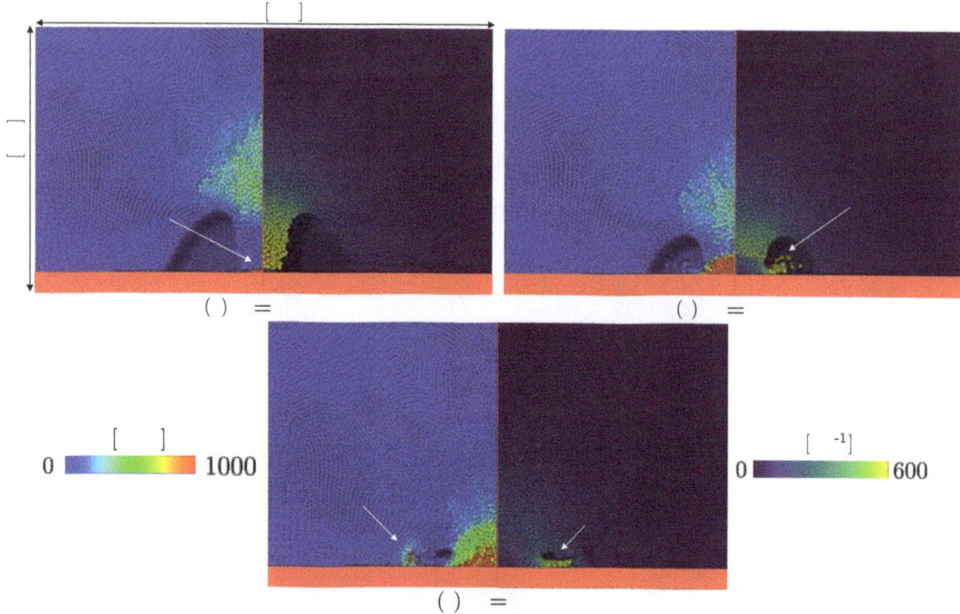

Figure 5. Pressure and velocity field in the domain represented in Figure 1 for different collapse snapshots in Region II ($dL/R_0 = 133$, $\gamma = 0.6$ & $P_\infty = 50$ [MPa]): (**a**) The jet impacts with the surface generating a water hammer impact. (**b**) The impact splits the jet in a high-speed/low pressure circular lateral jet. (**c**) The lateral jet impacts with the cavity side generating side shocks in the liquid phase.

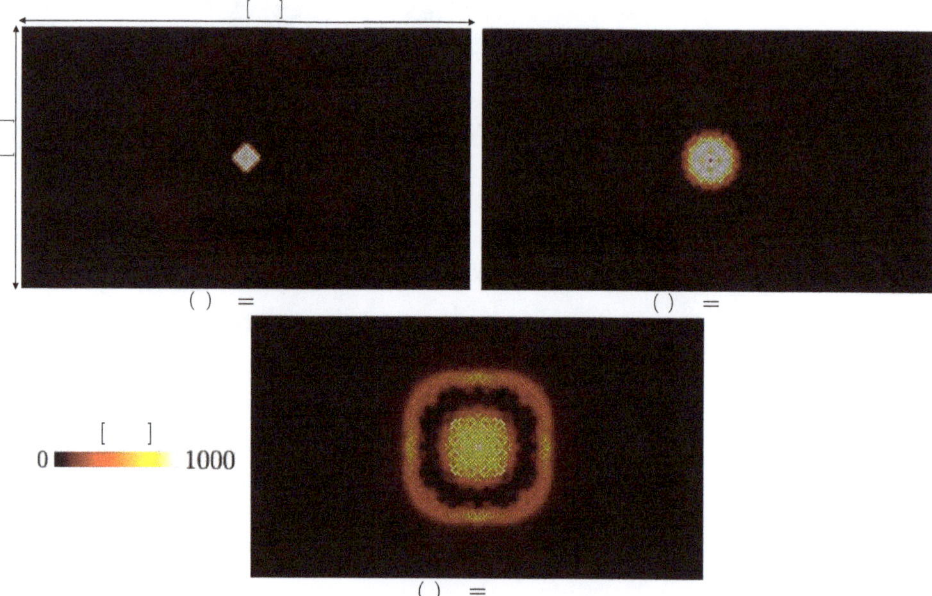

Figure 6. Pressure field over the surface for different collapse snapshots in Region II ($dL/R_0 = 133$, $\gamma = 0.6$ & $P_\infty = 50$ [MPa]): (**a**) Pressure field of the surface during the water hammer impact between the surface and the jet. (**b**) Pressure field of the surface during the jet split. (**c**) Pressure field of the surface at the impact between the lateral jet and the cavity side.

The pressure field shown in Figure 6 have non-intuitive square symmetry. This symmetry is forced by the topology of the particle distribution. However, these "artefacts" are temporary and only appears when the fluid instantly interacts with the surface plane and the circular symmetry is restored during the ring expansion, see Figure 7.

4.3. Region III: Ring Expansions

After the ring formation the collapse proceeds and, as the ring reduces its volume, a high-pressure shock is generated in the inner section of the ring (see Figure 8a). This produces a maximal pressure wave with the shape of circle, Figure 7a, whose pressure is higher than the pressure the centre of the plate. This shock will move to the centre of the surface, Figures 7b and 8b, generating the second pressure peak at the centre shown in Figure 3.

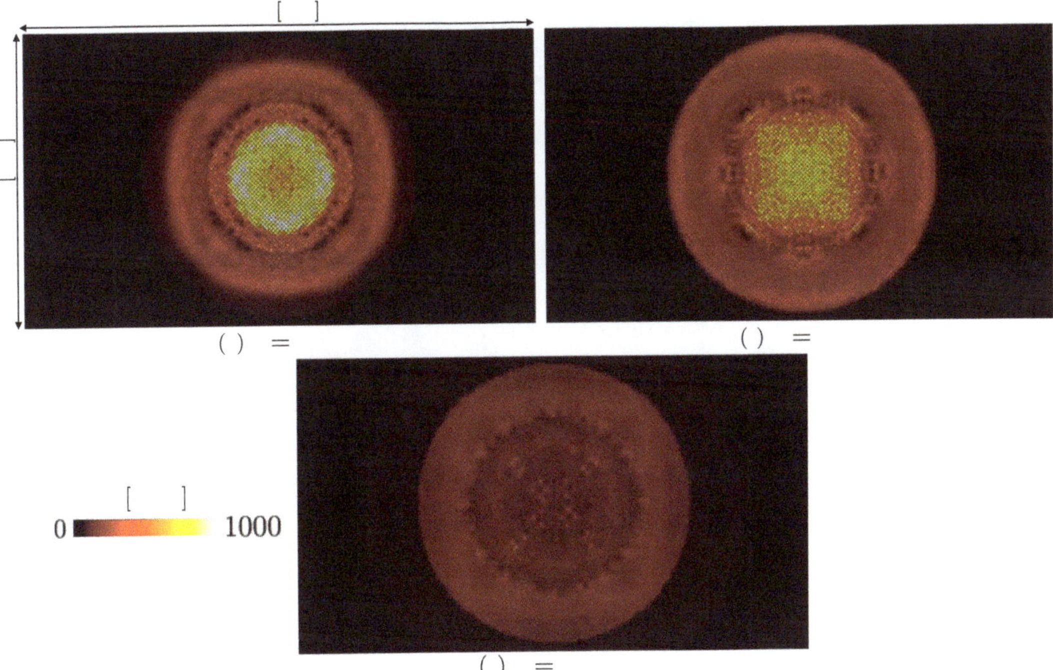

Figure 7. Pressure field over the surface for different collapse snapshots in Region III ($dL/R_0 = 133$, $\gamma = 0.6$ & $P_\omega = 50$ [MPa]); (**a**) Pressure field on the surface at the inner shock generation. (**b**) Pressure field of the surface at the second pressure peak. (**c**) Pressure field of the surface during the ring expansion.

Around the ring there is a non-uniform pressure field, see Figure 8c, that prevents a final "inertial collapse" but rather induce a rotation and an expansion of the ring to dissipate the energy gained during the collapse.

The ring behaves as a vortex ring, which dissipates the energy gained during the collapse by spinning and expanding around its central axis. This behaviour can be seen in the Videos S1–S3 available in the supplementary materials, while a schematic representation of the ring is shown in Figure 9.

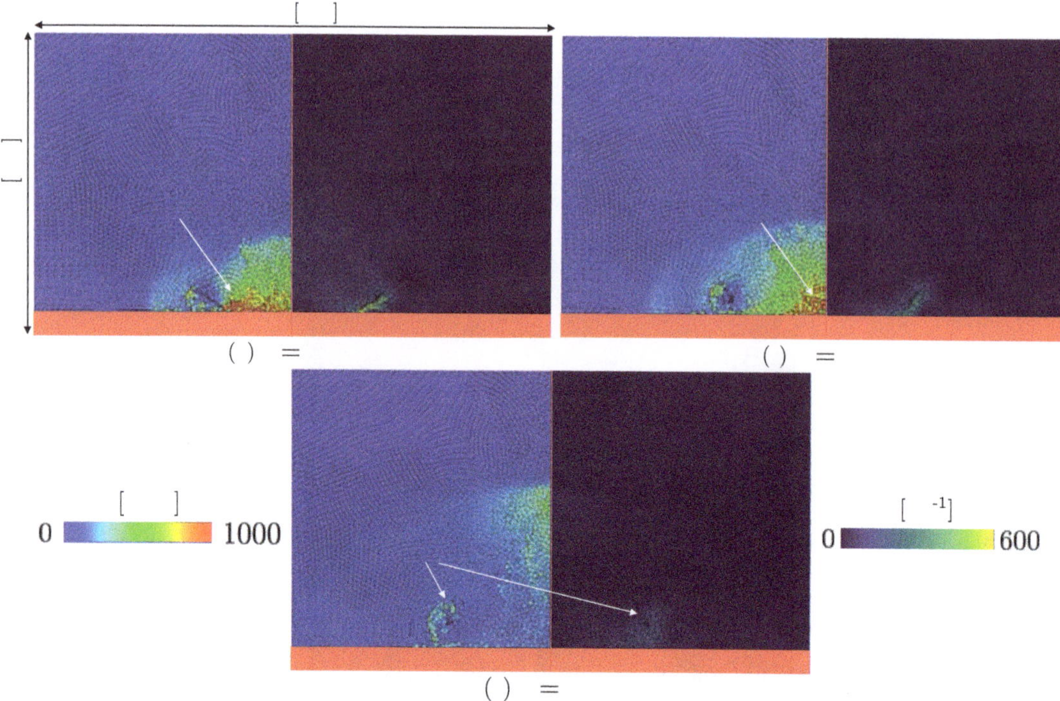

Figure 8. Pressure and velocity field in the domain represented in Figure 1 for different collapse snapshots in Region III ($dL/R_0 = 133$, $\gamma = 0.6$ & $P_\infty = 50$ [MPa]): (**a**) Inner high pressure shock generation. (**b**) Second pressure peak generated by the travelling inner shock. (**c**) Collapsing ring expansion.

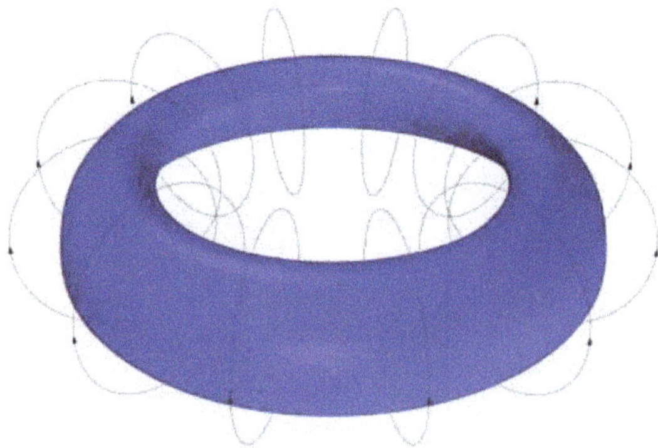

Figure 9. Schematic representation collapsing vortex ring: blue surface represents the collapsing ring, black arrows represent the liquid flow.

At the end of its motion the ring closes without generating shock waves, unlike noticed on the 2D model for the similar chase [1].

4.4. 3D and 2D Hydrodynamics Comparison

In this section we compare the hydrodynamics features and the pressures field between the 3D model and analogous 2D models [31–33] and, particularly, the 2D study made by Albano and Alexiadis [1]. We chose to compare 2D and 3D models using our previous (cartesian) 2D model [1] because we already have all the simulations available. Joshi et al. [33] carried out axisymmetric 2D simulations and the main features of the cartesian 2D and the axisymmetric 2D are the same. Our conclusions, therefore, can be generalised to both cartesian and axisymmetric 2D.

In Albano and Alexiadis [1] and here, we use exactly the same physical parameters for the simulation: driving pressure, initial radius, gamma, EOS, h, α and particle resolution. Figure 10 shows a side-by-side comparison of the last phase for a high-resolution collapse between the 2D model and the 3D model.

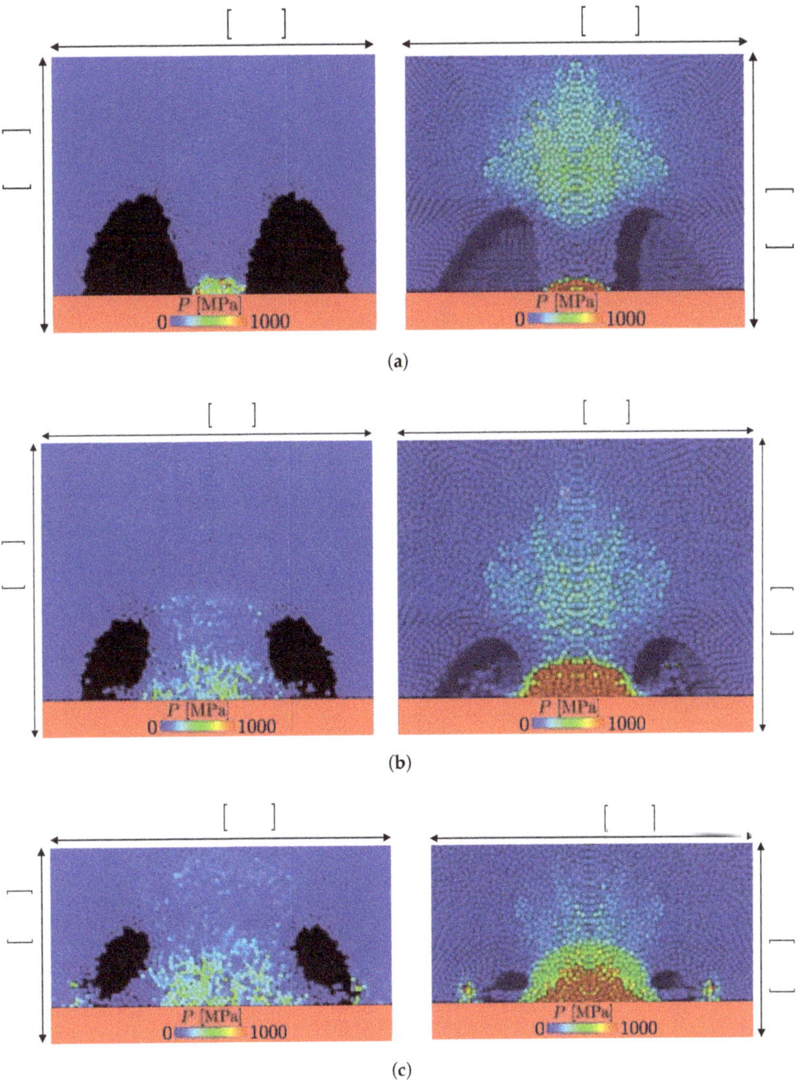

Figure 10. *Cont.*

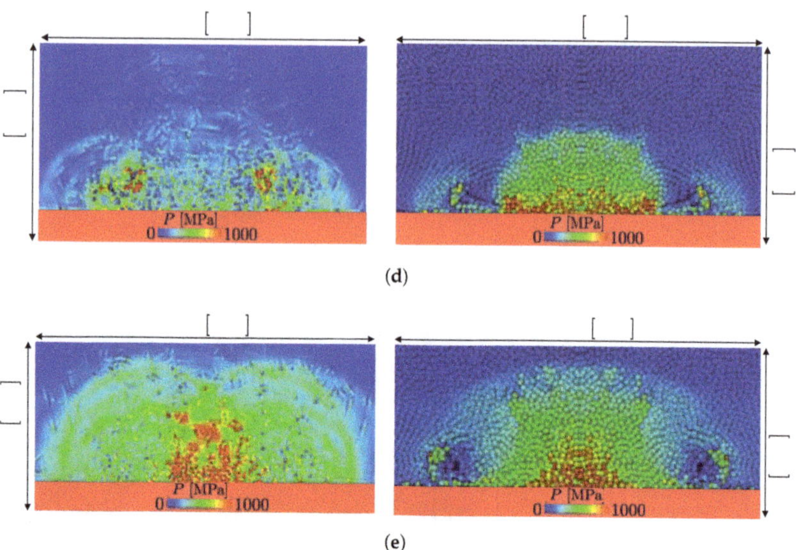

Figure 10. Pressure field for 2D [1] (left side) and 3D (right side) SPH model for a non symmetrical Rayleigh collapse ($dL/R_0 = 133$, $\gamma = 0.6$ & $P_\infty = 50$ MPa): (**a**) Water hammer impact of the jet with the surface. (**b**) Lateral jet formation. (**c**) Lateral jet impact with the cavity sides and side shock formation. (**d**) Inner high pressure shock generation. (**e**) Second pressure peak generated by the travelling inner shock.

In all phases of the collapse, the 3D model shows a smother pressure field. This shows that the 2D parameters were good enough to simulate the phenomenon, but the model induces additional constrains to keep the particle in the plane affecting the smoothness of the pressure field.

Overall, the 3D and the 2D models are both able to capture similar hydrodynamics features: (1) formation of re-entrant jet (2) formation of lateral jets (3) ring formation (corresponding to circle formation in 2D) (4) generation of side pressure waves. However, unlike the 3D model, in the 2D the new-formed circle does not dissipate their energy as vortex rings, but they generate a third pair of shock waves during the final phase of the collapse. This implies that 2D simulations of Rayleigh collapse can be considered accurate up to the final phase of the collapse where they show a third pressure peak that, which does not occur in 3D simulations.

This hydrodynamic difference also reflects on the pressure trend over the nearby solid. As can be seen in Figure 3, the second pressure peak is lower than the first peak while for the 2D the second is higher than the first [1]. This can be explained as follow:

- For the 3D: the energy gained by the fluid during the collapse, as a high speed jet, is partially used to generate a first maximal pressure peak shown in Figure 3 because of the water hammer impact between the jet and the surface. The residual energy is partially used to generate the second local maximal peak in region III and partially dissipate during the ring expansion.
- For the 2D: as before, the energy gained by the fluid is partially used to generate a first local maximal peak with the water hammer impact. The residual energy, since the vortex dissipation is not present, is totally dissipated by the circle collapse generating a second maximal pressure peak [1].

Another advantages of the 3D model over 2D models is for studying the pressure field over the nearby surface: with the 2D the layer of particles representing the surface is composed only by a single line and this makes complicate to study the spatial distribution and some noise is expected [1,34]. However, the advantage of the 2D model over the 3D is the computational cost of the simulation. A simulation with a total time of $t = 1$ μs with $N_{p,2D} = 9{,}575{,}802$ particles took 4 h of runtime on a supercomputer using 200 computational cores. On the other hand the 3D simulation with a total time of $t = 0.8$ μs with $N_{p,3D} = 20{,}292{,}752$ particles took 27 h of runtime using 200 computational cores.

5. Conclusions

In this work we proposed the first 3D SPH model to simulate a Rayleigh collapse. The Hydrodynamics of the model is validated for the symmetrical collapse case using the Rayleigh-Plesset Equation.

The model is then used to simulate a non symmetrical collapse of an empty cavity where a solid surface is acting as an anisotropic driver with $\gamma = 0.6$.

From the pressure trend over the surface is possible to identify three regions where the hydrodynamics of the collapsing cavity shows different features:

- Region *I*: The anisotropic pressure field in the liquid by the driver generates a high-speed jet from the top of the cavity that will impact on the surface.
- Region *II*:The jet impacts with the solid surface generating the first pressure peak and splitting in lateral a lateral jet that, by impacting with the cavity sides, makes the cavity assume the shape of a ring.
- Region *III*: The new formed ring reaches its minimal volume generating a shock wave an a second pressure peak and then the ring behaves as a vortex ring by dissipating its energy by rotating and expanding.

Finally, the model is compared with a similarly set up 2D model to investigate the difference between the two approaches: at cost of a great increase in computational cost, the 3D shows smoother pressure and velocity fields and also displays a different hydrodynamics behaviour in the last moment of the collapse with the formation of the vortex ring.

A natural evolution of the presented model is in the Discrete Multi-Physics framework [58]:

1. By substituting the SPH particle acting as a solid surface with Discrete Element Method or Peridynamic particles would be possible to study the erosion process of cavitation.
2. By filling the cavity with SPH particles, as done in previous work [31,32,34], following a gas EOS would be possible to study the temperature and pressure profile inside the cavity.
3. By enabling a multiphase energy exchange (gas-liquid, gas-solid and liquid-solid), as done in a previous work [34], would be possible to investigate the role of temperature in the hydrodynamic of the collapse and tin the erosion process within the same computational framework.

Supplementary Materials: The following are available online at https://www.mdpi.com/article/10.3390/chemengineering5030063/s1, Video S1: Non-symmetrical Rayleigh collapse ($\gamma = 0.6$) with pressure field in the liquid phase, Video S2: Non-symmetrical Rayleigh collapse ($\gamma = 0.6$) with velocity field in the liquid phase, Video S3: Pressure field over the surface generated during the Rayleigh non-symmetrical collapse ($\gamma = 0.6$).

Author Contributions: A.A. (Andrea Albano) and A.A. (Alessio Alexiadis) conceptualise the work; A.A. (Andrea Albano) designed the work and performed the simulations; A.A. (Andrea Albano) and A.A. (Alessio Alexiadis) contributed in writing/reviewing and editing the paper. Both authors have read and agreed to the published version of the manuscript.

Funding: This work was supported by the US Office of Naval Research Global (ONRG) under 256 NICOP Grant N62909-17-1-2051.

Data Availability Statement: All relevant data are within the manuscript and its Supporting Information files.

Conflicts of Interest: The authors declare no conflict of interest.

References

1. Albano, A.; Alexiadis, A. Non-Symmetrical Collapse of an Empty Cylindrical Cavity Studied with Smoothed Particle Hydrodynamics. *Appl. Sci.* **2021**, *11*, 3500. [CrossRef]
2. Rayleigh, L. VIII. On the pressure developed in a liquid during the collapse of a spherical cavity. *Lond. Edinb. Dublin Philos. Mag. J. Sci.* **1917**, *34*, 94–98. [CrossRef]
3. Plesset, M.S. The dynamics of cavitation bubbles. *J. Appl. Mech.* **1949**, *16*, 277–282. [CrossRef]
4. Benjamin, T.B. Pressure waves from collapsing cavities. In Proceedings of the 2nd Symposium on Naval Hydrodynamics, Washington, DC, USA, 25–29 August 1958; pp. 207–229.
5. Hickling, R.; Plesset, M.S. Collapse and rebound of a spherical bubble in water. *Phys. Fluids* **1964**, *7*, 7–14. [CrossRef]
6. Knapp, R.; Daily, J.; Hammitt, F. *Cavitation*; McGraw-Hill: New York, NY, USA, 1970; Volume 39.
7. Vogel, A.; Lauterborn, W.; Timm, R. Optical and acoustic investigations of the dynamics of laser-produced cavitation bubbles near a solid boundary. *J. Fluid Mech.* **1989**, *206*, 299–338. [CrossRef]
8. Philipp, A.; Lauterborn, W. Cavitation erosion by single laser-produced bubbles. *J. Fluid Mech.* **1998**, *361*, 75–116. [CrossRef]
9. Giannadakis, E.; Gavaises, M.; Arcoumanis, C. Modelling of cavitation in diesel injector nozzles. *J. Fluid Mech.* **2008**, *616*, 153–193. [CrossRef]
10. Hsiao, C.T.; Jayaprakash, A.; Kapahi, A.; Choi, J.K.; Chahine, G.L. Modelling of material pitting from cavitation bubble collapse. *J. Fluid Mech.* **2014**, *755*, 142–175. [CrossRef]
11. Sun, X.; Xuan, X.; Song, Y.; Jia, X.; Ji, L.; Zhao, S.; Yoon, J.Y.; Chen, S.; Liu, J.; Wang, G. Experimental and numerical studies on the cavitation in an advanced rotational hydrodynamic cavitation reactor for water treatment. *Ultrason. Sonochem.* **2021**, *70*, 105311. [CrossRef]
12. Besant, W.H. *A Treatise on Hydrostatics and Hydrodynamics*; Deighton, Bell: London, UK, 1859.
13. Kim, K.H.; Chahine, G.; Franc, J.P.; Karimi, A. *Advanced Experimental and Numerical Techniques for Cavitation Erosion Prediction*; Springer: Berlin/Heidelberg, Germany, 2014; Volume 106.
14. Karimi, A.; Martin, J. Cavitation erosion of materials. *Int. Met. Rev.* **1986**, *31*, 1–26. [CrossRef]
15. Brennen, C.E. *Cavitation and Bubble Dynamics*; Cambridge University Press: Cambridge, UK, 2014.
16. Turangan, C.; Ball, G.; Jamaluddin, A.; Leighton, T. Numerical studies of cavitation erosion on an elastic–plastic material caused by shock-induced bubble collapse. *Proc. R. Soc. A Math. Phys. Eng. Sci.* **2017**, *473*, 20170315. [CrossRef]
17. Plesset, M.S.; Prosperetti, A. Bubble dynamics and cavitation. *Annu. Rev. Fluid Mech.* **1977**, *9*, 145–185. [CrossRef]
18. Keller, J.B.; Miksis, M. Bubble oscillations of large amplitude. *J. Acoust. Soc. Am.* **1980**, *68*, 628–633. [CrossRef]
19. Tomita, Y.; Shima, A. Mechanisms of impulsive pressure generation and damage pit formation by bubble collapse. *J. Fluid Mech.* **1986**, *169*, 535–564. [CrossRef]
20. Kudryashov, N.A.; Sinelshchikov, D.I. Analytical solutions of the Rayleigh equation for empty and gas-filled bubble. *J. Phys. A Math. Theor.* **2014**, *47*, 405202. [CrossRef]
21. Kudryashov, N.A.; Sinelshchikov, D.I. Analytical solutions for problems of bubble dynamics. *Phys. Lett. A* **2015**, *379*, 798–802. [CrossRef]
22. Naude, C.F.; Ellis, A.T. On the Mechanism of Cavitation Damage by Nonhemispherical Cavities Collapsing in Contact with a Solid Boundary. *J. Basic Eng.* **1961**, *83*, 648–656. [CrossRef]
23. Kling, C.L.; Hammitt, F.G. A photographic Study of Spark-Induced Cavitation Bubble Collapse. *J. Basic Eng.* **1972**, *94*, 825–832. [CrossRef]
24. Lauterborn, W.; Bolle, H. Experimental investigations of cavitation-bubble collapse in the neighbourhood of a solid boundary. *J. Fluid Mech.* **1975**, *72*, 391–399. [CrossRef]
25. Flannigan, D.J.; Suslick, K.S. Inertially confined plasma in an imploding bubble. *Nat. Phys.* **2010**, *6*, 598–601. [CrossRef]
26. Plesset, M.S.; Chapman, R.B. Collapse of an initially spherical vapour cavity in the neighbourhood of a solid boundary. *J. Fluid Mech.* **1971**, *47*, 283–290. [CrossRef]
27. Blake, J.; Taib, B.; Doherty, G. Transient cavities near boundaries. Part 1. Rigid boundary. *J. Fluid Mech.* **1986**, *170*, 479–497. [CrossRef]
28. Johnsen, E.; Colonius, T. Numerical simulations of non-spherical bubble collapse. *J. Fluid Mech.* **2009**, *629*, 231–262. [CrossRef]
29. Beig, S.; Aboulhasanzadeh, B.; Johnsen, E. Temperatures produced by inertially collapsing bubbles near rigid surfaces. *J. Fluid Mech.* **2018**, *852*, 105–125. [CrossRef]
30. Supponen, O.; Obreschkow, D.; Tinguely, M.; Kobel, P.; Dorsaz, N.; Farhat, M. Scaling laws for jets of single cavitation bubbles. *J. Fluid Mech.* **2016**, *802*, 263–293. [CrossRef]
31. Nair, P.; Tomar, G. Simulations of gas-liquid compressible-incompressible systems using SPH. *Comput. Fluids* **2019**, *179*, 301–308. [CrossRef]
32. Pineda, S.; Marongiu, J.C.; Aubert, S.; Lance, M. Simulation of a gas bubble compression in water near a wall using the SPH-ALE method. *Comput. Fluids* **2019**, *179*, 459–475. [CrossRef]
33. Joshi, S.; Franc, J.P.; Ghigliotti, G.; Fivel, M. SPH modelling of a cavitation bubble collapse near an elasto-visco-plastic material. *J. Mech. Phys. Solids* **2019**, *125*, 420–439. [CrossRef]

34. Albano, A.; Alexiadis, A. A smoothed particle hydrodynamics study of the collapse for a cylindrical cavity. *PLoS ONE* **2020**, *15*, e0239830.
35. Liu, G.R.; Liu, M.B. *Smoothed Particle Hydrodynamics: A Meshfree Particle Method*; World Scientific: Singapore, 2003.
36. Liu, M.; Liu, G.; Lam, K. Investigations into water mitigation using a meshless particle method. *Shock Waves* **2002**, *12*, 181–195. [CrossRef]
37. Albano, A.; Alexiadis, A. Interaction of Shock Waves with Discrete Gas Inhomogeneities: A Smoothed Particle Hydrodynamics Approach. *Appl. Sci.* **2019**, *9*, 5435. [CrossRef]
38. Swegle, J.; Attaway, S. On the feasibility of using smoothed particle hydrodynamics for underwater explosion calculations. *Comput. Mech.* **1995**, *17*, 151–168. [CrossRef]
39. Liu, M.; Liu, G.; Lam, K.; Zong, Z. Smoothed particle hydrodynamics for numerical simulation of underwater explosion. *Comput. Mech.* **2003**, *30*, 106–118. [CrossRef]
40. Johnson, G.R.; Stryk, R.A.; Beissel, S.R. SPH for high velocity impact computations. *Comput. Methods Appl. Mech. Eng.* **1996**, *139*, 347–373. [CrossRef]
41. Lucy, L.B. A numerical approach to the testing of the fission hypothesis. *Astron. J.* **1977**, *82*, 1013–1024. [CrossRef]
42. Gingold, R.A.; Monaghan, J.J. Smoothed particle hydrodynamics: Theory and application to non-spherical stars. *Mon. Not. R. Astron. Soc.* **1977**, *181*, 375–389. [CrossRef]
43. Monaghan, J.; Gingold, R.A. Shock simulation by the particle method SPH. *J. Comput. Phys.* **1983**, *52*, 374–389. [CrossRef]
44. Morris, J.; Monaghan, J. A switch to reduce SPH viscosity. *J. Comput. Phys.* **1997**, *136*, 41–50. [CrossRef]
45. Monaghan, J. SPH and Riemann solvers. *J. Comput. Phys.* **1997**, *136*, 298–307. [CrossRef]
46. Sirotkin, F.V.; Yoh, J.J. A Smoothed Particle Hydrodynamics method with approximate Riemann solvers for simulation of strong explosions. *Comput. Fluids* **2013**, *88*, 418–429. [CrossRef]
47. Shao, S.; Lo, E.Y. Incompressible SPH method for simulating Newtonian and non-Newtonian flows with a free surface. *Adv. Water Resour.* **2003**, *26*, 787–800. [CrossRef]
48. Hosseini, S.; Manzari, M.; Hannani, S. A fully explicit three-step SPH algorithm for simulation of non-Newtonian fluid flow. *Int. J. Numer. Methods Heat Fluid Flow* **2007**, *17*, 715–735. [CrossRef]
49. Shadloo, M.S.; Oger, G.; Le Touzé, D. Smoothed particle hydrodynamics method for fluid flows, towards industrial applications: Motivations, current state, and challenges. *Comput. Fluids* **2016**, *136*, 11–34. [CrossRef]
50. Rahmat, A.; Yildiz, M. A multiphase ISPH method for simulation of droplet coalescence and electro-coalescence. *Int. J. Multiph. Flow* **2018**, *105*, 32–44. [CrossRef]
51. Hopp-Hirschler, M.; Shadloo, M.S.; Nieken, U. A smoothed particle hydrodynamics approach for thermo-capillary flows. *Comput. Fluids* **2018**, *176*, 1–19. [CrossRef]
52. Nasiri, H.; Jamalabadi, M.Y.A.; Sadeghi, R.; Safaei, M.R.; Nguyen, T.K.; Shadloo, M.S. A smoothed particle hydrodynamics approach for numerical simulation of nano-fluid flows. *J. Therm. Anal. Calorim.* **2019**, *135*, 1733–1741. [CrossRef]
53. Ng, K.; Ng, Y.; Sheu, T.; Alexiadis, A. Assessment of Smoothed Particle Hydrodynamics (SPH) models for predicting wall heat transfer rate at complex boundary. *Eng. Anal. Bound. Elem.* **2020**, *111*, 195–205. [CrossRef]
54. Plimpton, S. *Fast Parallel Algorithms for Short-Range Molecular Dynamics*; Technical Report; Sandia National Labs.: Albuquerque, NM, USA, 1993.
55. Ganzenmüller, G.C.; Steinhauser, M.O.; Van Liedekerke, P.; Leuven, K.U. The implementation of Smooth Particle Hydrodynamics in LAMMPS. *Paul Van Liedekerke Kathol. Univ. Leuven* **2011**, *1*, 1–26.
56. Albano, A.; Le Guillou, E.; Danzé, A.; Moulitsas, I.; Sahputra, I.H.; Rahmat, A.; Duque-Daza, C.A.; Shang, X.; Ng, K.C.; Ariane, M.; et al. How to modify LAMMPS: From the prospective of a Particle method researcher. *ChemEngineering* **2021**, *5*, 30. [CrossRef]
57. Stukowski, A. Visualization and analysis of atomistic simulation data with OVITO–the Open Visualization Tool. *Model. Simul. Mater. Sci. Eng.* **2009**, *18*, 015012. [CrossRef]
58. Alexiadis, A. The discrete multi-hybrid system for the simulation of solid-liquid flows. *PLoS ONE* **2015**, *10*, e0124678. [CrossRef] [PubMed]

MDPI
St. Alban-Anlage 66
4052 Basel
Switzerland
Tel. +41 61 683 77 34
Fax +41 61 302 89 18
www.mdpi.com

ChemEngineering Editorial Office
E-mail: chemengineering@mdpi.com
www.mdpi.com/journal/chemengineering